AF361736

Nutritional Components of Wheat Based Food: Composition, Properties and Uses

Nutritional Components of Wheat Based Food: Composition, Properties and Uses

Editors

Donatella Bianca Maria Ficco
Grazia Maria Borrelli

Basel • Beijing • Wuhan • Barcelona • Belgrade • Novi Sad • Cluj • Manchester

Editors
Donatella Bianca Maria Ficco
Centro di Ricerca
Cerealicoltura e Colture
Industriali (CREA-CI)
Foggia, Italy

Grazia Maria Borrelli
Centro di Ricerca
Cerealicoltura e Colture
Industriali (CREA-CI)
Foggia, Italy

Editorial Office
MDPI
St. Alban-Anlage 66
4052 Basel, Switzerland

This is a reprint of articles from the Special Issue published online in the open access journal *Foods* (ISSN 2304-8158) (available at: https://www.mdpi.com/journal/foods/special_issues/wheat_composition_properties_uses).

For citation purposes, cite each article independently as indicated on the article page online and as indicated below:

Lastname, A.A.; Lastname, B.B. Article Title. *Journal Name* **Year**, *Volume Number*, Page Range.

ISBN 978-3-0365-9508-5 (Hbk)
ISBN 978-3-0365-9509-2 (PDF)
doi.org/10.3390/books978-3-0365-9509-2

Contents

About the Editors

Donatella Bianca Maria Ficco

Donatella Bianca Maria Ficco, long-term researcher in the Council for Agricultural Research and Economics, Research Centre for Cereal and Industrial Crops of Foggia (Italy) since 2012. She has published 38 articles in scientific journals indexed with a relative quality index, in addition to 75 works accepted in different national and international congresses. All these works are framed within three research lines: (a) phenotyping of quality and nutraceutical characters and process operations for the development of functional foods, (b) development of genetic maps to study the genetic basis of agronomically important traits and (c) storage protein and celiac disease. She is co-inventor of the patent (PCT/IT2012/000287) on "Peptides having protective effect towards the inflammatory activity of peptide 31-43 of -gliadin in celiac disease". She is a member of the expert working group on 'Durum wheat genomics and breeding' within the Wheat Initiative.

Grazia Maria Borrelli

Grazia Maria Borrelli graduated in Biological Sciences at University of Pisa and obtained his PhD in Sustainable Agricultural Ecosystems from the University of Foggia. She is currently a researcher at Council for Agricultural Research and Economics, Research Centre for Cereal and Industrial Crops in Foggia.

Her research activity mainly concerns durum wheat and falls within three main topics: the study of lipoxygenase role in pasta color, cereal processing, pigmented wheats for functional foods; in vitro culture and genetic transformation, including transgenesis an New Breeding Technologies, for traits of agronomical interest; multiple plant responses to abiotic stresses.

She has contributed as author or co-author of 31 papers published in scientific indexed journals, 8 papers published in popular Journals, and 3 book chapters by international publishers, in addition to 34 proceedings accepted in different national and international congresses.

Preface

Functional cereal-based foods are receiving increasing attention by consumers as they are a good source of energy, minerals, vitamins and fiber. These products can be obtained by high-phytochemical wheat species to use as wholemeal or selected by-products. Therefore, one of the major emerging technologies is the extraction of bioactive compounds from waste of other plant organs and their application in the flour enrichment with multiple benefits for a wide range of consumers.

Donatella Bianca Maria Ficco and Grazia Maria Borrelli
Editors

 *foods*

Editorial

Nutritional Components of Wheat Based Food: Composition, Properties, and Uses

Donatella Bianca Maria Ficco * and Grazia Maria Borrelli

Consiglio per la Ricerca in Agricoltura e l'Analisi dell'Economia Agraria—Centro di Ricerca Cerealicoltura e Colture Industriali, 71122 Foggia, Italy; graziamaria.borrelli@crea.gov.it
* Correspondence: donatellabm.ficco@crea.gov.it

Citation: Ficco, D.B.M.; Borrelli, G.M. Nutritional Components of Wheat Based Food: Composition, Properties, and Uses. *Foods* **2023**, *12*, 4010. https://doi.org/10.3390/foods12214010

Received: 25 October 2023
Accepted: 30 October 2023
Published: 2 November 2023

Wheats (bread and durum wheats) and their main end-use products (particularly bread and pasta) have an important role in the Mediterranean diet as they substantially contribute to nutrient intake. Furthermore, whole grains are also a source of dietary fiber, minerals, vitamins, and phenolic compounds with beneficial effects on human health.

This Special Issue comprises a collection of ten peer-reviewed articles centered on the variability of the chemical, functional, and technological features of raw materials for the identification of those useful for the improvement of derived wheat based foods. In total, 66 authors contributed to this collection, which consists of nine research articles and one review article. Among the accepted submissions, the following topics were mainly covered: the exploitation of landraces, ancient and modern genotypes belonging to different cereal species, as a source of useful compounds to obtain staple foods with superior nutrition and functional properties (five articles); the use of functional/fortifying ingredients starting from raw materials (two articles) or by-products/wastes generated in the agribusiness industries (two articles) to improve the nutritional and antioxidant properties of bread and pasta, mainly; and finally, the optimization of the use of wheat aleurone for bakery products (one review article).

Recently, to cope with climate change and to meet the needs for more sustainable low-input grain production, the exploration of genetic diversity has become fundamental. Landraces, adapted to specific agro-climatic conditions, constitute a reservoir of genetic diversity for the development of new agricultural systems and new interesting products, with a higher content of micronutrients (minerals; vitamins) and bioactive components. In this perspective, pigmented landraces are an underexploited resource of phytochemicals. Ladhari et al. [1] investigated the compositional properties of two differently anthocyanin-rich Ethiopian durum wheat landraces in relation to protein content, dry gluten, ash, total polyphenols, anthocyanins, proanthocyanidins, and specific phenolic acids, confirming them to be a source of primary and secondary metabolites of interest for human health and nutrition.

Furthermore, both old varieties that have been replaced with more productive ones and some of the species that were cultivated in the past have been recovered during recent decades for regional crop production and gourmet foods as they have more favorable compositions in terms of health.

Among the ancient wheats, diploid einkorn (*T. monococcum*, L., AA genome), tetraploid emmer (*T. turgidum* L. subsp. *dicoccum* Thell., AABB genomes), and hexaploid spelt (*T. aestivum* L. subsp. *spelta* Thell, AABBDD genomes) represent interesting genetic materials.

Lovegrove et al. [2] analyzed three cultivars of emmer and five cultivars each of spelt and bread wheat, grown in two consecutive years at two nitrogen levels that reflect those of low-input and intensive farming systems, for their dietary fiber, polar metabolites, protein, phenolics, and mineral micronutrients. While the ranges of the components resulted to be similar among the three cereal types, an exception regarding mineral micronutrients (zinc and iron) which are prevalent in spelt and emmer was found. It follows that the consumption of these wheats could contribute to the ways to cope with micronutrient deficiencies, which affect billions of people worldwide.

The renewed interest in more natural foods promoted the use of these wheats as raw materials for the making of products such as bread, pasta, biscuits, pancakes, etc. In this regard, studies on processing quality have increased in importance. Brandolini et al. [3] studied the composition and technological characteristics of the flours and breads of two elite einkorn wheats in comparison with those of a high-quality bakery bread wheat, all cropped in four different environments. The einkorns confirmed better flour composition than that of bread wheat in terms of proteins, soluble pentosans, and yellow pigment. Technologically, they exhibit breadmaking behavior that is similar to that of bread wheat. Einkorn breads had a softer texture, are maintained for a longer shelf-life period, and have a slower retrogradation period than that of bread wheat. They demonstrate that the possibility of obtaining einkorn breads with technological properties comparable to those of breadmaking-quality bread wheat and with superior nutritional value is linked to the use of high-quality cultivars and to the optimization of processing conditions.

In a similar approach, Huertas-García et al. [4] evaluated a wide collection (eighty-eight) of Spanish traditional spelt accessions for grain/ flour quality, dough, and baking traits and compared them with those of ten modern wheat cultivars (nine bread wheats and one modern spelt cultivar). In comparison with wheats, Spanish spelt accessions showed softer grains, higher protein content and gluten extensibility, and lower gluten strength. When considering bread-making quality, two spelt genotypes were shown to have the highest loaf volume, the most important trait through which to assess wheat baking quality, and to have the possibility to be used for breeding purposes aimed at the improvement of bread making quality.

Finally, Borrelli et al. [5] compared old and modern cultivars of three cereal species, durum wheat, emmer wheat, and barley, in a 2-year evaluation, studying the composition of phenolics and carotenoids, and antioxidant activity in wholemeals and related biscuits, providing information on consumer acceptability. Barley showed the highest levels for all traits, followed by emmer, with the exception of carotenoid content, which was prevalent in the modern durum wheat cultivar. Generally, the highest values of phytochemicals were observed in the modern cultivars, while the old ones had the highest total phenolic content, according to the literature. The same trend was observed in biscuits. Emmer showed the best overall acceptability among consumers with even higher scores than those of commercial biscuits, while barley biscuits, albeit ranking the highest in terms of phytochemicals, were shown to be less appreciated. New insights into consumer-led food product development are required to find optimal formulations for better nutritional, sensorial, and health-related end-products. Once identified, the genetic materials deemed of interest can be recommended for supply chain studies or employed to transfer qualitative and bioactive traits into elite cultivars.

With growing demands for functional, natural, and low-calorie ingredients, extensive research is being carried out to identify novel ingredients that can be useful in the enrichment of formulations for various cereal-based products. To this aim, the partial replacement of wheat flour with phytochemical-rich, functional plant-based flours or powders is an interesting strategy. For instance, Cornelian cherries (CC; *Cornus mas* L.) are edible fruits with a color ranging from red to purple, that are rich in vitamin C and polyphenols. CC-derived ingredients have been used in many formulations to produce enhanced food products. Šimora et al. [6] investigated cornelian cherry powder (CCP) as a functional ingredient for bread production, via the incorporation of five CCP levels (0, 1, 2, 5, and 10% w/w) in flour. Overall, the results showed that replacing wheat flour with up to 5% of CCP produced the most suitable formulation with which to yield bread loaves that are richer in antioxidants, without negatively impacting the physical properties and sensory attributes of the bakery.

In a similar way, hemp can be used to enrich functional pasta. Hemp seeds, mainly used as animal feed, are receiving great attention in relation to their use by humans as a source of nutrients, in particular for their high levels in polyunsaturated fatty acids.

Bonacci et al. [7] incorporated hemp seed flour, at different percentages (5%, 7.5%, and 10%) and with two different particle sizes, as a fortifying ingredient in the production of pasta, studying the effect on its rheological and chemical characteristics. The minor particle size and 7.5% substitution level have been identified as the optimal processing conditions in which to produce cooked fortified pasta of high quality, that is nutritionally rich (with improved minerals, and an improved amino acid and polyunsaturated fatty acid content and profile), and has good organoleptic properties.

Functional ingredients may also be derived from by-products or food processing waste, with the aim of sustainably using natural resources, providing additional economic benefits to food businesses.

An example of the circular bioeconomy comes from brewers' spent grain (BSG), the main by-product generated from the brewing industry. BSG is rich in antioxidant compounds such as phenolic acids, flavonoids, tannins, and proanthocyanidins, amino phenolics, and fiber (β-glucans and arabinoxylans). Baiano et al. [8] considered BSG derived from the brewing of Belgian-style white beers a functional ingredient in bread making. Three different mixtures of barley malt and of unmalted bread wheat/durum wheat/emmer wheat, cropped in two geographical areas, were used for bread making, at two different percentages (20 and 25%) of substitution of wheat flour type 0, with the aim to evaluate the effects of such replacements on overall bread quality and functional characteristics. The authors stated that supplementation with the highest percentage of BSGs exerted a positive influence on the contents of phenolics and dietary fibers with good structural and sensory attributes, owing to the simultaneous addition of gluten. Bread supplemented with bread or durum wheat spent grains resulted in a good compromise between the content of nutraceuticals and overall sensory quality.

Further, given the growing interest of consumers in so-called "superfoods", the use of the avocado fruit is seeing great relevance owing to its great potential, conferred by its antioxidant, fiber, and low sugar content. Avocados are mostly consumed fresh, but they can also be processed, and hence, several components of the fruit, including the peel and seeds, these being rich in protein, fiber, and numerous bioactive compounds, are wasted. This waste may constitute a suitable ingredient in the form of powders for pasta, breads, dry soups, and other food recipes. In this regard, Viola et al. [9] aimed to recycle avocado waste through dehydration and milling. The resulting powder was used as a functional ingredient for the processing of sourdough semolina bread, a product consumed daily in Southern Italy. The authors demonstrated that avocado waste products can be successfully incorporated into leavened baked products to improve their characteristics, particularly in terms of antioxidant content, also resulting in sensorially appreciation due to aroma and color.

Owing to the rising demand for novel formulations, combining the right ingredient with an efficient processing method has become a critical endeavor.

Although health benefits might be achieved through the use of wholemeal for cereal-based products due to its richness in bioactive compounds, it noted that the use of wholemeal, containing bran, negatively affects technological properties during food processing and the sensory acceptance of final products. A compromise could be achieved via the use of aleurone, which is considered by millers as a bran layer rich in dietary fiber, proteins, and bioactive compounds. Unfortunately, its use is limited, most likely due to issues related to its extraction and to its negative rheological and technological effects on dough, which are mainly ascribable to its content of dietary fiber, particularly the water-insoluble type. Therefore, specific modifications to aleurone's components, performed without losing any of the health benefits prior to its incorporation into a food process, are welcomed.

Lebert et al. [10] reviewed different extraction techniques to test the potentiality of this tissue as a nutrient-rich ingredient in bread-making. The authors found an improved nutritional profile in aleurone-rich flour including dietary fiber and protein, minerals, antioxidants, phytoestrogens, and sterols. Despite these beneficial nutritional properties, changes during dough development could affect pasting properties and dough rheology, depending on the percentage of aleurone in the flour. Considering that the aleurone layer is tightly bound to seed coats, it is rather difficult to separate it from the endosperm and from the rest of the bran during conventional milling. Most existing processes for the retrieval of the aleurone layer begin with bran material and provide for the main mechanical, physical and chemical procedures, with varying aleurone purities and yields. Moreover, reducing the particle size of an aleurone material can also lead to beneficial effects in terms of the bioaccessibility of antioxidant compounds. Unfortunately, there is no an univocal method related to its extraction, and the latter also depends on the natural variability of the grain's constituents, constantly requiring the adaptation of the process itself to each raw material. Further, for the wide use of aleurone as an ingredient, extraction methods should be standardized, easy to apply, and not too costly.

The articles compiled in this Special Issue demonstrate a research trajectory in which natural variation in the grain composition of wheats and related cereals, the use of functional food ingredients, and the process used should be considered when studying the wheat supply chain.

Funding: This study includes results obtained within the project entitled "Nobili Cereali", funded by PSR 2014–2020 Regione Campania—Misura 16.1.1–Azione 2 "Sostegno ai POI".

Conflicts of Interest: The authors declare no conflict of interest.

References

1. Ladhari, A.; Corrado, G.; Rouphael, Y.; Carella, F.; Nappo, G.R.; Di Marino, C.; De Marco, A.; Palatucci, D. Chemical, Functional, and Technological Features of Grains, Brans, and Semolina from Purple and Red Durum Wheat Landraces. *Foods* **2022**, *11*, 1545. [CrossRef]
2. Lovegrove, A.; Dunn, J.; Pellny, T.K.; Hood, J.; Burridge, A.J.; America, A.H.P.; Gilissen, L.; Timmer, R.; Proos-Huijsmans, Z.A.M.; van Straaten, J.P.; et al. Comparative Compositions of Grain of Bread Wheat, Emmer and Spelt Grown with Different Levels of Nitrogen Fertilisation. *Foods* **2023**, *12*, 843. [CrossRef]
3. Brandolini, A.; Lucisano, M.; Mariotti, M.; Estivi, L.; Hidalgo, A. Breadmaking Performance of Elite Einkorn (*Triticum monococcum* L. subsp. monococcum) Lines: Evaluation of Flour, Dough and Bread Characteristics. *Foods* **2023**, *12*, 1610. [CrossRef] [PubMed]
4. Huertas-García, A.B.; Guzmán, C.; Ibba, M.I.; Rakszegi, M.; Sillero, J.C.; Alvarez, J.B. Processing and Bread-Making Quality Profile of Spanish Spelt Wheat. *Foods* **2023**, *12*, 2996. [CrossRef] [PubMed]
5. Borrelli, G.M.; Menga, V.; Giovanniello, V.; Ficco, D.B.M. Antioxidants and Phenolic Acid Composition of Wholemeal and Refined-Flour, and Related Biscuits in Old and Modern Cultivars Belonging to Three Cereal Species. *Foods* **2023**, *12*, 2551. [CrossRef] [PubMed]
6. Šimora, V.; Ďúranová, H.; Brindza, J.; Moncada, M.; Ivanišová, E.; Joanidis, P.; Straka, D.; Gabríny, L.; Kačániová, M. Cornelian Cherry (*Cornus mas*) Powder as a Functional Ingredient for the Formulation of Bread Loaves: Physical Properties, Nutritional Value, Phytochemical Composition, and Sensory Attributes. *Foods* **2023**, *12*, 593. [CrossRef] [PubMed]
7. Bonacci, S.; Di Stefano, V.; Sciacca, F.; Buzzanca, C.; Virzì, N.; Argento, S.; Melilli, M.G. Hemp Flour Particle Size Affects the Quality and Nutritional Profile of the Enriched Functional Pasta. *Foods* **2023**, *12*, 774. [CrossRef] [PubMed]
8. Baiano, A.; la Gatta, B.; Rutigliano, M.; Fiore, A. Functional Bread Produced in a Circular Economy Perspective: The Use of Brewers' Spent Grain. *Foods* **2023**, *12*, 834. [CrossRef] [PubMed]
9. Viola, E.; Buzzanca, C.; Tinebra, I.; Settanni, L.; Farina, V.; Gaglio, R.; Di Stefano, V. A Functional End-Use of Avocado (cv. Hass) Waste through Traditional Semolina Sourdough Bread Production. *Foods* **2023**, *12*, 3743. [CrossRef]
10. Lebert, L.; Buche, F.; Sorin, A.; Aussenac, T. The Wheat Aleurone Layer: Optimisation of Its Benefits and Application to Bakery Products. *Foods* **2022**, *11*, 3552. [CrossRef]

Article

Chemical, Functional, and Technological Features of Grains, Brans, and Semolina from Purple and Red Durum Wheat Landraces

Afef Ladhari [1], Giandomenico Corrado [2], Youssef Rouphael [2], Francesca Carella [3], Giuseppina Rita Nappo [4], Cinzia Di Marino [5], Anna De Marco [6] and Domenico Palatucci [3,*]

[1] Laboratoire GREEN TEAM (LR17AGR01), Institut National Agronomique de Tunisie (INAT), Université de Carthage, 43 Avenue Charles Nicolle, Tunis 1082, Tunisia; afef.ladh@yahoo.fr

[2] Department of Agricultural Sciences, University of Naples Federico II, Via Università 100, 80055 Portici, Italy; giandomenico.corrado@unina.it (G.C.); youssef.rouphael@unina.it (Y.R.)

[3] Department of Biology, University of Naples Federico II, Via Cinthia 21, 80126 Naples, Italy; francesca.carella@unina.it

[4] Sannio Tech Consortium, Piazza San G. Moscati 5, 82030 Apollosa, Italy; pinanappo@gmail.com

[5] Department of Chemical Sciences, University of Naples Federico II, Via Cinthia 26, 80126 Naples, Italy; cdimarin@unina.it

[6] Department of Pharmacy, University of Naples Federico II, Via Montesano 49, 80131 Naples, Italy; anna.demarco@unina.it

* Correspondence: domenicopalatucci@libero.it

Citation: Ladhari, A.; Corrado, G.; Rouphael, Y.; Carella, F.; Nappo, G.R.; Di Marino, C.; De Marco, A.; Palatucci, D. Chemical, Functional, and Technological Features of Grains, Brans, and Semolina from Purple and Red Durum Wheat Landraces. *Foods* **2022**, *11*, 1545. https://doi.org/10.3390/foods11111545

Academic Editors: Donatella Bianca Maria Ficco and Grazia Maria Borrelli

Received: 13 April 2022
Accepted: 20 May 2022
Published: 25 May 2022

Abstract: A main reason of the increasing interest in cereal landraces is their potential to offer more diversified and functional staple food. For instance, landraces are an underexploited resource of pigmented varieties, appreciated for the high accumulation of phytochemicals with known health benefits. This study characterized the chemical, functional, and technological features of the bran, semolina, and grains of two durum wheat (*Triticum turgidum* L. subsp. *durum*, Desf.) landraces, named 'Purple' and 'Red' for their grain color, collected in Ethiopia and grown and sold in southern Italy as a niche product. Specifically, we analyzed the protein content, dry gluten, ash, total polyphenols, anthocyanins, proanthocyanidins, and specific phenolic acids. We also evaluated the antioxidant activity using DPPH- and ABTS-based methods. The two landraces had positive nutritional features, such as a high protein content, a rich and composite range of secondary metabolites (which include specific phenolic acids and anthocyanins), and antioxidant activities in all the fractions analyzed. The germplasm under investigation therefore has a well-justified potential to yield functional products and to diversify durum wheat-based foods.

Keywords: durum wheat; diversity; pigmented cereals; phytochemicals; anthocyanins; antioxidant activity; protein; gluten

1. Introduction

Wheat is one of the first domesticated cereal plants and it has been globally cultivated for its grains since the dawn of civilization. In the last decade, its world production has increased, currently reaching 750 million tons [1], while the sowing area has fluctuated around 220 million hectares. China (17%) and India (12%) are the top producers and collectively the European Union produces around 15% of the world's total. In Italy, about 2 million hectares are cultivated, prevalently with durum wheat, for a production of 8 million tons [2]. The yield increase in the last century is the joint result of different factors, with plant breeding having a significant role in shaping the morphological and technological features of contemporary varieties [3]. These are characterized by a reduced height, a more efficient assimilate partitioning, diminished sensitivity to photoperiod, adaptability to certain agronomic conditions, and resistance to specific races of fungal pathogens [3], traits that are expected to be absent in the old varieties of wheats [4].

Durum wheat (*Triticum turgidum* L. subsp. *durum*, Desf.) is a tetraploid species (AABB; 2n = 4x = 28) with better tolerance to drought and heat than the hexaploid common wheat (*Triticum aestivum* L; AABBDD; 2n = 6x = 42). Durum wheat, also known as hard, pasta, or macaroni wheat, is mainly cultivated in Mediterranean countries, North America, Argentina, and eastern Europe. In Italy, the production is principally located in southern regions such as Apulia and Sicily. Durum wheat is central to the gastronomy of Mediterranean countries because it is employed to produce pasta and couscous, as well as bulgur, puddings, pastries, freekeh, kishk, and other traditional dishes. An important aesthetic and commercial feature of durum wheat semolina is the color. The typical yellow-amber pigmentation is predominantly due to lipophilic carotenoids within the kernel, mainly lutein [5]. Nonetheless, anthocyanins are also another class of pigmented phytochemicals that can be present in high amount in the grains of some wheat varieties. According to the quantity and type, this class of water-soluble pigments can give rise to wheat grains with colors ranging from red to purple [5].

In recent years, the scientific interest and appreciation of the quality of traditional wheat varieties has increased for a more sustainable low-input production of grain, as germplasm with an enhanced phytochemical profile, and as a source of adaptive traits in the face of climate change [6,7]. For instance, considering that wheat is a staple food in several countries, anthocyanin-rich grains can be used to produce a wide range of foods with enhanced nutraceutical and pharmaceutical properties. Old varieties are also gaining popularity to satisfy consumer demand for regional crop production and food manufacturing, to diversify the dietary basket, and to provide commercially novel products richer in health-promoting ingredients [7,8]. Regrettably, the compositional properties of anthocyanin-rich grains of landraces, as well as old durum wheat varieties, have not yet been fully acknowledged, not only if compared with soft wheat varieties, but also with old species such as einkorn, emmer, and spelt [5,9–12].

The aim of this research was to explore the chemical, functional, and technological features (such as the contents of proteins; dry gluten; gluten index; ash; total polyphenols; antioxidant activity) of two differently pigmented *Triticum turgidum* landraces. Moreover, we quantified major anthocyanins (Delphinidin 3-glucoside, Delphinidin 3-rutinoside, Cyanidin 3-glucoside, Petunidin 3-glucoside, Peonidin 3-glucoside, and Malvidin 3-glucoside), anthocyanidins (i.e., cyanidin, delphinidin, malvidin, peonidin, and petunidin), and specific phenolic acids (ferulic acid, *p*-hydroxybenzoic acid, vanillic acid, and *p*-coumaric acid). These two landraces were named 'Purple' and 'Red' according to the color of the grain and were originally collected from the Oromya region, one of regional states of Ethiopia. Cereal landraces are mainly evaluated as a source of inheritable traits that may favor local adaptation and productivity in sustainable agriculture [13], but they also have desirable characteristics related to food quality and nutritional benefits. Therefore, our detailed characterization contributes to demonstrate the value of durum wheat landraces as a rich source of primary and secondary metabolites of interest for human health and nutrition.

2. Materials and Methods

2.1. Materials

We analyzed two durum wheat (*Triticum turgidum* L. subsp. *durum*, Desf.) landraces named 'Purple' and 'Red'. They are maintained and multiplied by Agrismarter (Foggia, Italy), marketed by the company Granomischio (Foggia, Italy), and cultivated in areas near the Daunian Mountains (Apulia region, Italy). Grains (i.e., intact caryopses), bran (i.e., kernel components except the flour fraction at the given extraction rate, ~65–75%), and semolina were provided also by the Granomischio company in September 2021. For their analysis, grains were ground to a fine powder using a blender stored at −20 °C. Semolina and bran were processed by a professional milling company with a multi-pass roller system. Briefly, grains were purified (mainly to remove low-density particles), surface cleaned (to remove impurities and possible abiotic and biotic contamination), pearled/scoured, milled, and purified/sieved to semolina.

Chemicals, analytical grade reagents, and standards of phenolic compounds were obtained from Sigma-Aldrich (Milan, Italy).

2.2. Determination of Dry Gluten

Gluten was extracted from semolina ($n = 5$) according to previously described procedure [14]. Twenty grams of semolina were suspended in 12.5 mL of a 4% monosodium/disodium phosphate buffer at pH of 6.8, diluted to a ratio of 1:40 with a 2% NaCl solution. After 30 min, the dough was washed using a Glutomatic System 2200 (Perkin Elmer, Turin, Italy) with the NaCl solution to remove soluble proteins and starch. The pure gluten obtained was dried in an oven (Heraeus T6200, Progitec, Sabaudia, LT, Italy). The gluten content was expressed as dry gluten per 100 g of material.

2.3. Determination of Gluten Index

The determination of the gluten index comprised three steps: (i) the gluten extraction and quantification of the wet gluten; (ii) the centrifugation of the wet gluten; and (iii) the calculation of the gluten index, according to standard procedures [14] using a Gluten Index 2100 centrifuge (Bastak, Ankara, Turkey). The gluten index is the percentage ratio of the wet gluten remaining on the sieve (after centrifugation) to the total wet gluten.

2.4. Determination of Total Protein Content

Total proteins were quantified using the Kjeldahl method. Briefly, samples (1 g) were digested in 15 mL of 98% H_2SO_4 in the presence of a catalyst (K_2SO_4:$CUSO_4$, 9:1 *w/w*). Then, 50 mL of 40% NaOH (*w/v*) was added to covert the released ammonium into ammonia, which was distilled and collected in a flask containing a known amount of excess acid (0.1 M HCl). The excess acid was back-titrated with 0.1 M NaOH. The protein content refers to 100 g of substance and is calculated using the following formula:

$$\text{Proteins (\%)} = [100 \times (V_{NaOH} \times C_{NaOH} - V_{HCl} \times C_{HCl}) \times 14.0067 \times 5.70]/g$$

where V_{NaOH} are the liters of NaOH; C_{NaOH} is the molar concentration of NaOH; V_{HCl} are the liters of HCl; C_{HCl} is the molar concentration of HCl; 14.0067 is the atomic weight of nitrogen; 5.70 is the conversion factor for proteins; and g is the weight of the sample in grams [15].

2.5. Determination of Ash

The quantification of ash was performed essentially as described [16]. Briefly, a weighed sample of semolina (5–10 g) was placed in a platinum capsule and heated at 550–590 °C in the muffle furnace Srefo R-1905 (Zhuhai Refine Zhizao, Guangdong, China) until light gray ash was obtained (5 h). The weight of the ash refers to 100 g of dry matter. The measurement was carried out on five replicates.

2.6. Phenolic Extraction

The phenolic extracts were prepared as reported [17]. The steps of the approach are summarized in Figure 1. About 1.5 g of wheat material was pulverized and suspended in 30 mL of a methanol/hydrochloric acid solution (99:1, *v/v*). The mixture was stirred for 30 min at room temperature. The suspension was then centrifuged (PK121R Multispeed, ALC International, Milan, Italy) at 10,000 rpm for 5 min at room temperature. The sediment was extracted five times.

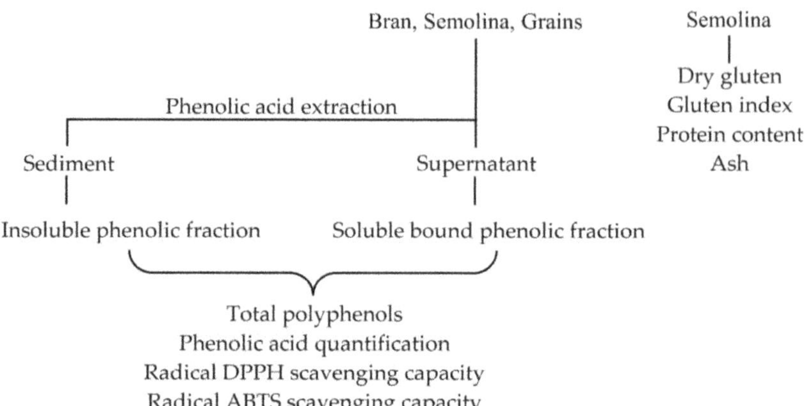

Figure 1. Flowchart of the process used for the extraction and analysis of the wheat material.

2.7. Soluble Phenolic Fraction

Two milliliters of the acidified methanolic extract were dried using a rotavapor (Rotavapor RE 111, Buchi, Switzerland) at 30 °C. The dried material was then suspended in 20 mL of 3M KOH and stirred for about three hours at room temperature. The solution was acidified with 1M HCl to pH 3 and extracted three times with a 1:1 (v/v) mixture of petroleum ether/ethyl acetate. Extracts were treated with anhydrous sodium sulfate, filtered on a Whatman paper Grade 1, dried under a light stream of nitrogen, and resuspended in methanol.

2.8. Insoluble Bound Phenolic Fraction

The sediment (Figure 1) was suspended in 20 mL of 3M KOH at room temperature for about three hours, under continuous stirring, acidified with 1M HCl to pH 3, and extracted three times with a 1:1 (v/v) petroleum ether/methylene chloride solution. Extracts were treated as in paragraph 2.2 and resuspended in methanol. All extractions were performed avoiding direct light to minimize photo-oxidation.

2.9. Total Polyphenols

The total phenol content in the soluble and insoluble bound extracts and in the different organic extracts (Figure 1) was quantified with the Folin–Ciocalteu method using gallic acid as a standard as previously reported [18]. The extracts were solubilized in 5 mL of methanol, with the aid of a sonicator. One milliliter was taken and filtered on a Phenex filter (0.45 μm) and diluted to a 10 mL final volume. Samples ($n = 5$) and solutions of the standard were tested with the colorimetric method and absorbance was read at 765 nm [19]. Specifically, 0.2 mL of the solution of the sample to be analyzed (or of the standard, or Milli Q water, in the case of the blank) was added with 0.8 mL of Milli Q water and 0.2 mL of the Folin–Ciocalteu reagent. The solution was incubated for 5 min and then another 0.8 mL of Milli Q water and 2 mL of an aqueous 8% Na_2CO_3 solution were added. A standard curve was built with gallic acid at the 4.8, 9.6, 48, 96, 240, and 480 μg/mL concentrations. The minimum threshold for accepting the calibration curve was r = 0.97. Results were expressed as milligrams of gallic acid equivalents.

2.10. Phenolic Acid Quantification by HPLC

Both phenolic extracts (Figure 1) were analyzed by HPLC in accordance with the already published protocols [20,21]. A Shimadzu LC-8A HPLC instrument (Shimadzu, Milan, Italy) was used with a 2.6 mm 100 Å (100 × 4.6 mm) Kinetex reverse phase column. The eluent phase consisted of a mixture of A (2% AcOH in water, v/v) and B (methanol), with a constant flow of 1.2 mL/min and a wavelength of the UV detector set at 280 nm.

The gradient, in terms of eluent B, was 10% at time 0, 20% at 10 min, 25% at 15 min, and 30% at 30 min. The presence of ferulic acids, *p*-hydroxybenzoic acid, and vanillic and coumaric acids was measured through the corresponding calibration lines obtained from the corresponding standard samples commercially available from Sigma-Aldrich. The calibration curves were linear in the concentration intervals considered. In particular, the detection limits were equal to 12, 0.08, 0.1, and 1.11 µg/mL, respectively, for the four aforementioned phenolic acids. The analyses were performed five times; the results were expressed as micrograms per kilogram of wheat on a dry matter (DM) basis.

2.11. Determination of Anthocyanins Content

Total anthocyanin extract was prepared essentially as reported [22] and quantification was carried out with a spectrophotometric method [23], using catechin as standard with concentration ranging from 2 to 200 µg/mL. Results were expressed as µg catechin equivalents/g dry weight material. The equation obtained from the standard curve is y = 0.0018x + 0.0146, where y is absorbance at 535 nm and x is concentration of catechin standard.

2.12. Determination of Proanthocyanidin Content

Proanthocyanidin extract was prepared using previously published procedures [23] and the proanthocyanidin content was determined also as already described [24], using catechin as standard with concentrations ranging from 100 to 1000 µg/mL. Results were expressed as µg catechin equivalents/g dry weight material. The equation obtained from the standard curve is y = 0.0023x + 0.0187, where y is absorbance at 510 nm and x is concentration of catechin standard.

2.13. Radical DPPH Scavenging Capacity

The DPPH· antioxidant activity of the material under investigation (Figure 1) was evaluated using already published procedures [25] with minor modifications. One milliliter of extraction solvent with different extract dilutions was added to two mL of DPPH in methanol (5×10^{-5} M). The reaction was carried out at 25 °C for 30 min. After half an hour, the absorbance value reached a constant value, which was used to calculate the percentage of residual DPPH. Radical reduction by antioxidants was monitored by measuring the absorbance at 517 nm using a Perkin Elmer Lambda 7 spectrophotometer (Beckman, Brea, CA, USA). Five extracts were analyzed for each sample, each with four different dilutions. A regression line was also calculated for the reference antioxidant Trolox, with concentrations ranging from 3 to 50 µM. The antioxidant activity was expressed as the ratio between the I_{50} of Trolox and the I_{50} of the sample, that is, micromoles of Trolox equivalent (TE) per gram of DM.

2.14. Radical ABTS Scavenging Capacity

The ABTS antioxidant activity (Figure 1) was evaluated as already published [26], quantifying the ability of natural extracts to convert the radical cation $ABTS^{+}$, generated from the corresponding acid using as oxidizing agent sodium or potassium persulfate ($K_2S_2O_8$ or $Na_2S_2O_8$) in its neutral form. For the ABTS assay, the antioxidant capacity was also expressed as the Trolox equivalent antioxidant capacity (TEAC), a unit of measurement defined as the quantity of Trolox needed to obtain the same antioxidant activity as the sample (micromoles of TE per g of sample).

2.15. Statistical Analysis

Data are reported as mean value ± standard deviation (SD). The normality of the data distribution was assessed by the Shapiro-Wilk test. The independent Student's *t*-test was employed for mean separation considering as the threshold of statistical significance a *p*-value lower than 0.05. Calculations were performed with the SigmaPlot 12.2 software (Systat Software, San Jose, CA, USA).

3. Results and Discussion

3.1. Analysis of Dry Gluten, Gluten Index, Protein Content, and Ash

The two wheat varieties analyzed had a similar percentage of dry gluten, around 12.5–12.8% (Table 1). Nonetheless, *T. durum* 'Purple' had a higher gluten index (GI of 37 against 31 of the 'Red'), values that can be considered low in durum wheat [27–29]. The GI is a widely accepted parameter to express gluten strength, and it is considered highly inheritable [30]. Hence, the observed difference reveals distinct technological features of the two landraces. Nonetheless, the relationship of the GI with the protein content should not be thought as linear [29,31]. For both varieties, the latter was not far from the upper limit for durum wheat, usually from 7% to 18%, with an average of 12%. The protein content and the gluten strength are the main features of the starting material that determine the quality of pasta [28]. Even so, the protein content is often considered more important than the strength of the gluten, being more closely correlated with the positive features of dried pasta, although this relationship can be affected by the processing method and the gluten composition [32]. The semolina of 'Purple' and 'Red' had ash contents of 1.25 and 1.35%, respectively. The ash content in the grain is under genetic and environmental control. It is typically influenced by mineral fertilization and positively correlated with the protein content [33]. A low ash content in milled durum grains is considered a quality feature, although it does not significantly affect the industrial performance of the semolina. In Italy, as in other countries, the ash content in common and durum wheat is regulated by law (D.P.R. 9 February 2001, n. 187). The observed values exceed the threshold for "*semola*" and are within limits for "*semolato*". It should be added that the ash content lowers with sequential debrannings. For instance, in a group of 11 Italian varieties encompassing traditional and contemporary durum wheat cultivars, the ash content reached an acceptable level for "*semola*" after five successive debranning treatments [34].

Table 1. Protein, dry gluten and ash content and gluten index in semolina of durum wheat 'Purple' and 'Red' landraces. Values are reported as mean ± SD ($n = 5$). Significant differences between the two varieties for bran, semolina, or grain are indicated by lowercase letters (t-test, $p < 0.05$).

Parameters (%)	'Purple'	'Red'
Protein	15.20 ± 0.10 b	16.80 ± 0.20 a
Dry gluten	12.80 ± 0.88	12.50 ± 0.06
Ash	1.25 ± 0.02 b	1.35 ± 0.01 a
Gluten index	37.00 ± 0.20 a	31.00 ± 0.60 b

3.2. Polyphenol Content

The grain, semolina, and bran of both 'Purple' and 'Red' were subjected to an evaluation of the total polyphenol content because these are a wide group of chemically diverse secondary metabolites that are generally known for their positive effects on human health [35,36]. Cereals, as many other plant species, contain different phenolic compounds, and the most important are the derivatives of the cinnamic acid (i.e., coumaric, caffeic, and ferulic acids), flavonoids, and lignans [37]. It is believed that their function is predominantly non-nutritive [37].

The quantification of total polyphenols indicated that the bran of the 'Purple' landrace had a significantly higher amount (almost double) of both phenolic fractions, while differences were not significant for the insoluble fractions of the semolina and grains (Table 2).

Table 2. Soluble and insoluble polyphenolic fraction in 'Purple' and 'Red' landraces. Values are reported as mean ± SD (n = 5). Significant differences between the two varieties for bran, semolina, or grain are indicated by lowercase letters (t-test, $p < 0.05$).

Polyphenols (µg/g DM Gallic Acid Equivalent)	Bran		Semolina		Grain	
	'Purple'	'Red'	'Purple'	'Red'	'Purple'	'Red'
Soluble phenolic fraction	112 ± 21 a	64 ± 4 b	31 ± 4 b	49 ± 3 a	103 ± 15 a	65 ± 5 b
Insoluble bound phenolic fraction	2391 ± 471 a	1430 ± 284 b	751 ± 105	849 ± 112	2380 ± 455	1840 ± 329

Specifically, 'Purple' bran and grains had an almost double phenolic content (112 and 103 µg/g DM of the equivalents of gallic acid, respectively) compared to those of the 'Red' (64 and 65 µg/g DM of gallic acid equivalents, respectively), while the semolina content of 'Red' (49 µg/g DM of gallic acid equivalents) was significantly higher than that of the semolina of 'Purple' (31 µg/g DM of gallic acid equivalents). The same was true for the insoluble bound phenolic fractions. In 'Purple', the phenolic content in the bran and grain was 40% and 23% higher compared to 'Red'. This can be explained considering that the phenolic content in wheat grains is mainly concentrated in the bran with a limited contribution of the organs and tissue originating the milled semolina. Nonetheless, the two varieties differed in the variation of the ratio insoluble/soluble polyphenols across the analyzed material. This parameter little varied in 'Purple' (coefficient of variation, CV: 7.9%), while the CV was 24.3% for 'Red', whose semolina had the lowest insoluble/soluble polyphenol ratio.

3.3. Determination of Phenolic Acids

HPLC analysis of the semolina revealed several phenolic derivatives, with four compounds (i.e., ferulic, p-hydroxybenzoic, vanillic, and p-coumaric acids) being the most abundant. The analysis of the two landraces indicated a similar profile in terms of quantity and rank of the chemical compounds, with the ferulic acid always present in a predominant amount, followed, for both varieties, by the vanillic acid or the p-coumaric acid in the soluble or insoluble fraction, respectively (Table 3).

Table 3. Identification and quantification of the acids in the soluble and insoluble bound phenolic fraction in semolina of 'Purple' and 'Red' based on retention times and calibration lines for comparison with the corresponding commercial products. Values are reported as mean ± SD (n = 5). Mean values between the two varieties were not statistically different (t-test, $p \geq 0.05$).

Soluble Hpenolic Fraction (mg/kg DM)	'Purple'	'Red'
Ferulic acid	17.9 ± 1.10	18.3 ± 1.05
p-Hydroxybenzoic acid	3.1 ± 0.28	3.3 ± 0.27
Vanillic acid	7.2 ± 0.33	7.4 ± 0.38
p-Coumaric acid	2.2 ± 0.32	2.5 ± 0.30
Insoluble Bound Phenolic Fraction (mg/kg DM)		
Ferulic acid	625 ± 39.3	639 ± 41.4
p-Hydroxybenzoic acid	2.1 ± 0.29	2.3 ± 0.28
Vanillic acid	3.1 ± 0.35	3.4 ± 0.29
p-Coumaric acid	18.5 ± 2.35	19.2 ± 2.26

In the 'Red' variety, the ferulic acid was 639 mg/kg DM and 18.3 mg/kg DM in the insoluble fraction and in the soluble fraction, respectively (Table 3). The ferulic acid is considered the predominant free- and bound-form of polyphenols in cereals, especially (brown) rice and corn [38]. In durum wheat, the ferulic acid contents are slightly influenced by the environment in normal agronomic conditions, and it is mostly influenced by the genotype and altered by abiotic stress [39]. In the soluble fraction, of the three other quantified phenolic acids, the most abundant was vanillic acid, with concentrations in the range of 7.2–7.4 µg/kg DM, followed by *p*-hydroxybenzoic acid, with concentrations in the range of 3.1–3.3 µg/kg DM. The *p*-coumaric acid was the less abundant constituent, with concentrations in the range of 2.2–2.5 µg/kg DM. In the insoluble fraction of the three other quantified phenolic acids, the most abundant was *p*-coumaric acid, with concentrations in the range of 18.5–19.2 µg/kg DM, and with much lower concentrations of vanillic acid (3.1–3.4 µg/kg DM) and *p*-hydroxybenzoic acid (2.1–2.3 mg/kg DM). Overall, the *p*-hydroxybenzoic and the vanillic acids were present in higher quantities in the soluble fractions of the semolina, while the amount of ferulic and *p*-coumaric acids was larger than insoluble fraction. Finally, we did not observe differences between the two varieties in the acids in the soluble and insoluble bound phenolic fraction of the semolina.

3.4. Anthocyanins and Proanthocyanidins Content

Anthocyanins and proanthocyanidins (expressed as the micrograms of catechin equivalents per gram of DM) were also quantified in the grain, semolina, and bran of 'Purple' and 'Red' (Table 4).

Table 4. Anthocyanins and proanthocyanidins in the 'Purple' and 'Red' landraces. Values are reported as mean $\pm$ SD ($n = 5$). Significant differences between the two varieties for bran, semolina, or grain are indicated by lowercase letters (t-test, $p < 0.05$).

Phenolic Components	Bran		Semolina		Grain	
(µg/g DM Catechin Equivalent)	'Purple'	'Red'	'Purple'	'Red'	'Purple'	'Red'
Anthocyanins	72.9 $\pm$ 1.1 a	36.3 $\pm$ 0.4 b	12.5 $\pm$ 0.8 b	16.8 $\pm$ 0.1 a	116.6 $\pm$ 1.9 a	39.2 $\pm$ 0.8 b
Proanthocyanidins	1530 $\pm$ 133 a	1031 $\pm$ 33 b	244 $\pm$ 22 b	466 $\pm$ 111 a	3437 $\pm$ 377 a	1807 $\pm$ 166 b

The content of anthocyanins in the grain of 'Purple' was higher than in 'Red', almost double in the bran (72.9 vs. 36.3 µg/g DM) and triple in the grain (116.6 vs. 39.2 µg/g DM) (Table 4). In contrast, the semolina of 'Red' contained 25% more anthocyanin than 'Purple' (16.8 vs. 12.5 µg/g DM, respectively). A similar trend was observed for proanthocyanidins (polymers or oligomers of anthocyanidin). Their content was 33% higher in the bran of the 'Purple' compared to that of the 'Red' (1530 vs. 1031 µg/g DM of catechin equivalents, respectively) and 47% more in the grain of the 'Purple' than that of the 'Red' (3437 vs. 1807 µg/g DM of catechin equivalents, respectively). On the other hand, the amount of proanthocyanidins in the semolina was almost double (+47%) in the 'Red' compared to the 'Purple' (466 vs. 244 µg/g DM of catechin equivalents, respectively).

Anthocyanins are a class of water-soluble pigments belonging to the flavonoid family. They show a range of pharmacological activities because of their antioxidant and anti-inflammatory properties, with potential therapeutic benefits [40]. In durum wheat, anthocyanins mainly accumulate in the pericarp and aleurone [5]. This may explain why the less colored 'Red' variety yielded semolina with a significantly higher content of anthocyanins and proanthocyanidins. Anthocyanin-rich grains can be then used to produce functional foods and considering that these compounds are in the less noble coat of the grain, it has been also proposed that grains can be also exploited as a natural source to extract these pigments [5].

The determination of the anthocyanin compounds by HPLC revealed the presence of significant differences between the material and the variety (Table 5). Overall, the cyanidin 3-glucoside was the most abundant compound, followed by peonidin 3-glucoside. Quantitative differences between the varieties were most pronounced for the bran, with the 'Purple' having on average a threefold higher amount than 'Red' of the detected molecules. Comparing the material, as expected, anthocyanins were in much lower quantities in the semolina, with the two major anthocyanins detected only in the 'Purple' variety. Interestingly, the analysis of the grain revealed both qualitative and quantitative differences. Specifically, the differences in relative terms between the two varieties were more limited compared to the bran. Moreover, only the grain of the 'Red' variety contained detectable amounts of delphinidin 3-glucoside and delphinidin 3-rutinoside.

Table 5. Anthocyanin composition in the 'Purple' and 'Red' landraces (mean $\pm$ SD; n = 5) determined by HPLC. Values are reported as mean $\pm$ SD (n = 5). Significant differences between the two varieties for bran, semolina, or grain are indicated by lowercase letters (t-test, $p < 0.05$).

Anthocyanins (µg/kg DM)	Bran		Semolina		Grain	
	'Purple'	'Red'	'Purple'	'Red'	'Purple'	'Red'
Delphinidin 3-glucoside	-	-	-	-	-	1.20 $\pm$ 0.10
Delphinidin 3-rutinoside	-	-	-	-	-	2.66 $\pm$ 0.14
Cyanidin 3-glucoside	5.09 $\pm$ 0.11 a	1.15 $\pm$ 0.05 b	2.05 $\pm$ 0.15	-	7.32 $\pm$ 0.48 a	2.39 $\pm$ 0.03 b
Petunidin 3-glucoside	1.53 $\pm$ 0.11 a	0.57 $\pm$ 0.07 b	-	-	1.55 $\pm$ 0.05 a	1.29 $\pm$ 0.01 b
Peonidin 3-glucoside	2.05 $\pm$ 0.05 a	0.86 $\pm$ 0.06 b	0.89 $\pm$ 0.05	-	3.35 $\pm$ 0.15 a	1.68 $\pm$ 0.04 b
Malvidin 3-glucoside	-	-	-	-	-	-

Cyanidin 3-glucoside is often the most abundant anthocyanins in colored cereals (e.g., rice and corn), as well as in most of the plants [40]. For purple common wheat, cyanidin 3-glucoside, peonidin 3-glucoside, and cyanidin 3-galactoside have been described as the most abundant compounds [41,42]. In blue common wheat, cyanidin 3-glucoside is predominant, with pelargonidin 3-glucoside and cyanidin 3-galactoside present in lower amounts [43]. Although intra-varietal differences in colors are sufficiently explained by variation in anthocyanins [44], it is not straightforward to correlate the color of a variety with the type, number, and quantity of pigments, also considering that the influence of the anthocyanins on the plant tissue hue (and tint) is determined by various factors besides their total amount and ratio [45,46]. It is therefore interesting that the grain of the 'Red' variety contained delphinidins in a low amount compared to other pigments, while these were not detected in the 'Purple'. These anthocyanins are typically associated with dark grains [5]. For instance, in a blue common wheat, delphinidin 3-glucoside and delphinidin 3-rutinoside accounted for 69.3% of the detected anthocyanins [42]. Nonetheless, they were also not found in a purple common wheat [42]. In addition to wheats also in purple rye grains cyanidin-3-glucoside is the predominant anthocyanin, followed by peonidin-3-glucoside [47,48] as in our 'Purple'.

We also quantified major anthocyanidins (i.e., cyanidin, delphinidin, malvidin, peonidin, and petunidin) in the 'Purple' and 'Red' bran, semolina, and grain (Table 6) because these are typical of colored wheat grains [10]. In grains, cyanidin was the most abundant aglycones for all the material, followed by delphinidins. Other works indicated that cyanidin also the main aglycone in purple common and durum wheat varieties, but it was followed by peonidin [49]. Petunidin was detected in smaller quantities only in the two brans, while peonidin and malvidin only in the grains of the 'Purple'.

Table 6. Anthocyanidin composition in the 'Purple' and 'Red' (mean $\pm$ SD; n = 5) determined by HPLC. Values are reported as mean $\pm$ SD (n = 5). Significant differences between the two varieties for bran, semolina, or grain are indicated by lowercase letters (t-test, $p < 0.05$).

Anthocyanidins (µg/kg DM)	Bran		Semolina		Grain	
	'Purple'	'Red'	'Purple'	'Red'	'Purple'	'Red'
Delphinidin	1.75 $\pm$ 0.05 a	0.91 $\pm$ 0.01 b	1.15 $\pm$ 0.05	-	3.15 $\pm$ 0.45 a	0.87 $\pm$ 0.03 b
Cyanidin	1.87 $\pm$ 0.03	1.60 $\pm$ 0.20	1.27 $\pm$ 0.07 a	0.56 $\pm$ 0.14 b	5.45 $\pm$ 0.35 a	1.47 $\pm$ 0.03 b
Petunidin	-	-	-	-	1.68 $\pm$ 0.08 a	0.45 $\pm$ 0.05 b
Peonidin	-	-	-	-	1.35 $\pm$ 0.05	-
Malvidin	-	-	-	-	1.15 $\pm$ 0.75	-
Total anthocyanins (Anthocyanins + Proanthocyanidins)	12.29 $\pm$ 0.15 a	5.09 $\pm$ 0.37 b	5.36 $\pm$ 0.32 a	0.56 $\pm$ 0.14 b	25.00 $\pm$ 0.40 a	12.01 $\pm$ 0.09 b

The total anthocyanin content was higher in the 'Purple' than in the 'Red'; precisely, it was just over double of that contained in the bran (25 vs. 12.01 µg/kg DM), more than double of that contained in the grain (12.29 vs. 5.09 µg/kg DM), and almost 10 times that contained in the semolina (5.36 vs. 0.56 µg/kg DM).

3.5. Antioxidant Activity

Tests for the evaluation of antioxidant activity show the highest values for the grain in both varieties compared to bran and semolina. Moreover, in both assays the highest values were observed for the 'Purple' variety compared to that 'Red' (Table 7). Mean values were not significantly different between the semolina of the two varieties. In the brans, were higher values in the DPPH test (respectively, ABTS test) were recorded for the 'Purple' (resp. 'Red'). As previously reported [50], the highest antioxidant capacity was observed in whole grains, which contain more bioactive compounds of health interest, such as insoluble fiber, phenolic acids, and alkylresorcinols [51].

Table 7. Antioxidant activity (µmol Trolox equivalent/g DM) in 'Purple' and 'Red' landraces. Values are reported as mean $\pm$ SD ($n = 5$). Significant differences between the two varieties for bran, semolina, or grain are indicated by lowercase letters (t-test, $p < 0.05$).

Assay	Bran		Semolina		Grain	
	'Purple'	'Red'	'Purple'	'Red'	'Purple'	'Red'
DPPH	9.95 ± 0.30 a	8.47 ± 0.44 b	8.14 ± 0.04	8.03 ± 0.01	11.07 ± 0.59 a	9.85 ± 0.01 b
ABTS	1.75 ± 0.30 b	2.70 ± 0.12 a	1.67 ± 0.24	1.68 ± 0.10	5.29 ± 0.25 a	4.33 ± 0.40 b

4. Conclusions

The characterization of *Triticum durum* 'Purple' and 'Red', two varieties imported from Ethiopia and grown in southern Italy, highlighted their distinctive features such as an above-average protein content, which should positively influence the pasta-making quality. The ferulic acid was particularly abundant among phenolic acids in both the soluble and insoluble phenolic fractions of the grain, semolina, and bran. Even if related to the analysis of two landraces, our work also revealed the qualitative and quantitative diversity in anthocyanin content in durum wheat. It should be added that the material under investigation derives from a non-mass-production system in which grains are processed with a multi-pass methodology using a rolling mill. These systems generate less heat during grinding also because multiple passes allow the achievement of size reduction more gradually [52]. In the future, it may be worth investigating to what extent the high antioxidant activities in the semolina may be affected by the milling method. The lower gluten index, high protein level, rich and composite range of secondary metabolites, along with the antioxidant activities, indicate that the germplasm under investigation has interesting features for the niche market of functional durum wheat products in specific geographical areas as an alternative to the mass-produced Italian goods required for international markets.

Author Contributions: Conceptualization, D.P. and A.D.M.; methodology, D.P., C.D.M. and A.L.; formal analysis, G.R.N., F.C., C.D.M. and D.P.; investigation, F.C. and D.P.; data curation, Y.R., G.C., D.P. and A.D.M.; designed the research study, supervision, writing—original draft preparation and the last version of the manuscript, Y.R., G.C., A.D.M. and D.P. All authors have read and agreed to the published version of the manuscript.

Funding: This research received no external funding.

Data Availability Statement: The data are contained within the article.

Acknowledgments: We thank the non-profit organization "Associazione Italiana per la Promozione delle Ricerche sull'Ambiente e la Salute umana" (AIPRAS) for its support.

Conflicts of Interest: The authors declare no conflict of interest.

References

1. Hasanuzzaman, M.; Nahar, K.; Hossain, M.A. *Wheat Production in Changing Environments*; Spinger: Dordrecht, The Netherlands, 2019.
2. Senatore, M.; Ward, T.; Cappelletti, E.; Beccari, G.; McCormick, S.; Busman, M.; Laraba, I.; O'Donnell, K.; Prodi, A. Species diversity and mycotoxin production by members of the *Fusarium tricinctum* species complex associated with *Fusarium* head blight of wheat and barley in Italy. *Int. J. Food Microbiol.* **2021**, *358*, 109298. [CrossRef]
3. Lupton, F. *Wheat Vreeding: Its Scientific Basis*; Spinger: Dordrecht, The Netherlands, 2014.
4. Royo, C.; Nachit, M.M.; Di Fonzo, N.; Araus, J.L.; Pfeiffer, W.H.; Slafer, G.A. *Durum Wheat Breeding: Current Approaches and Future Strategies, Volumes 1 and 2*; Food Products Press: Binghamton, NY, USA, 2005; p. 525.
5. Ficco, D.B.; Mastrangelo, A.M.; Trono, D.; Borrelli, G.M.; De Vita, P.; Fares, C.; Beleggia, R.; Platani, C.; Papa, R. The colours of durum wheat: A review. *Crop Pasture Sci.* **2014**, *65*, 1–15. [CrossRef]
6. Arzani, A.; Ashraf, M. Cultivated ancient wheats (*Triticum* spp.): A potential source of health-beneficial food products. *Comp. Rev. Food Sci. Food Safety* **2017**, *16*, 477–488. [CrossRef] [PubMed]
7. Cooper, R. Re-discovering ancient wheat varieties as functional foods. *J. Tradit. Complementary Med.* **2015**, *5*, 138–143. [CrossRef] [PubMed]
8. Dinu, M.; Whittaker, A.; Pagliai, G.; Benedettelli, S.; Sofi, F. Ancient wheat species and human health: Biochemical and clinical implications. *J. Nutr. Biochem.* **2018**, *52*, 1–9. [CrossRef]
9. Shoeva, O.Y.; Gordeeva, E.I.; Khlestkina, E.K. The regulation of anthocyanin synthesis in the wheat pericarp. *Molecules* **2014**, *19*, 20266–20279. [CrossRef]
10. Lachman, J.; Martinek, P.; Kotíková, Z.; Orsák, M.; Šulc, M. Genetics and chemistry of pigments in wheat grain–A review. *J. Cereal Sci.* **2017**, *74*, 145–154. [CrossRef]
11. Brandolini, A.; Hidalgo, A.; Gabriele, S.; Heun, M. Chemical composition of wild and feral diploid wheats and their bearing on domesticated wheats. *J. Cereal Sci.* **2015**, *63*, 122–127. [CrossRef]
12. Yilmaz, V.A.; Brandolini, A.; Hidalgo, A. Phenolic acids and antioxidant activity of wild, feral and domesticated diploid wheats. *J. Cereal Sci.* **2015**, *64*, 168–175. [CrossRef]
13. Newton, A.C.; Akar, T.; Baresel, J.P.; Bebeli, P.J.; Bettencourt, E.; Bladenopoulos, K.V.; Czembor, J.H.; Fasoula, D.A.; Katsiotis, A.; Koutis, K. Cereal landraces for sustainable agriculture. *Sustain. Agric.* **2011**, *2*, 147–186.
14. AACC. *Method 38-12.02*, 11th ed.; AACC International, Inc.: St. Paul, MN, USA, 2005.
15. Hayes, M. Measuring protein content in food: An overview of methods. *Foods* **2020**, *9*, 1340. [CrossRef]
16. Ismail, B.P. Ash content determination. In *Food Analysis Laboratory Manual*; Spinger: Dordrecht, Netherlands, 2017; pp. 117–119.
17. Abdel-Aal, E.-S.; Hucl, P.; Sosulski, F.; Graf, R.; Gillott, C.; Pietrzak, L. Screening spring wheat for midge resistance in relation to ferulic acid content. *J. Agric. Food Chem.* **2001**, *49*, 3559–3566. [CrossRef]
18. Singleton, V.L.; Orthofer, R.; Lamuela-Raventós, R.M. [14] Analysis of total phenols and other oxidation substrates and antioxidants by means of folin-ciocalteu reagent. In *Methods Enzymol.*; Elsevier: Amsterdam, The Netherlands, 1999; Volume 299, pp. 152–178.
19. Bastola, K.P.; Guragain, Y.N.; Bhadriraju, V.; Vadlani, P.V. Evaluation of standards and interfering compounds in the determination of phenolics by Folin-Ciocalteu assay method for effective bioprocessing of biomass. *Am. J. Anal. Chem.* **2017**, *8*, 416–431. [CrossRef]
20. Medina, M.B. Simple and rapid method for the analysis of phenolic compounds in beverages and grains. *J. Agric. Food Chem.* **2011**, *59*, 1565–1571. [CrossRef]
21. Dellagreca, M.; Fiorentino, A.; Izzo, A.; Napoli, F.; Purcaro, R.; Zarrelli, A. Phytotoxicity of secondary metabolites from *Aptenia cordifolia*. *Chem. Biodivers.* **2007**, *4*, 118–128. [CrossRef]
22. Hosseinian, F.S.; Li, W.; Beta, T. Measurement of anthocyanins and other phytochemicals in purple wheat. *Food Chem.* **2008**, *109*, 916–924. [CrossRef]
23. Siebenhandl, S.; Grausgruber, H.; Pellegrini, N.; Del Rio, D.; Fogliano, V.; Pernice, R.; Berghofer, E. Phytochemical profile of main antioxidants in different fractions of purple and blue wheat, and black barley. *J. Agric. Food Chem.* **2007**, *55*, 8541–8547. [CrossRef]
24. Sun, B.; Ricardo-da-Silva, J.M.; Spranger, I. Critical factors of vanillin assay for catechins and proanthocyanidins. *J. Agric. Food Chem.* **1998**, *46*, 4267–4274. [CrossRef]
25. Lavelli, V.; Hidalgo, A.; Pompei, C.; Brandolini, A. Radical scavenging activity of einkorn (*Triticum monococcum* L. subsp. monococcum) wholemeal flour and its relationship to soluble phenolic and lipophilic antioxidant content. *J. Cereal Sc.* **2009**, *49*, 319–321.
26. Re, R.; Pellegrini, N.; Proteggente, A.; Pannala, A.; Yang, M.; Rice-Evans, C. Antioxidant activity applying an improved ABTS radical cation decolorization assay. *Free Radic. Biol. Med.* **1999**, *26*, 1231–1237. [CrossRef]
27. Vida, G.; Szunics, L.; Veisz, O.; Bedő, Z.; Láng, L.; Árendás, T.; Bónis, P.; Rakszegi, M. Effect of genotypic, meteorological and agronomic factors on the gluten index of winter durum wheat. *Euphytica* **2014**, *197*, 61–71. [CrossRef]
28. Ames, N.; Clarke, J.; Marchylo, B.; Dexter, J.; Woods, S. Effect of environment and genotype on durum wheat gluten strength and pasta viscoelasticity. *Cereal Chem.* **1999**, *76*, 582–586. [CrossRef]
29. Oikonomou, N.; Bakalis, S.; Rahman, M.; Krokida, M. Gluten index for wheat products: Main variables in affecting the value and nonlinear regression model. *Int. J. Food Prop.* **2015**, *18*, 1–11. [CrossRef]

30. Clarke, F.; Clarke, J.; Pozniak, C.; Knox, R.; McCaig, T. Protein concentration inheritance and selection in durum wheat. *Can. J. Plant Sci.* **2009**, *89*, 601–612. [CrossRef]
31. Cecchini, C.; Bresciani, A.; Menesatti, P.; Pagani, M.A.; Marti, A. Assessing the rheological properties of durum wheat semolina: A Review. *Foods* **2021**, *10*, 2947. [CrossRef]
32. Ames, N.; Clarke, J.; Marchylo, B.; Dexter, J.; Schlichting, L.; Woods, S. The effect of extra-strong gluten on quality parameters in durum wheat. *Can. J. Plant Sci.* **2003**, *83*, 525–532. [CrossRef]
33. Peterson, C. Influence of cultivar and environment on mineral and protein concentrations of wheat flour, bran, and grain. *Cereal Chem.* **1986**, *63*, 118–186.
34. Fares, C.; Troccoli, A.; Di Fonzo, N. Use of friction debranning to evaluate ash distribution in Italian durum wheat cultivars. *Cereal Chem.* **1996**, *73*, 232–234.
35. Dykes, L.; Rooney, L. Phenolic compounds in cereal grains and their health benefits. *Cereal Foods World* **2007**, *52*, 105–111. [CrossRef]
36. Hodzic, Z.; Pasalic, H.; Memisevic, A.; Srabovic, M.; Saletovic, M.; Poljakovic, M. The influence of total phenols content on antioxidant capacity in the whole grain extracts. *Eur. J. Sci. Res.* **2009**, *28*, 471–477.
37. Salunkhe, D.; Jadhav, S.; Kadam, S.; Chavan, J.; Luh, B. Chemical, biochemical, and biological significance of polyphenols in cereals and legumes. *Crit. Rev. Food Sci. Nutr.* **1983**, *17*, 277–305. [CrossRef]
38. Van Hung, P. Phenolic compounds of cereals and their antioxidant capacity. *Crit. Rev. Food Sci. Nutr.* **2016**, *56*, 25–35. [CrossRef]
39. Lafiandra, D.; Masci, S.; Sissons, M.; Dornez, E.; Delcour, J.; Courtin, C.; Caboni, M.F. Kernel components of technological value. In *Durum Wheat Chemistry and Technology*, 2nd ed.; Sissons, M., Marchylo, B., Abecassis, J., Eds.; AACC International Inc.: St. Paul, MN, USA, 2012; pp. 85–124.
40. Khoo, H.E.; Azlan, A.; Tang, S.T.; Lim, S.M. Anthocyanidins and anthocyanins: Colored pigments as food, pharmaceutical ingredients, and the potential health benefits. *Food Nutr. Res.* **2017**, *61*, 1361779. [CrossRef]
41. Abdel-Aal, E.-S.M.; Hucl, P. Composition and stability of anthocyanins in blue-grained wheat. *J. Agric. Food Chem.* **2003**, *51*, 2174–2180. [CrossRef]
42. Abdel-Aal, E.-S.M.; Young, J.C.; Rabalski, I. Anthocyanin composition in black, blue, pink, purple, and red cereal grains. *J. Agric. Food Chem.* **2006**, *54*, 4696–4704. [CrossRef]
43. Hu, C.; Cai, Y.-Z.; Li, W.; Corke, H.; Kitts, D.D. Anthocyanin characterization and bioactivity assessment of a dark blue grained wheat (*Triticum aestivum* L. cv. Hedong Wumai) extract. *Food Chem.* **2007**, *104*, 955–961. [CrossRef]
44. Biolley, J.; Jay, M. Anthocyanins in modern roses: Chemical and colorimetric features in relation to the colour range. *J. Exp. Bot.* **1993**, *44*, 1725–1734. [CrossRef]
45. Fossen, T.; Cabrita, L.; Andersen, O.M. Colour and stability of pure anthocyanins influenced by pH including the alkaline region. *Food Chem.* **1998**, *63*, 435–440. [CrossRef]
46. Torskangerpoll, K.; Andersen, Ø.M. Colour stability of anthocyanins in aqueous solutions at various pH values. *Food Chem.* **2005**, *89*, 427–440. [CrossRef]
47. Dedio, W.; Hill, R.; Evans, L. Anthocyanins in the pericarp and coleoptiles of purple wheat. *Can. J. Plant Sc.* **1972**, *52*, 977–980. [CrossRef]
48. Dedio, W.; Hill, R.; Evans, L. Anthocyanins in the pericarp and coleoptiles of purple-seeded rye. *Can. J. Plant Sc.* **1972**, *52*, 981–983. [CrossRef]
49. Abdel-Aal, E.S.M.; Hucl, P.; Shipp, J.; Rabalski, I. Compositional differences in anthocyanins from blue-and purple-grained spring wheat grown in four environments in Central Saskatchewan. *Cereal Chem.* **2016**, *93*, 32–38. [CrossRef]
50. Cammerata, A.; Laddomada, B.; Milano, F.; Camerlengo, F.; Bonarrigo, M.; Masci, S.; Sestili, F. Qualitative characterization of unrefined durum wheat air-classified fractions. *Foods* **2021**, *10*, 2817. [CrossRef]
51. Hemery, Y.; Rouau, X.; Lullien-Pellerin, V.; Barron, C.; Abecassis, J. Dry processes to develop wheat fractions and products with enhanced nutritional quality. *J. Cereal Sci.* **2007**, *46*, 327–347. [CrossRef]
52. Prabhasankar, P.; Haridas Rao, P. Effect of different milling methods on chemical composition of whole wheat flour. *Eur. Food Res. Technol.* **2001**, *213*, 465–469. [CrossRef]

 *foods*

Article

Comparative Compositions of Grain of Bread Wheat, Emmer and Spelt Grown with Different Levels of Nitrogen Fertilisation

Alison Lovegrove [1], Jack Dunn [1], Till K. Pellny [1], Jessica Hood [1], Amanda J. Burridge [2], Antoine H. P. America [3], Luud Gilissen [4], Ruud Timmer [5], Zsuzsan A. M. Proos-Huijsmans [6], Jan Philip van Straaten [6], Daisy Jonkers [7], Jane L. Ward [1], Fred Brouns [8] and Peter R. Shewry [1,*]

1 Rothamsted Research, Harpenden AL5 2JQ, UK
2 Life Sciences, University of Bristol, 24 Tyndall Avenue, Bristol BS8 1TQ, UK
3 BU Bioscience, Plant Sciences Group, Wageningen University and Research, 6700 AA Wageningen, The Netherlands
4 Plant Breeding, Wageningen University & Research, 6700 AJ Wageningen, The Netherlands
5 Business Unit Field Crops, Wageningen University & Research, 8200 AK Lelystad, The Netherlands
6 Nederlands Bakkerij Centrum, Agro Business Park 75-83, 6708 PV Wageningen, The Netherlands
7 Division Gastroenterology-Hepatology, Department of Internal Medicine and School for Nutrition and Translational Research in Metabolism (NUTRIM), Maastricht University, 6200 MD Maastricht, The Netherlands
8 Department of Human Biology, Faculty of Health, Medicine and Life Sciences, School for Nutrition and Translational Research in Metabolism (NUTRIM), Maastricht University, 6700 MD Maastricht, The Netherlands
* Correspondence: peter.shewry@rothamsted.ac.uk; Tel.: +44-1582-938195

Abstract: Five cultivars of bread wheat and spelt and three of emmer were grown in replicate randomised field trials on two sites for two years with 100 and 200 kg nitrogen fertiliser per hectare, reflecting low input and intensive farming systems. Wholemeal flours were analysed for components that are suggested to contribute to a healthy diet. The ranges of all components overlapped between the three cereal types, reflecting the effects of both genotype and environment. Nevertheless, statistically significant differences in the contents of some components were observed. Notably, emmer and spelt had higher contents of protein, iron, zinc, magnesium, choline and glycine betaine, but also of asparagine (the precursor of acrylamide) and raffinose. By contrast, bread wheat had higher contents of the two major types of fibre, arabinoxylan (AX) and β-glucan, than emmer and a higher AX content than spelt. Although such differences in composition may be suggested to result in effects on metabolic parameters and health when studied in isolation, the final effects will depend on the quantity consumed and the composition of the overall diet.

Keywords: bread wheat; emmer; spelt; fibre; metabolites; minerals; phenolics; fertilisation; health benefits

Citation: Lovegrove, A.; Dunn, J.; Pellny, T.K.; Hood, J.; Burridge, A.J.; America, A.H.P.; Gilissen, L.; Timmer, R.; Proos-Huijsmans, Z.A.M.; van Straaten, J.P.; et al. Comparative Compositions of Grain of Bread Wheat, Emmer and Spelt Grown with Different Levels of Nitrogen Fertilisation. *Foods* **2023**, *12*, 843. https://doi.org/10.3390/foods12040843

Academic Editors: Donatella Bianca Maria Ficco, Grazia Maria Borrelli and Grant Campbell

Received: 16 January 2023
Revised: 7 February 2023
Accepted: 13 February 2023
Published: 16 February 2023

1. Introduction

Wheat is the major staple crop in temperate countries, with annual global yields exceeding 700 million tonnes. About 95% of the total production is hexaploid bread wheat (*Triticum aestivum* L. subsp. *aestivum*, genome constitution AABBDD) which originated about 10,000 years ago, with most of the remaining 5% production being tetraploid pasta wheat (*Triticum turgidum* L. subsp. *durum*) (AABB genomes). Bread and durum wheats are "free threshing" (the hulls being readily separated from the grain during harvest), which is regarded as an advanced trait. However, both bread and pasta wheats have been subjected to intensive breeding, focusing on improving their agronomic performance and increasing their yield and quality (for making bread and pasta, respectively). Hence, although there is wide genetic variation in both species, modern cultivars (those developed by scientific breeding during the last few decades) tend to be less genetically diverse than older cultivars and traditional types of wheat dating from before the application of breeding (called land races) [1].

Ancient wheats are diploid einkorn (*T. monococcum*, L., AA genome), tetraploid emmer (*T. turgidum* L. subsp. *dicoccum* Thell., AABB genomes) and hexaploid spelt (*T. aestivum* L. subsp. *spelta* Thell, AABBDD genomes) and are generally hulled as opposed to free threshing. Although the term "ancient" is taken to imply that the genotypes grown today are similar to those grown in antiquity, this is certainly not the case, at least for spelt and emmer. Einkorn is a distinct species, which includes cultivated and wild forms, while emmer and spelt are subspecies of *T. turgidum* and *T. aestivum*, respectively. Furthermore, modern commercial cultivars of spelt may have introgressions (transfer of genetic information) from bread wheat due to cross-breeding, while all types of wheat cultivated today have been grown and hence selected (either unconsciously or deliberately) over thousands of years [2].

The composition of wheat grain is determined by the genotype, the environment, the farming system and the interactions between the genotype and these factors. Environmental factors are particularly important when comparing ancient and modern wheats, as the former are often grown in organic or low input systems with low nitrogen application to avoid lodging (bending of the plant at or near ground level, making harvest difficult and often leading to premature germination of grains) while modern semi-dwarf wheats are more usually grown in intensive high input systems with high nitrogen fertilisation [3].

The breeding and selection of modern bread wheats have focused on increasing yield and improving breadmaking quality, which is largely determined by the content and composition of gluten proteins. It has therefore been suggested that this has led to modern wheats having lower contents of micronutrients (minerals, vitamins) and bioactive components (phytochemicals) and higher contents of proteins that may lead to adverse reactions and diseases such as coeliac disease, wheat allergy and non-coeliac wheat sensitivity [4,5]. Hence, there has been increased interest in ancient wheats, which are assumed to have more favourable compositions for health [2].

We have therefore carried out detailed analyses of grain samples of three commercial cultivars of emmer and five cultivars each of spelt and bread wheat. All of the cultivars are adapted to Northern Europe and corresponded, with one substitution due to unavailability, to those selected to compare the effects of breads on health as part of the "Well-on-Wheat?" research consortium programme (https://www.wellonwheat.org, accessed on 1 February 2023). These samples were grown in replicate field trials in two Northern European countries (the UK and the Netherlands) for two years with low (100 kg/Ha) and high (200 kg/ha) applications of nitrogen fertilisation to reflect the different inputs used for ancient and modern wheats. Wholemeal samples were analysed for the major types of dietary fibre in white flour (arabinoxylan and β-glucan), polar metabolites, protein (as nitrogen), phenolics and mineral micronutrients to identify differences in composition.

2. Materials and Methods

2.1. Grain Samples

Commercial samples of five cultivars each of bread wheat (RAGT Reform, Capo, Bernstein, Kometus, Akteur) and spelt (Comburger, Zollernspelz, Attergauer, Bauländer Spelz, Franckenkorn) and three cultivars of emmer (Ramses, Roter Heidfelder, Späths Albjuwel) were grown in two years (2017–2018 and 2018–2019) at two nitrogen levels (100 and 200 kg N/Ha) in Flevoland (WUR Field crops, Lelystad, Flevoland, Netherlands, 52°53′94.69″ N, 5°56′56.77″ E) in three randomised replicate 6 × 1.5 m plots and at Rothamsted Research (Harpenden, Hertfordshire, UK, 51°48′19.79″ N 0°21′11.39″ W) in three randomised replicate plots of 1 × 1 m. All trials were autumn sown, but sowing, fertiliser application and harvest dates varied between sites and years, according to local conditions. Standard agronomic practices for the two sites were used. Grain samples of emmer and spelt were mechanically dehulled. Whole grain samples at about 14% water content were milled in two stages: firstly, a Retsch ZM 200 Model Ultra-Centrifugal Mill (Retsch Gmbh, Dusselgorf, Germany) using a 0.5 mm ring sieve and then a Glen Creston Ball Mill Retsch

Gmbh, Dusseldorf, Germany) using 5 ball bearings in a 5 cm diameter canister for 4 min for each sample.

2.2. Genotyping

The samples were genotyped using the Axiom 35k Wheat Breeders Genotyping Array (Thermo Fisher Scientific, Inc., Waltham, MA, USA) using the Affymetrix GeneTitan (Thermo Fisher Scientific, Inc.) [6]. Alleles were identified using the Affymetrix proprietary software package Axiom Analysis Suite V4.0.3.3 (Thermo Fisher Scientific, Inc.) and prior model 'Axiom_WhtBrd-1.r3'. A Dish QC threshold of 0.8 and call rate cut-off of 90% were used to adjust for hybridisation rates of spelt and emmer to the array. A distance matrix was generated from the scores using R package SNPRelate (Bioconductor Open Source, Harvard, MA, USA) [7]. The first two Principal Components accounting for over 25% of the variance (PC1:19.76%; PC2:5.53%) were plotted as a PCA plot.

2.3. Enzyme Fingerprinting of Arabinoxylan and β-Glucan

Three technical replicates of flour were digested with endoxylanase and lichenase (β-glucanase) to release arabinoxylan oligosaccharides (AXOS) from arabinoxylan (AX) and gluco-oligosaccharides (GOS) from β-glucan, respectively [8]. The oligosaccharides were separated using a 2 mm × 250 mm Carbopac PA-1 (Dionex) column [8] (dx.doi. org/10.17504/protocols.io.babriam6, accessed on 1 February 2023). The areas under the oligosaccharide peaks were combined to give total AX and total β-glucan (expressed in arbitrary units), respectively.

2.4. NMR Spectroscopy of Polar Metabolites

Sample preparation for ^{1}H-NMR was carried out as described by [9]. Signal intensities for characteristic spectral regions for 29 major metabolites were compared with a library of spectra of standards analysed under the same conditions.

2.5. Mineral Analysis

Nitrogen was determined on each biological replicate by Dumas combustion, using a Leco combustion analyzer (Leco Corp., St. Paul, MN, USA). Iron and zinc were determined by Optima 7300 DV Inductively Coupled Plasma–Optical Emission Spectrometer (ICP–OES) (Perkin Elmer, Waltham, MA, USA) after digestion with nitric and perchloric acids. Certified external standards and in-house standards were used to monitor performance using Shewhart Control Charts.

2.6. Total Phenolics

Total phenolics were determined based on [10]. Triplicate 75 mg samples of each biological replicate were vortexed with 1.5 mL acidified methanol and then mixed at 850 rpm on an Eppendorf Thermomixer (Eppendorf Ltd., Stevenage, UK) for 2 h at 23 °C. After centrifugation (Eppendorf Ltd., Stevenage, UK) at 5000× *g* for 10 min, 1 mL of supernatant was removed into a fresh Eppendorf tube and 200 µL aliquots mixed with 1.5 mL of ×10 diluted Folin–Ciocalteu reagent (Sigma-Aldrich, St. Louis, MO, USA) and left to stand for 5 min. 1.5 mL of 6% (*w/v*) aqueous sodium carbonate solution was added, mixed and stood at room temperature for 90 min. The absorbance at 725 nm was then measured (Jenway 6715 UV/Vis spectrophotometer, Cole-Parmer, St Neots, UK), and the concentration of phenolics was calculated using ferulic acid as a standard (Sigma-Aldrich) and a standard curve from 20, 40, 100, 150 and 200 µg/mL with three technical replicates of each point.

2.7. Statistical Analysis

All data were analysed using analysis of variance (ANOVA) in Genstat 21 (VSN International, Hemel Hempstead, UK). The block structure was Trial/Block/Subblock, where Trial captures the location and year of each trial. There are 3 blocks within each trial and

2 sub-blocks within each block to which the Nlevel treatment was applied. Lines were considered to be applied to plots within each sub-block. The treatment structure was Nlevel*(Grain/(cultivarBreadwheat + cultivarEmmer + cultivarSpelt)) where the factor Grain tests for differences between the three grain types. The nested factors cultivarBreadwheat, cultivarEmmer and cultivarSpelt test for differences between the lines within each grain type. The Nlevel factor tests for differences between the two Nlevels and their interactions with grain type and lines are also included. Some variables were transformed in order to meet the normality and homoscedasticity assumptions of the analysis. Means and 95% confidence intervals are given in Tables 1 and 3 while the transformations used are indicated in Supplementary Tables of means and *p* values.

Principal component analysis (PCA) and orthogonal partial least squares-discrimination analysis (OPLS-DA) were carried out using SIMCA-P software (version 13, MKS Umetrics) (Sartorius UK Ltd., Epsom, UK).

3. Results

Five cultivars each of bread wheat (RAGT Reform, Capo, Bernstein, Kometus, Akteur) and spelt (Comburger, Zollernspelz, Attergauer, Bauländer Spelz, Franckenkorn) and three cultivars of emmer (Ramses, Roter Heidfelder, Späths Albjuwel) were selected, all of which have been grown commercially in Northern Europe. The genomic relationships between the cultivars were determined using the Axiom 35k Wheat Breeders Genotyping Array, comprising 35,143 single nucleotide polymorphism (SNP) markers. Principal component analysis (PCA) showed a clear separation of the three cereal types, confirming that the spelt lines used did not have recent introgressions from bread wheat (Figure 1).

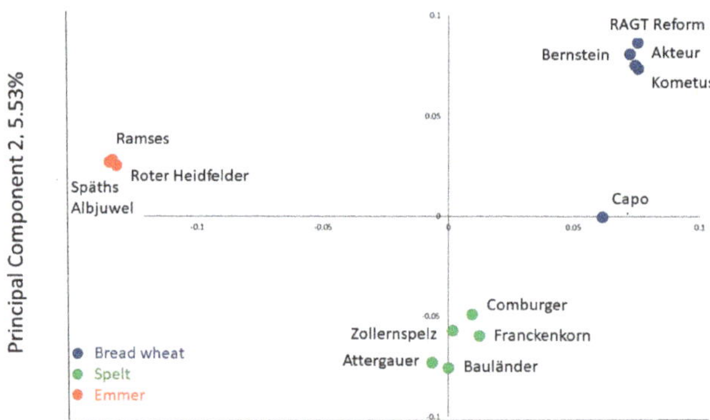

Figure 1. Genomic relationships of the 13 genotypes, illustrated by Principal Component Analysis of markers determined using the Axiom HD Genotyping Array (comprising 819,571 SNP markers).

3.1. Grain Composition

In order to determine the variation in composition within and between bread wheat, spelt and emmer, the four environments (site × year combinations) were treated as blocks in the statistical analyses. In addition, to compare the effects of nitrogen fertilisation on grain composition, the data for the 100 kg/Ha and 200 kg/Ha treatments were analysed separately and compared formally by including nitrogen as a factor in the statistical analyses. The ranges of contents are illustrated in Figures 2, 3 and 5, while Tables 1–4 and Supplementary Tables S1–S8 present means, 95% confidence intervals, SEMs and observed statistical significance determined by ANOVA.

The groups of components are discussed below.

3.2. Protein and Minerals

Grain protein content (determined as N × 5.7) (Figure 2A, Table 1) overlapped in range between bread wheat, spelt and emmer but was highest in spelt and lowest in bread wheat. It was also higher at 200 kg/Ha than at 100 kg/Ha. ANOVA (Table 2) showed statistically significant differences in protein content between bread wheat, emmer and spelt and statistically significant effects of nitrogen on their protein contents. By contrast, the three cereal types did not differ significantly in their response to nitrogen, and there was little variation in the effects of nitrogen between the cultivars within a single cereal type.

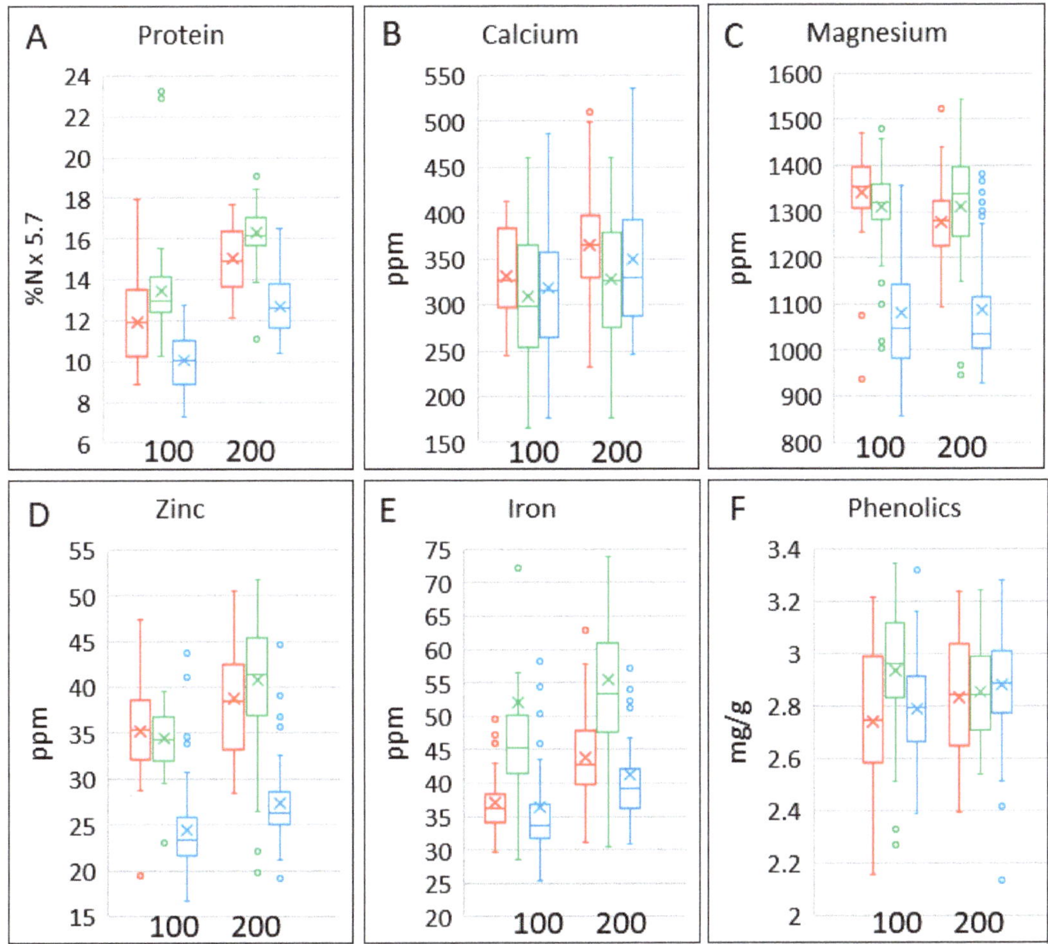

Figure 2. Contents of protein, minerals and total phenolics in grains of the three cereal types grown in four environments. Colour code: red, emmer; green, spelt; blue, bread wheat. The bar shows the range of the whole dataset. The box shows the middle two quartiles, separated by the horizontal line, which is the median, and the vertical lines are the upper and lower quartiles, respectively. Outliers are shown as circles. The x is the mean average. All analyses are expressed on a dry weight basis.

Table 1. Means and 95% confidence intervals (in parentheses) of contents of selected minerals, metabolites and groups of metabolites in grain of the three types of wheat grown with 100 and 200 kg N/Ha.

	100 kg N/Ha			200 kg N/Ha		
	Bread Wheat	**Emmer**	**Spelt**	**Bread Wheat**	**Emmer**	**Spelt**
%N	1.764 (1.68, 1.847)	2.095 (2.001, 2.189)	2.354 (2.27, 2.437)	2.23 (2.146, 2.313)	2.636 (2.542, 2.73)	2.857 (2.773, 2.94)
Ca	311.6 (300.2, 323.5)	327.7 (312.3, 343.8)	301.8 (290.7, 313.3)	341.6 (329.1, 354.6)	358.5 (341.7, 376.1)	320.2 (308.5, 332.4)
Fe	35.33 (33.36, 37.42)	36.65 (34.01, 39.5)	47.65 (44.98, 50.46)	40.04 (37.8, 42.4)	43.22 (40.11, 46.58)	54.01 (51, 57.21)
Mg	1081 (1054, 1108)	1341 (1309, 1374)	1311 (1284, 1338)	1087 (1060, 1114)	1278 (1246, 1311)	1312 (1285, 1339)
Zn	24.41 (22.78, 26.04)	35.22 (33.34, 37.11)	34.37 (32.74, 36)	27.27 (25.64, 28.9)	38.76 (36.88, 40.65)	40.78 (39.15, 42.41)
total phenolics	2786 (2732, 2840)	2742 (2677, 2807)	2934 (2880, 2987)	2879 (2826, 2933)	2830 (2765, 2894)	2854 (2800, 2908)
raffinose	5.95 (5.819, 6.082)	7.077 (6.908, 7.245)	6.916 (6.785, 7.048)	5.944 (5.813, 6.076)	6.465 (6.296, 6.633)	6.743 (6.611, 6.874)
asparagine	0.4907 (0.4681, 0.5144)	0.6588 (0.6195, 0.7007)	0.6981 (0.666, 0.7318)	0.5664 (0.5403, 0.5937)	0.7803 (0.7337, 0.8298)	0.8648 (0.825, 0.9065)
glycine betaine	1.311 (1.262, 1.362)	1.49 (1.432, 1.548)	1.606 (1.551, 1.662)	1.307 (1.257, 1.357)	1.466 (1.41, 1.525)	1.537 (1.483, 1.591)
choline	0.1748 (0.1711, 0.1785)	0.2148 (0.2101, 0.2195)	0.2137 (0.21, 0.2174)	0.1854 (0.1817, 0.1891)	0.2207 (0.216, 0.2254)	0.2229 (0.2192, 0.2266)
galactinol	0.2224 (0.2122, 0.2328)	0.3976 (0.3795, 0.4161)	0.2692 (0.2579, 0.2806)	0.2223 (0.2121, 0.2327)	0.3657 (0.3484, 0.3834)	0.2751 (0.2637, 0.2867)
inositol	3.08 (3.024, 3.136)	3.216 (3.141, 3.292)	3.421 (3.365, 3.477)	3.097 (3.041, 3.153)	3.096 (3.021, 3.171)	3.25 (3.194, 3.306)
total amino acids	5.753 (5.618, 5.891)	6.522 (6.331, 6.719)	6.607 (6.452, 6.766)	6.084 (5.941, 6.23)	6.724 (6.528, 6.927)	6.951 (6.788, 7.118)
total organic acids	2.014 (1.934, 2.093)	2.247 (2.156, 2.337)	2.133 (2.053, 2.213)	2.048 (1.968, 2.128)	2.321 (2.231, 2.411)	2.245 (2.166, 2.325)
total methyl donors	1.482 (1.436, 1.529)	1.702 (1.644, 1.761)	1.814 (1.758, 1.871)	1.486 (1.44, 1.533)	1.685 (1.628, 1.744)	1.758 (1.703, 1.813)
total sugars	29.47 (28.78, 30.18)	35.79 (34.71, 36.9)	29.88 (29.18, 30.6)	28.75 (28.07, 29.44)	33.62 (32.61, 34.67)	28.8 (28.13, 29.5)

The box plots show that the contents of calcium (Figure 2B) and magnesium (Figure 2C) were not affected by nitrogen, but the content of magnesium was lower in bread wheat grain than in grains of spelt or emmer. The contents of iron and zinc were lowest in bread wheat grain (Figure 2D,E), while the content of iron was lower in emmer than in spelt grain. The contents of both minerals were also higher at 200 kg N/Ha than at 100 kg N/Ha. However, ANOVA showed significant differences between the contents of all minerals in the grain types and significant effects of nitrogen fertilisation on the contents of all minerals except magnesium, which showed a significant interaction due to a difference between nitrogen fertilisation in emmer only (Table 2). There were also significant differences between the contents of all minerals (except zinc in spelt and iron in all grain types) between the cultivars within each type, but no differences in the effects of nitrogen between the cultivars within the types (except for iron in spelt) (Table 2 and Table S1).

3.3. Total Phenolics

Phenolics are the most abundant phytochemicals in wheat grain [11]. The contents of total phenolics varied widely (Figure 2F), with significant differences between cereal types and cultivars within types (Table 2). There was an interaction between cereal type and nitrogen, with total phenolics being lower at 100 kg N/Ha for all cereal types apart from spelt where total phenolics were higher at 100 kg N/Ha than at 200 kg N/Ha.

3.4. Polar Metabolites

The contents of polar metabolites in the samples were determined by ^{1}H NMR spectroscopy. This allowed the quantification of monosaccharide (glucose, fructose, galactose), disaccharide (maltose, sucrose) and trisaccharide (raffinose) sugars, organic acids (malic, acetic, fumaric), the sugar alcohols inositol and galactinol, the "methyl donors" choline and betaine and thirteen amino acids (alanine, aspartic acid, asparagine, glycine, glutamic acid, glutamine, γ-amino butyric acid (GABA), isoleucine, leucine, phenylalanine, tyrosine, tryptophan and valine). Data for all components are given in Supplementary Tables S2–S4, while selected components and groups of components are shown in Tables 1 and 2 and Figure 3.

The contents of all metabolites and groups of metabolites overlapped between the cereal types, but differences between the ranges in the types are observed (Figure 3). Asparagine is of particular interest because it is the limiting factor for the formation of acrylamide during processing [12,13]. The contents of total amino acids (Figure 3A) and of asparagine (Figure 3B) were significantly lower in bread wheat and higher in spelt, with a significant effect of nitrogen. However, there were no differences in the effects of nitrogen between and within the three cereal types.

The contents of total sugars (mono-, di- and trisaccharides) and total organic acids varied widely (Figure 3C,D), but both were significantly higher in emmer. Total organic acids were also significantly lower in bread wheat (Table 2). The contents of sugars were significantly affected by nitrogen, with no differences between the effects of nitrogen on the different cereal types or cultivars within the types (Table 2). The concentration of glycine betaine was about 10-fold greater than that of choline, which is typical for wheat [14]. Although the ranges overlapped (Figure 3E,F), they were significantly higher in spelt and lower in bread wheat, with little effect of nitrogen or variation within the types (Table 2).

Raffinose (the trisaccharide galactose, glucose, fructose) is a non-digestible and fermentable carbohydrate (being part of the FODMAP (fermentable oligosaccharides, disaccharides, monosaccharides and polyols) fraction) while galactinol (1-alpha-D-Galactosyl-myo-inositol) and inositol ((1R,2S,3r,4R,5S,6s)-Cyclohexane-1,2,3,4,5,6-hexol) are precursors in raffinose synthesis [15]. Raffinose (Figure 3I) accounted for about a quarter of the total sugars (Figure 3C) and was significantly lower in bread wheat than in emmer or spelt. The contents of inositol were about half of those of raffinose and were higher in spelt (Figure 3G). Galactinol was present at much lower concentrations and was significantly higher in emmer and lower in bread wheat than in spelt (Figure 3H). The contents of raffinose and inositol were affected by nitrogen level and, with the exception of galactinol in spelt, varied significantly between cultivars of the three cereal types (Table 2).

PCA analysis of the metabolite dataset showed partial separation of the three cereal types, based on 48.9% of the total variance (Figure 4A). To improve the discrimination between cereal types, we repeated the analysis using supervised multivariate analysis (orthogonal partial least squares discrimination analysis, OPLS-DA), selecting for differences between the cereal types (Figure 4C). This gave clear separation between the three types with the loadings plot (Figure 4F) showing that emmer was characterised by high contents of maltose, galactinol and glucose and spelt by higher levels of amino acids (glycine, asparagine, leucine, isoleucine and valine) compared with bread wheat. OPLS-DA was also used to separate the samples based on nitrogen level (Figure 4B), the loadings plot (Figure 4E) showing higher contents of amino acids at 200 kg N/Ha. These differences are also illustrated by the difference plots in Supplementary Figure S1.

Table 2. p values from ANOVA of minerals and selected metabolites in the three cereal types and cultivars.

	NLevel	Grain	Nlevel. Grain	Grain. Cultivar. Bread Wheat	Grain. Cultivar. Emmer	Grain. Cultivar. Spelt	Nlevel. Grain. Cultivar Bread Wheat	Nlevel. Grain. Cultivar Emmer	Nlevel. Grain. Cultivar Spelt
%N	<0.001	<0.001	0.475	<0.001	0.098	<0.001	0.704	0.212	0.642
log(Ca)	<0.001	<0.001	0.639	<0.001	<0.001	<0.001	0.672	0.195	0.495
log(Fe)	<0.001	<0.001	0.815	0.566	0.957	0.064	0.375	0.936	**0.023**
Mg	0.46	<0.001	**0.016**	<0.001	**0.03**	<0.001	0.228	0.351	0.749
Zn	<0.001	<0.001	**0.008**	**0.012**	<0.001	0.556	0.409	0.834	0.994
Total phenolics	0.407	<0.001	<0.001	0.002	<0.001	<0.001	0.253	0.516	0.153
raffinose	**0.005**	<0.001	<0.001	<0.001	<0.001	**0.001**	0.988	0.271	0.709
Log(asparagine)	<0.001	<0.001	0.352	<0.001	**0.022**	<0.001	0.624	0.34	0.78
Sqrt (glycine betaine)	0.358	<0.001	0.087	<0.001	<0.001	<0.001	**0.014**	0.951	0.305
choline	<0.001	<0.001	0.535	<0.001	**0.005**	<0.001	0.852	0.109	1
Sqrt (galactinol)	0.399	<0.001	0.061	<0.001	<0.001	0.605	0.935	0.146	0.951
inositol	**0.002**	<0.001	**0.009**	<0.001	<0.001	<0.001	0.883	0.06	0.678
$\log_e$(total amino acids)	**0.002**	<0.001	0.598	0.099	<0.001	**0.001**	0.095	0.672	0.464
total organic acids	0.163	<0.001	0.338	<0.001	<0.001	<0.001	0.159	0.771	0.551
$\log_e$ (total methyl donors)	0.529	<0.001	0.141	<0.001	<0.001	<0.001	**0.011**	0.905	0.395
$\log_e$(total sugars)	**0.005**	<0.001	0.397	<0.001	<0.001	<0.001	0.726	0.23	0.726

Statistically significant values ($p < 0.05$) are given in bold. Some variables required transformation, square root (Sqrt) or $\log_e$, to meet the assumptions of the analysis.

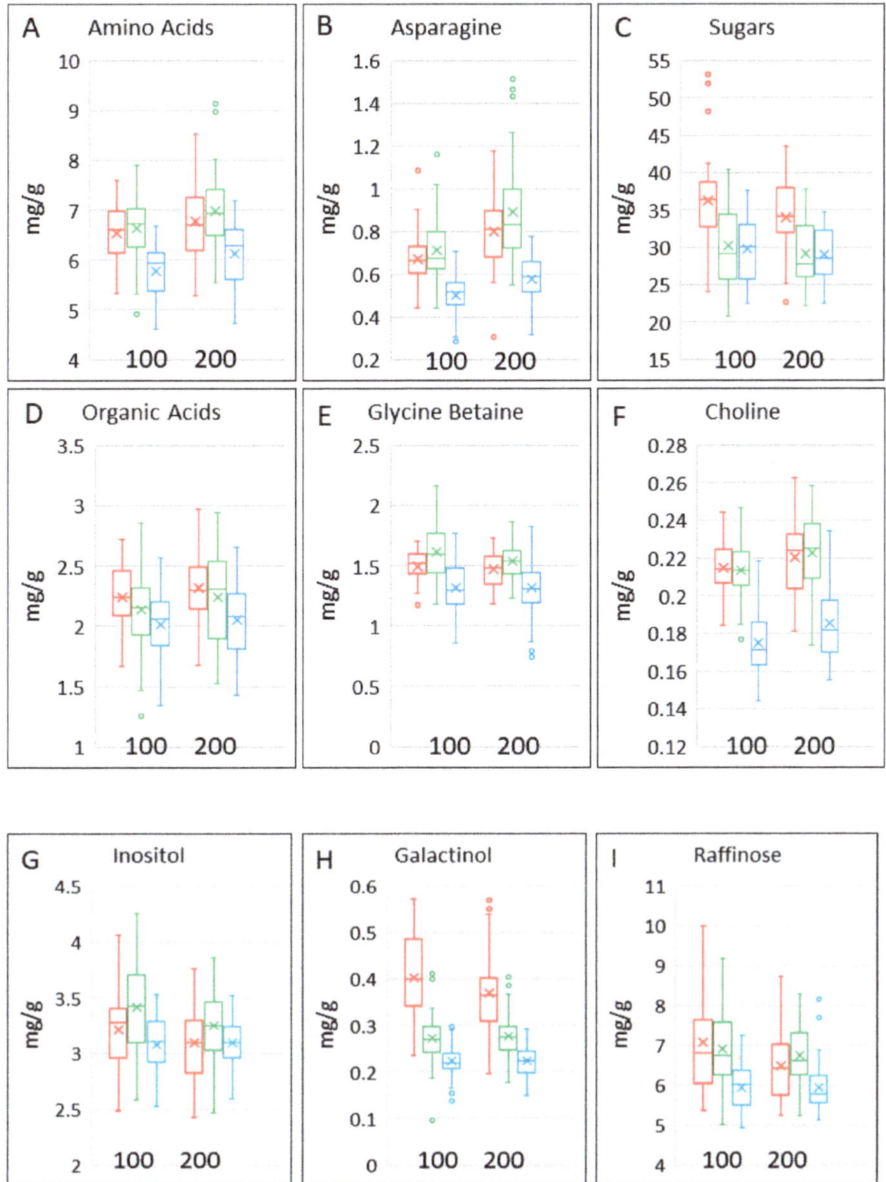

Figure 3. Contents of selected polar metabolites and groups of metabolites in grains of the three cereal types grown in four environments. Colour code: red, emmer; green, spelt; blue, bread wheat. The bar shows the range of the whole data set. The box shows the middle two quartiles, separated by the horizontal line, which is the median, and the vertical lines are the upper and lower quartiles, respectively. Outliers are shown as circles. The x is the mean average. All analyses are expressed on a dry weight basis.

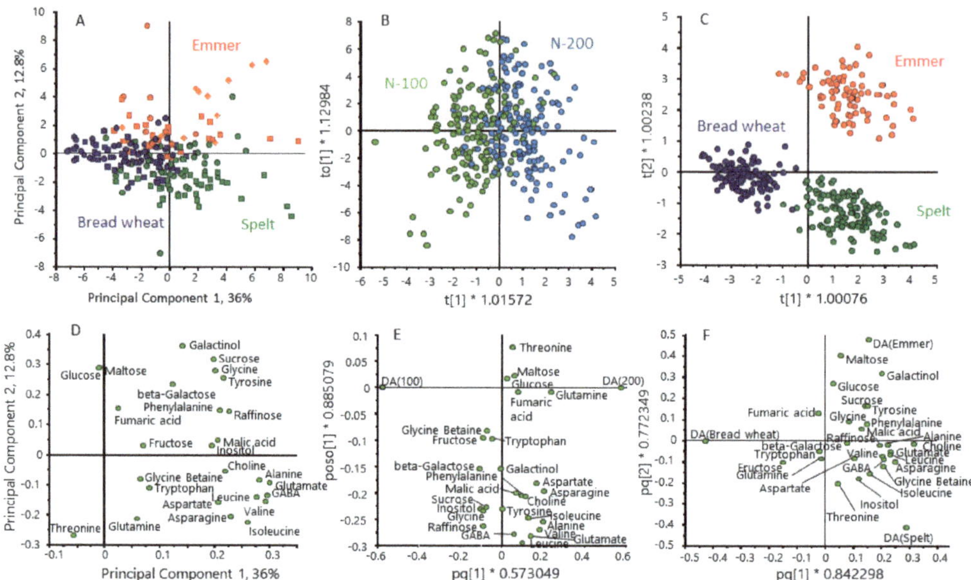

Figure 4. Multivariate analysis of the contents of polar metabolites in grains of the three cereal types grown in four environments. Principal component analysis PCA (**A**) and orthogonal partial least squares discrimination analysis (OPLS-DA) (**B,C**) selecting for differences between nitrogen treatments (**B**) and cereal types (**C**). (**D–F**) loading plots for (**A–C**), respectively. Colour code: red, emmer; green, spelt; blue, bread wheat.

3.5. Dietary Fibre

The contents of AX and β-glucan in the grain types are shown in Figure 5A,B and in Table 3. The contents of both AX and β-glucan were lower in emmer grain, with AX being highest in bread wheat grain and β-glucan highest in spelt grain. Hence, the ratio of AX: β-glucan was lower in spelt grain than in the other cereals (Figure 5D). The combined contents of these two components were also highest in bread wheat grain (Figure 5C), reflecting the fact that the content of AX was about three- to four-fold greater than that of β-glucan. There was significant variation in the contents of AXOS between the cultivars of the three types of wheat and of β-glucan between the cultivars of bread wheat and emmer, but no significant effects of nitrogen fertilisation (Table 4).

AX and β-glucan were determined by enzyme fingerprinting, which uses enzymes (endoxylanase and lichenase, respectively) to digest the polymers to release oligosaccharides separated and quantified by HP-AEC. The oligosaccharides released have defined structures, and their proportions, therefore, provide information on the structures of the polymers. In the case of AX, the oligosaccharides (AXOS) comprise chains of one to five xylose residues, one or more of which may be substituted with either one or two arabinose residues. The ratio of monosubstituted to disubstituted AX may affect the properties of the molecules and is generally higher in spelt and bread wheat than in emmer (Figure 6F). β-glucan comprises linear chains of glucose molecules linked predominantly by β(1-4) bonds. However, these β(1-4) bonds are interspersed with β(1-3) bonds that generally occur every three to four glucose residues, although some longer stretches of β(1-4) linked glucose residues (up to 14) have been reported. The distribution of β(1-3) bonds results in conformational changes in the linear glucan molecules, which affect their solubility and viscosity. Lichenase is a type of β-glucanase that releases mainly gluco-oligosaccharides (GOS) of three or four glucose residues (called G3 and G4), reflecting the relative abundances of β(1-3) and β(1-4) bonds. The ratio of G3:G4 GOS is higher in emmer and lower in spelt than in bread wheat (Figure 5E, Table 3).

Table 3. Means and 95% confidence intervals (in parentheses) of contents and compositions of arabinoxylan and β-glucan in grain of the three types of wheat grown with 100 and 200 kg N/Ha.

	100 kg N/Ha			200 kg N/Ha		
	Bread	Emmer	Spelt	Bread	Emmer	Spelt
TOT-AX	29.06 (28.18, 29.96)	20.17 (19.43, 20.93)	27.33 (26.51, 28.18)	29.61 (28.72, 30.53)	19.88 (19.16, 20.64)	27.26 (26.44, 28.11)
TOT-BG	8.146 (7.869, 8.432)	5.617 (5.386, 5.859)	9.256 (8.942, 9.581)	7.63 (7.371, 7.898)	5.142 (4.93, 5.363)	8.521 (8.232, 8.82)
ratio G3:G4 GOS	2.429 (2.395, 2.463)	2.54 (2.497, 2.583)	2.268 (2.234, 2.302)	2.381 (2.347, 2.415)	2.477 (2.434, 2.52)	2.252 (2.218, 2.286)
ratio TOT-AXOS: TOT-BG	3.596 (3.516, 3.677)	3.619 (3.516, 3.723)	2.968 (2.887, 3.048)	3.915 (3.834, 3.996)	3.895 (3.792, 3.999)	3.212 (3.131, 3.293)
ratio M:D AXOS	2.061 (2.035, 2.088)	1.865 (1.834, 1.897)	2.095 (2.068, 2.122)	2.032 (2.005, 2.058)	1.85 (1.818, 1.881)	2.071 (2.044, 2.097)

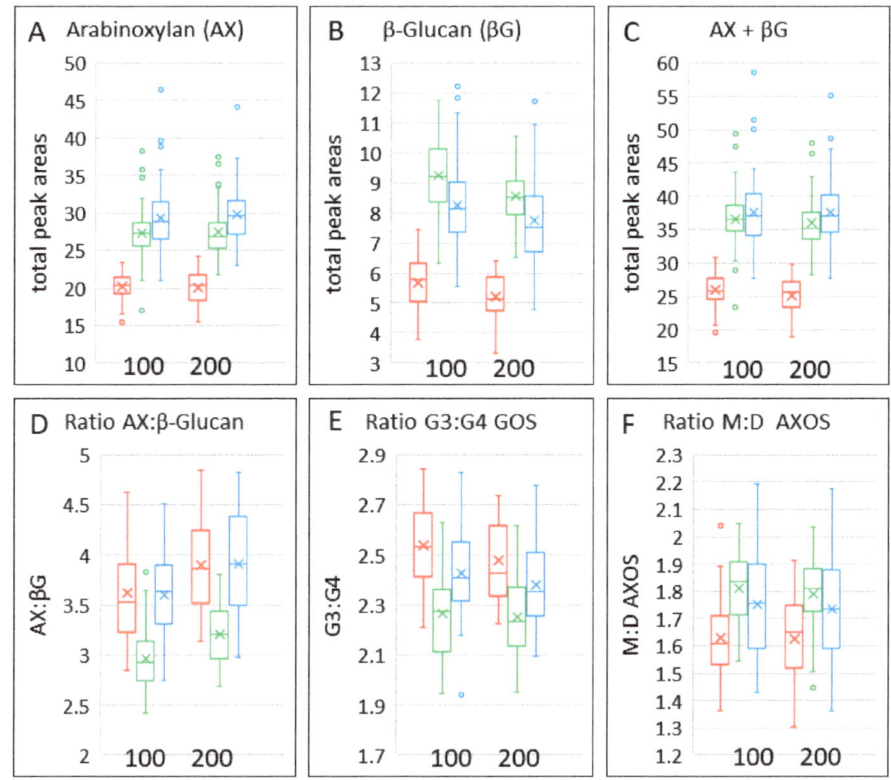

Figure 5. Contents, ratios and structures of arabinoxylan and β-glucan in grains of the three cereal types grown in four environments. Colour code: red, emmer; green, spelt; blue, bread wheat. The box shows the middle two quartiles, separated by the horizontal line, which is the median, and the vertical lines are the upper and lower quartiles, respectively. Outliers are shown as circles. The x is the mean average.

Differences in the structures of AX and β-glucan in the cereal types are illustrated by the multivariate analysis in Figure 6. OPLS-DA confirmed that there were no effects of nitrogen on AX and β-glucan structure (Figure 6B) but gave clear separation of the cereal types (Figure 6C), with the loadings plot (Figure 6F) showing that spelt differed from emmer and bread wheat in having higher proportions of G3 and G4 GOS and bread wheat higher proportions of substituted AXOS These differences are also illustrated by the difference plots in Supplementary Figure S2

Table 4. p-values from ANOVA of the proportions of AXOS and GOS in the three cereal types and cultivars.

	Nlevel	Grain	Nlevel. Grain	Grain. Cultivar. Bread wheat	Grain. Cultivar. Emmer	Grain. Cultivar. Spelt	Nlevel. Grain. Cultivar. Bread wheat	Nlevel. Grain. Cultivar. Emmer	Nlevel. Grain. Cultivar. Spelt
$\log_e$ (TOT-AXOS)	0.855	**<0.001**	0.514	**<0.001**	**<0.001**	**<0.001**	0.42	0.848	0.926
$\log_e$ (TOT-BG)	**0.002**	**<0.001**	0.758	**<0.001**	**<0.001**	0.085	0.659	0.501	0.76
ratio G3:G4 GOS	**0.032**	**<0.001**	0.427	0.509	0.197	**<0.001**	0.711	0.513	0.423
ratio TOT -AXOS: TOT-BG	**<0.001**	**<0.001**	0.655	**<0.001**	0.121	**<0.001**	0.715	0.218	0.564
Sqrt (ratio M:D AXOS)	0.093	**<0.001**	0.899	**<0.001**	**<0.001**	**<0.001**	0.249	0.775	0.902

Statistically significant values ($p < 0.05$) are given in bold. Some variables required transformation, square root (Sqrt) or $\log_e$, to meet the assumptions of the analysis.

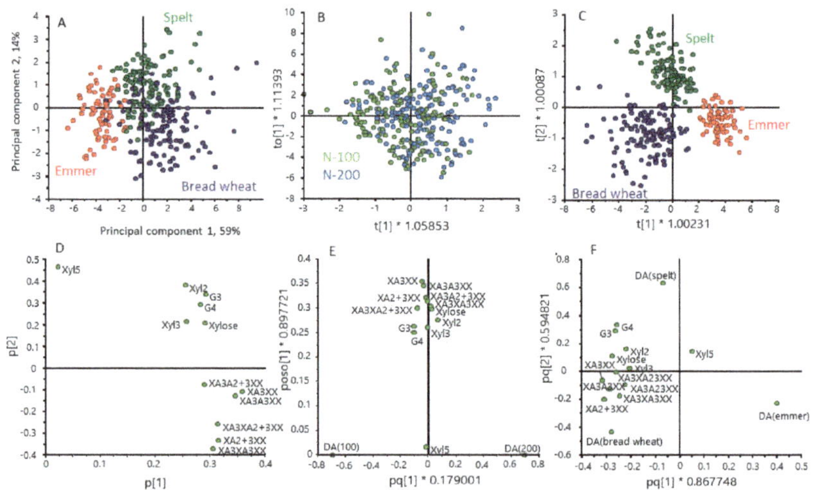

Figure 6. Multivariate analysis of the contents of AXOS and GOS in grains of the three cereal types grown in four environments. Principal component analysis PCA (**A**) and orthogonal partial least squares discrimination analysis (OPLS-DA) (**B,C**) selecting for differences between nitrogen treatments (**B**) and cereal types (**C**). (**D–F**) loading plots for (**A–C**), respectively. Colour code: red, emmer; green, spelt; blue, bread wheat.

4. Discussion

We have carried out comparative analyses of the grain of five cultivars each of bread wheat and spelt and three cultivars of emmer, focusing on components that may contribute to health effects. The cultivars selected were all commercially available in Germany at the time of the study, and analyses of flours blended from commercial grain samples and of doughs and breads produced from the blended flours using sourdough and yeast-based systems have been reported elsewhere [16]. The cultivars were grown in replicated randomised field trials at sites in the UK and Netherlands for two years, giving four environments (sites and years). Furthermore, two levels of nitrogen fertiliser were applied to reflect the commercial use of low and high input systems for ancient wheats (emmer, spelt) and modern bread wheat, respectively.

Because only small numbers of cultivars were compared, the results cannot be taken to represent the full range of diversity within the three types of wheat. Nevertheless, wide variation within each type was observed, resulting from the effects of genotype, environment and nitrogen fertilisation. However, because only two sites and years were compared, it was not possible to calculate the separate contributions of genotype, environment and G x E interactions, and the year/site combinations were, therefore, treated as "environments".

The variation in composition within emmer, spelt and bread wheat resulted in overlapping ranges in the contents of all components determined: protein, minerals, polar metabolites and AX and β-glucan fibre. Nevertheless, some statistically significant differences between the three cereal types were observed.

The lower protein content of modern bread wheats observed here is well-established and considered to result from "yield dilution"; the higher yields of modern wheats result from an increased accumulation of starch that dilutes other components [17]. Similarly, modern semi-dwarf bread wheats are known to have lower contents of Fe, Zn and Mg. This may result from the effects of the semi-dwarf phenotype, perhaps combined with some yield dilution [18,19]. The modern bread wheats were also significantly lower in asparagine, the precursor of acrylamide, which formed during processing [12] but also had lower contents of glycine betaine and choline, which have benefits for cardiovascular health by reducing the concentration of homocysteine in blood [20,21]. Hence, it is not possible to conclude that any of the three types of cereal is consistently "better" in terms of its content of polar metabolites.

Published values for the contents of total dietary fibre in whole grains of bread wheat range between 11.5–15.5% dry weight, of which about half (5.5–7.4% dry weight) is AX, with a lower content of β-glucan (0.51–0.96%) [22]. White flour contains significantly less fibre (due to the removal of the bran), about 4% dry weight, with AX and β-glucan accounting for about 70% and 20% of the total, respectively [23].

In the present study, bread wheat had higher contents of AX and β-glucan than emmer and higher AX content than spelt. Although spelt was higher in β-glucan, this component was present in lower concentrations than AX, and hence, the sum of the two types of fibre was highest in bread wheat. A meta-analysis of dietary fibre components in whole grain also showed slightly lower contents of AX in spelt than in bread wheat, with a wider range [24]. Some differences in fibre structure were also observed, with spelt having a lower ratio of G3:G4 GOS released from β-glucan while emmer had a lower ratio of monosubstituted:disubstituted AXOS released from AX. The significance of these differences for the behaviour of the AX and β-glucan fractions in foods and in the gastrointestinal tract is not known.

The bread wheats also had lower contents of raffinose, which is not absorbed in the human small intestine but rapidly fermented in the colon, forming part of the FODMAP fraction, which may contribute to discomfort due to gas production in individuals suffering from non-celiac wheat sensitivity and irritable bowel syndrome (IBS) [25]. However, the relevance of this to symptom control is limited as the major FODMAP fraction in wheat is fructans, which were not measured in the present work.

The effects of nitrogen fertilisation were also determined as it is usual to grown ancient and modern wheats under low input and intensive production systems, respectively. High nitrogen resulted in higher contents of minerals (iron, zinc and magnesium), as reported in a number of studies [26]. Similarly, a significant positive relationship between free asparagine content and total grain protein content has been reported [27,28]. Although only small effects of nitrogen on other components were observed, ANOVA showed that these varied between cereal types. ANOVA also showed interactions between N level and the proportions of AXOS and GOS released in all three types of wheat but no interactions with cultivars within the types.

A number of other studies have compared genotypes of modern and ancient wheats. For example, a series of studies compared the agronomic performance, yield, grain quality traits and contents of a range of "bioactive" components in 15 cultivars each of einkorn, emmer, spelt, bread wheat and durum wheat grown on four sites, with nitrogen fertilisation levels varying between wheat types to reflect commercial practice [29–31]. However, the relevance of such reported differences in composition to human health remains unclear.

This is, at least in part, due to the fact that the relevance of the parameters measured to human health has not been established. For example, although differences between the in vitro antioxidant capacity of cereal flours have been reported [32–34], these cannot be generalized to imply health benefits in humans in vivo [35]. Similarly, glucose released during the digestion of flour in vitro cannot be used to predict glycaemic response in vivo, which is determined by the competing effects of the appearance of glucose and the disappearance (cellular uptake) of glucose in blood [36–38]. As a result, although the glycaemic index (GI) calculated based on in vitro digestion may correlate with that determined in vivo, the absolute values may differ [39]. Such calculations based on in vitro digestion have been reported to result in over-estimation of the in vivo GI by 22% to 50% [40]. Differences in food structure resulting from processing will also affect oral mastication and gastrointestinal digestion. For example, the high density of pasta results in a significantly lower glycaemic response than those of flour and bread [41].

Finally, statistically significant differences in the compositions of cereal flours may not represent significant biological differences when considered in the context of processed food consumed as part of a mixed meal and with other foods consumed over a 24 h intake period.

Comparisons are also often reported using analytical data from different studies and/or data from published food composition tables (for example, [42,43]). This is clearly not valid as the cultivars, growth environments, agronomic practices and methods used for sample preparation and analysis will affect the results obtained.

Taking these factors into account, it is not surprising that different conclusions have been drawn on the health benefits of ancient compared with modern wheats. Thus, whereas Shewry and Hey [24] concluded that based on a comparison of grain samples grown and analysed under the same conditions, there is little evidence that ancient wheats are more "healthy" than modern wheats, Serban et al. [43] suggested that ancient wheat species have health benefits in relation to their nutraceutical composition.

In the context of the data presented here, the small differences in mean compositions of bread wheat, emmer or spelt and their overlapping quantitative ranges are unlikely to result in significant differences in health outcomes, with a possible exception being mineral micronutrients (zinc and iron, which are subject to low intakes in many population groups, with wheat being a significant dietary source [44].

Supplementary Materials: The following supporting information can be downloaded at: https://www.mdpi.com/article/10.3390/foods12040843/s1, Figure S1. Difference plots showing the metabolites contributing to the separations in Figure 4B (A) and Figure 4C (B, C, D). Figure S2. Difference plots showing the AXOS and GOS contributing to the separations in Figure 6A (A) and Figure 6B (B, C). Table S1. Means and SEMs of the contents of minerals and metabolites in grains of individual cultivars of the three cereal types grown in four environments. Table S2. Means and SEMs of minerals and metabolites in cultivars of the three cereal types grown at 100 and 200 kg N/Ha. Table S3. Means and SEMs of contents of polar metabolites in grains of the three cereal types grown in four environments at two Nlevels. Table S4. *p* values from ANOVA of the contents of polar metabolites in grains of the three cereal types grown in four environments at two Nlevels. Table S5. Means and SEMs for AXOS and GOS in grains of the three cereal types grown at 100 and 200 kg N/Ha. Table S6. *p* values from ANOVA of AXOS and GOS in the three types and cultivars of wheat. Statistically significant values (*p* < 0.05) are highlighted. Table S7. Means and SEMs of AXOS in the cultivars of the three cereals grown in four environments. Some variables required transformation, square root (Sqrt) or $\log_e$ (log), to meet the assumptions of the analysis. Table S8. Means and SEMs of AXOS in the cultivars of the three cereals grown in four environments with 100 and 200 kg N/Ha.

Author Contributions: Conceptualization, F.B., P.R.S., J.P.v.S., L.G., A.L. and D.J.; methodology, A.L., J.D. and J.L.W.; formal analysis, J.D., J.L.W., A.J.B. and J.H.; investigation, J.P.v.S., J.L.W., T.K.P., J.L.W., A.H.P.A., R.T. and Z.A.M.P.-H.; resources, P.R.S. and J.P.v.S.; data curation, J.D., J.H., J.L.W. and Z.A.M.P.-H.; writing—original draft preparation, F.B. and P.R.S.; writing—review and dditing, T.K.P., J.P.v.S., J.L.W., A.L., L.G., A.H.P.A. and D.J.; visualization, supervision, P.R.S., A.L., J.L.W. and D.J.; project administration, F.B. and D.J.; funding acquisition, F.B., P.R.S. and J.P.v.S. All authors have read and agreed to the published version of the manuscript.

Funding: "Well on Wheat?" is financed by a grant of the Dutch Government—"TKI- Top Knowledge Institute" and a wide range of additional partners* from the Agri-Food chain who donated unrestricted research grants. *These partners are AB-Mauri bakery Ingredients, Made, Netherlands; Borgesius Holding BV-Albert Heijn, Stadskanaal Netherlands; CSM innovation Bakery Center, Bingen, Germany; CYMMIT, Texcoco, Mexico; DSM Food Specialties, Delft, Netherlands; Fazer Bakeries Oy, Helsinki, Finland; Health Grain Forum, Vienna, Austria; ICC- Intl. Vienna, Austria; IWGA, Kansas 66210, USA; Lantmännen EK, Stockholm, Sweden; Mondelez, Saclay, France; Nederlands Bakkerij Centrum, Wageningen, Netherlands; Baking Industry Research Trust, Wellington, New Zealand; Nutrition et Sante, Revel, France; Puratos BV, Groot Bijgaarden, Belgium; Rademaker BV-Bakery equipments, Culemborg, Netherlands; Sonneveld Group BV, Papendrecht, Netherlands; Zeelandia-HJ Doeleman BV, Zierikzee, Netherlands. The project is coordinated and executed by an academic research consortium team (ARCT). Research specialists from funding partners, represented in a Research Steering Team (RST), are entitled to give scientific inputs to ARCT, which on its sole decision, may or may not take inputs into consideration. Decisions on studies set-up, execution and data interpretation of data are exclusively taken by ARCT. Scientific output communications are exclusively organized by ARCT without the involvement of funding partners. Rothamsted Research receives strategic funding from the Biotechnology and Biological Sciences Research Council (BBSRC) and the work forms part of the Designing Future Wheat strategic programme (BB/P016855/1).

Data Availability Statement: Full datasets are available from the corresponding author on request.

Acknowledgments: We thank Friedrich Longin (University of Hohenheim, Germany) for helping with the provision of grain samples and for dehulling emmer and spelt samples prior to analysis.

Conflicts of Interest: Part of the work of DJ outside the submitted work has been financed by public-private partnership grants of Top Knowledge Institute (TKI) Agri&Food and Health Holland, the NWO Carbohydrate Competence Center, by Organic A2BV/Mothersfinest BV, EU/FP7 Sysmed-IBD/305564, BIOM/305479 and Character/305676 and H2020 DISCOvERIE/848228. The authors declare no other conflict of interest.

References

1. Pont, C.; Leroy, T.; Seidel, M.; Tondelli, A.; Duchemin, W.; Armisen, D.; Lang, D.; Bustos-Korts, D.; Goué, N.; Balfourier, F. Tracing the ancestry of modern bread wheats. *Nat. Genet.* **2019**, *51*, 905–911. [CrossRef] [PubMed]
2. Gilissen, L.J.W.J.; Smulders, M.J.M. Gluten Quality and Quantity in Wheat and in Wheat-Derived Products. In *Biotechnological Strategies for the Treatment of Gluten Intolerances*; Rossi, M., Ed.; Academic Press-Elsevier: London, UK, 2021; pp. 79–129.
3. Shewry, P.R. What is gluten—Why is it special. *Front. Nutr.* **2019**, *6*, 101. [CrossRef] [PubMed]
4. Morris, C.E.; Sands, D.C. The breeder's dilemma–yield or nutrition? *Nat. Biotechnol.* **2003**, *24*, 1078–1080. [CrossRef] [PubMed]
5. Sands, D.C.; Morris, C.E.; Dratz, E.A.; Pilgeram, A.L. Elevating optimal human nutrition to a central goal of plant breeding and production of plant-based foods. *Plant Sci.* **2009**, *177*, 377–389. [CrossRef]
6. Burridge, A.J.; Winfield, M.O.; Allen, A.M.; Wilkinson, P.A.; Barker, G.L.A.; Coghill, J.; Waterfall, C.; Edwards, K.J. High-density SNP genotyping array for hexaploidy wheat and its relatives. In *Wheat Biotechnology: Methods and Protocols*; Bhalla, P.L., Singh, M.B., Eds.; Humana Press: New York, NY, USA, 2017; pp. 293–306. [CrossRef]
7. Zheng, X.; Levine, D.; Shen, J.; Gogarten, S.M.; Laurie, C.; Weir, B.S. A high performance computing toolset for relatedness and principal component analysis of SNP data. *Bioinformatics* **2012**, *28*, 3326–3328. [CrossRef]
8. Lovegrove, A.; Wilkinson, M.D.; Freeman, J.; Pellny, T.; Tosi, P.; Saulnier, L.; Shewry, P.R.; Mitchell, R.A.C. RNA interference suppression of genes in glycosyl transferase families 43 and 47 in wheat starchy endosperm causes large decreases in arabinoxylan content. *Plant Physiol.* **2013**, *163*, 95–107. [CrossRef]
9. Shewry, P.R.; Corol, D.; Jones, H.D.; Beale, M.H.; Ward, J.L. (2017) Defining genetic and chemical diversity in wheat grain by 1H-NMR spectroscopy of polar metabolites. *Mol. Nutr. Food Res.* **2017**, *61*, 1600807. [CrossRef]

10. Gao, L.; Wang, S.; Oomah, B.D.; Mazza, G. Wheat quality: Antioxidant activity of wheat millstreams. In *Wheat Quality Elucidation*; Ng, P., Wrigley, C.W., Eds.; AACC International: St. Paul, MN, USA, 2002; pp. 219–233.
11. Ward, J.L.; Poutanen, K.; Gebruers, K.; Piironen, V.; Lampi, A.-M.; Nyström, L.; Andersson, A.A.M.; Åman, P.; Boros, D.; Rakszegi, M.; et al. The HEALTHGRAIN cereal diversity screen: Concept, results and prospects. *J. Agric. Food Chem.* **2008**, *56*, 9699–9709. [CrossRef]
12. Mottram, D.S.; Wedzicha, B.L.; Dodson, A.T. Acrylamide is formed in the Maillard reaction. *Nature* **2002**, *419*, 448–449. [CrossRef]
13. Muttucumaru, N.; Elmore, J.S.; Curtis, T.; Mottram, D.S.; Parry, M.A.J.; Halford, N.G. Reducing acrylamide precursors in raw materials derived from wheat and potato. *J. Agric. Food Chem.* **2008**, *56*, 6167–6172. [CrossRef]
14. Corol, D.I.; Ravel, C.; Raksegi, M.; Bedo, Z.; Charmet, G.; Beale, M.H.; Shewry, P.R.; Ward, J.L. Effects of genotype and environment on the contents of betaine, choline and trigonelline in cereal grains. *J. Agric. Food Chem.* **2012**, *60*, 5471–5481. [CrossRef] [PubMed]
15. Sengupta, S.; Mukherjee, S.; Basak, P.; Majumder, A.L. Significance of galactinol and raffinose family oligosaccharide synthesis in plants. *Front. Plant Sci.* **2015**, *6*, 656. [CrossRef] [PubMed]
16. Shewry, P.R.; America, A.H.P.; Lovegrove, A.; Wood, A.J.; Plummer, A.; Evans, J.; van den Broeck, H.C.; Gilissen, L.; Mumm, R.; Ward, J.L.; et al. Comparative compositions of metabolites and dietary fibre components in doughs and breads produced from bread wheat, emmer and spelt and using yeast and sourdough processes. *Food Chem.* **2022**, *374*, 131710. [CrossRef] [PubMed]
17. Lovegrove, A.; Pellny, T.K.; Hassall, K.; Plummer, A.; Wood, A.; Bellisai, A.; Przewieslik-Allen, A.; Burridge, A.L.; Ward, J.L.; Shewry, P.R.; et al. Historical changes in the contents and compositions of fibre components and polar metabolites in white flour. *Sci. Rep.* **2020**, *10*, 5920. [CrossRef] [PubMed]
18. Fan, M.; Zhao, F.-J.; Fairweather-Tait, S.J.; Poulton, P.R.; Dunham, S.J.; McGrath, S.P. Evidence of decreasing mineral density in wheat grain over the last 160 years. *J. Trace Elem. Med. Biol.* **2018**, *22*, 315–324. [CrossRef] [PubMed]
19. Shewry, P.R.; Pellny, T.K.; Lovegrove, A. Is modern wheat bad for health? *Nat. Plants* **2016**, *2*, 16097. [CrossRef] [PubMed]
20. Obeid, R. The metabolic burden of methyl donor deficiency with focus on the betaine homocysteine methyltransferase pathway. *Nutrients* **2013**, *5*, 3481–3495. [CrossRef]
21. Zeisel, S.H.; Blusztajn, J.K. Choline and human nutrition. *Annu. Rev. Nutr.* **1994**, *14*, 269–296. [CrossRef]
22. Andersson, A.A.M.; Andersson, R.; Piironen, V.; Lampi, A.-M.; Nyström, L.; Boros, D.; Fraś, A.; Gebruers, K.; Courtin, C.M.; Delcour, J.; et al. Contents of dietary fibre components and their relation to associated bioactive components in whole grain wheat samples from the HEALTHGRAIN diversity screen. *Food Chem.* **2013**, *136*, 1243–1248. [CrossRef] [PubMed]
23. Mares, D.J.; Stone, B.A. Studies on wheat endosperm. I. Chemical composition and ultrastructure of the cell walls. *Aus. J. Biol. Sci.* **1973**, *26*, 793–812.
24. Shewry, P.R.; Hey, S. Do "ancient" wheat species differ from modern bread wheat in their contents of bioactive components? *J. Cereal Sci.* **2015**, *65*, 236–243. [CrossRef]
25. Barrett, J.S.; Gibson, P.R. Fermentable oligosaccharides, disaccharides, monosaccharides and polyols (FODMAPs) and nonallergic food intolerance: FODMAPs or food chemicals. *Ther. Adv. Gastroenterol.* **2012**, *5*, 261–268. [CrossRef] [PubMed]
26. Zebarth, B.J.; Warren, C.J.; Sheard, R.W. Influence of the rate of nitrogen fertilisation on the mineral content of winter wheat in Ontario. *J. Agric. Food Chem.* **1992**, *40*, 1526–1530. [CrossRef]
27. Shewry, P.R. Effects of nitrogen and sulfur nutrition on grain composition and properties of wheat and related cereals. In *The Molecular and Physiological Basis of Nutrient Use Efficiency in Crops*; Hawkesford, M.J., Barraclough, P., Eds.; Wiley-Blackwell: Chichester, UK, 2011; pp. 103–120.
28. Oddy, J.; Raffan, S.; Wilkinson, M.D.; Elmore, J.S.; Halford, N.G. Understanding the Relationships between Free Asparagine in Grain and Other Traits to Breed Low-Asparagine Wheat. *Plants* **2022**, *11*, 669. [CrossRef] [PubMed]
29. Longin, C.F.H.; Ziegler, J.U.; Schweiggert, R.M.; Koehler, P.; Carle, R.; Würschum, T. Comparative study of hulled (einkorn, emmer and spelt) and naked wheats (durum and bread wheat): Agronomic performance and quality traits. *Crop Sci.* **2016**, *56*, 302–311. [CrossRef]
30. Ziegler, J.U.; Schweiggert, R.M.; Würschum, T.; Longin, C.F.H.; Carle, R. Lipophilic antioxidants in wheat (*Triticum* spp.): A target for breeding new varieties for future functional cereal products. *J. Funct. Foods* **2016**, *20*, 594–605. [CrossRef]
31. Ziegler, J.U.; Steiner, D.; Longin, C.F.H.; Würschum, T.; Schweiggert, R.M.; Carle, R. Wheat and the irritable bowel syndrome—FODMAP levels of modern and ancient species and their retention during bread making. *J. Funct. Foods* **2016**, *25*, 257–266. [CrossRef]
32. Masisi, K.; Beta, T.; Moghadasian, M.H. Antioxidant properties of diverse cereal grains: A review on in vitro and in vivo studies. *Food Chem.* **2016**, *196*, 90–97. [CrossRef]
33. Leváková, L.; Lacko-Bartošová, M. Phenolic acids and antioxidant activity of wheat species: A review. *Agriculture* **2017**, *63*, 92–101. [CrossRef]
34. Kwiatkowski, C.A.; Harasim, E.; Feledyn-Szewczyk, B.; Joniec, J. The antioxidant potential of grains in selected cereals grown in an organic and conventional system. *Agriculture* **2022**, *12*, 1485. [CrossRef]
35. Pompella, A.; Sies, H.; Wacker, R.; Brouns, F.; Grune, T.; Biesalski, H.K.; Frank, J. The use of total antioxidant capacity as surrogate marker for food quality and its effect on health is to be discouraged. *Nutrition* **2014**, *30*, 791–793. [CrossRef] [PubMed]
36. Borczak, B.; Marek, M.; Sikora, E.; Dobosz, A.; Kapusta-Duch, J. Glycaemic index of white bread. *Starch* **2018**, *70*, 17022. [CrossRef]
37. Schenk, S.; Davidson, C.J.; Zderic, T.W.; Byerley, L.O.; Coyle, E.F. Different glycemic indexes of breakfast cereals are not due to glucose entry into blood but to glucose removal by tissue. *Am. J. Clin. Nutr.* **2003**, *78*, 742–748. [CrossRef]

38. Eelderink, C.; Moerdijk-Poortvliet, T.C.; Wang, H.; Schepers, M.; Preston, T.; Boer, T.; Vonk, R.J.; Schierbeek, H.; Priebe, M.G. The glycemic response does not reflect the in vivo starch digestibility of fiber-rich wheat products in healthy men. *J. Nutr.* **2012**, *142*, 258–263. [CrossRef] [PubMed]

39. Rojas-Bonzi, P.; Vangsøe, C.T.; Nielsen, K.L.; Lærke, H.N.; Hedemann, M.S.; Knudsen, K.E.B. The relationship between In vitro and In vivo starch digestion kinetics of breads varying in dietary fibre. *Foods* **2020**, *9*, 1337. [CrossRef]

40. Dodd, H.; Williams, S.; Brown, R.; Venn, B. Calculating meal glycemic index by using measured and published food values compared with directly measured meal glycemic index. *Am. J. Clin. Nutr.* **2011**, *94*, 992–996. [CrossRef]

41. Vanhatalo, S.; Dall'Asta, M.; Cossu, M.; Chiavaroli, L.; Francinelli, V.; Pede, G.D.; Dodi, R.; Närväinen, J.; Antonini, M.; Goldoni, M.; et al. Pasta structure affects mastication, bolus properties, and postprandial glucose and insulin metabolism in healthy adults. *J. Nutr.* **2022**, *152*, 994–1005. [CrossRef]

42. Călinoiu, L.F.; Vodnar, D.C. Whole Grains and Phenolic Acids: A Review on Bioactivity, Functionality, Health Benefits and Bioavailability. *Nutrients* **2018**, *10*, 1615. [CrossRef]

43. Serban, L.R.; Paucean, A.; Man, S.M.; Chis, M.S.; Muresan, V. Ancient wheat species: Biochemical profile and impact on sourdough bread characteristics—A review. *Processes* **2021**, *9*, 2008. [CrossRef]

44. Balk, J.; Connorton, J.M.; Wan, Y.; Lovegrove, A.; Moore, K.L.; Uauy, C.; Sharp, P.; Shewry, P.R. Improving wheat as a source of iron and zinc for global nutrition. *Nutr. Bull.* **2018**, *44*, 53–59. [CrossRef]

 *foods*

Article

Breadmaking Performance of Elite Einkorn (*Triticum monococcum* L. subsp. *monococcum*) Lines: Evaluation of Flour, Dough and Bread Characteristics

Andrea Brandolini [1,*], Mara Lucisano [2], Manuela Mariotti [2], Lorenzo Estivi [2] and Alyssa Hidalgo [2]

[1] Consiglio per la Ricerca in Agricoltura e l'Analisi dell'Economia Agraria—Unità di Ricerca per la Zootecnia e l'Acquacoltura (CREA-ZA), Via Piacenza 29, 26900 Lodi, Italy

[2] Dipartimento di Scienze per gli Alimenti, la Nutrizione e l'Ambiente (DeFENS), Università degli Studi di Milano, Via G. Celoria 2, 20133 Milan, Italy; mara.lucisano@unimi.it (M.L.); mariotti.manu@gmail.com (M.M.); lorenzo.estivi@unimi.it (L.E.); alyssa.hidalgovidal@unimi.it (A.H.)

* Correspondence: andrea.brandolini@crea.gov.it; Tel.: +39-0371-404750

Abstract: Einkorn flour, rich in proteins, carotenoids, and other antioxidants, generally has poor breadmaking value. In this research, the composition and technological characteristics of the flours and breads of two elite einkorns (Monlis and ID331) and a bread wheat (Blasco), cropped in four different environments, were evaluated. The einkorns confirmed better flour composition than bread wheat for proteins (on average, 16.5 vs. 10.5 g/100 g), soluble pentosans (1.03 vs. 0.85 g/100 g), and yellow pigment (10.0 vs. 1.0 mg/kg). Technologically, they had better SDS sedimentation values (89 vs. 66 mL), lower farinographic water absorption (52.6 vs. 58.8%), and a similar development time, stability, and degree of softening. Viscoelasticity tests showed lower storage and loss moduli and more prevalent elastic behaviour for Blasco, while rheofermentographic tests showed an anticipated development time (120.8 vs. 175.0 min), higher maximum height (73.0 vs. 63.0 mm), and superior retention coefficient (99.1 vs. 88.7%), but a lower CO_2 total (1152 vs. 1713 mL) for einkorn doughs. Einkorn breads were bigger than the control (736 vs. 671 cm^3); crumb pores percentage was similar, but medium-size pores were scarcer. Finally, a 52-h shelf-life trial demonstrated that einkorn bread had a softer texture, maintained for a longer time, and a slower retrogradation than the control. Therefore, choice of appropriate varieties and process optimisation allows the production of excellent einkorn breads with a superior nutritional value and longer shelf life.

Keywords: colour; farinograph; rheofermentograph; viscoelastic behaviour; bread shelf life; crumb porosity

Citation: Brandolini, A.; Lucisano, M.; Mariotti, M.; Estivi, L.; Hidalgo, A. Breadmaking Performance of Elite Einkorn (*Triticum monococcum* L. subsp. *monococcum*) Lines: Evaluation of Flour, Dough and Bread Characteristics. *Foods* **2023**, *12*, 1610. https://doi.org/10.3390/foods12081610

Academic Editor: Saroat Rawdkuen

Received: 8 March 2023
Revised: 28 March 2023
Accepted: 3 April 2023
Published: 10 April 2023

1. Introduction

Einkorn (*Triticum monococcum* L. subsp. *monococcum*), a diploid hulled wheat (2 n = 2 x = 14) with high protein [1–3], lutein, and antioxidants content [4,5], is considered a cereal with a poor baking attitude and a dough characterized by excessive stickiness. The farinographic tests often show evidence of scarce stability and a high degree of softening, while breadmaking yields small volume loaves, due to reduced leavening and easy collapse of the dough [6,7]. However, a great variation for breadmaking quality exists within the einkorn gene pool, and selected ecotypes with high SDS sedimentation values (>70 mL), good farinographic stability (360–720 s), and a limited degree of softening (20–50 UB) yielding breads with volumes similar or even higher than wheat breads have been identified [6–8]. Additionally, because their doughs present poor tolerance to mechanical processing and prolonged fermentation, gentle processing at a low speed and for a short duration (3–4 min) have been proposed to improve loaves' volume [9]. The use of sourdough fermentation has also been suggested to improve texture, volume, and shelf life of einkorn bread [10]. Crust shape and colour are similar to those of wheat loaves, but the crumb shows an enticing yellow tinge [11] for the high carotenoids content [5]. Furthermore, the lower endogenous enzymatic activity [12,13] allows for better preservation of antioxidants during storage [14,15] and processing [16,17], and limits heat damage [18], thus safeguarding the favourable nutritional characteristics of einkorn-derived products.

The potential health benefits of einkorn foods have been the subject of recent investigations. Although einkorn, like other wheats, such as barley and rye, is not suitable for consumption by people with celiac disease [19–21], einkorn may be better suited than other *Triticum* species for patients with chronic dysmetabolic diseases. For example, einkorn bread consumption leads to more favorable metabolic responses and greater satiety compared with standard wheat breads [22]. Investigating the effects of einkorn bread on the intestinal physiology and metabolism of pigs, Barone et al. [23] observed a lower postprandial insulin rise after einkorn consumption compared with bread wheat consumption; furthermore, the intestinal ecosystem was enriched in health-promoting bacteria. Einkorn's anti-inflammatory effects were also recorded in cultured cells [24].

Although einkorn breads are often characterized by an inferior technological quality compared with wheat breads, their enticing taste, aroma, attractive color, and health-promoting properties suggest that they may be a worthy addition to the increasing assortment of available products. Therefore, the aim of this research was to evaluate several breadmaking quality facets of two elite einkorn wheats (ID331 and Monlis) and, as control, of one bread wheat cultivar (Blasco); the two einkorn accessions were selected because in previous unpublished tests repeatedly showed the best breadmaking attitude. Hence, all-around information about the breadmaking properties of these two elite einkorns will allow the baking industry to develop new leavened products through adopting the most appropriate processing approaches. To this end, the composition and technological and rheological properties of their flours, doughs, and breads were assessed throughout the breadmaking process; furthermore, the shelf life of the bread loaves was investigated.

2. Materials and Methods

2.1. Materials

Two breadmaking-quality einkorns (ID331 and Monlis) and a breadmaking-quality bread wheat (cv. Blasco) were cropped in four different environments, two under conventional management (Sant'Angelo Lodigiano, LO, and Montelibretti, Rome, labelled as SAL and ROMA) and two under organic management (Sant'Angelo Lodigiano, LO, and Leno, BS, i.e., SALbio and LENObio). Some relevant agronomic parameters of the four environments are summarized in Table S1. The accessions were planted in 10 m^2 plots according to a Randomized Complete Block Design with three replications. For weed control, the herbicide Ariane II (Clopiralid 1.8% + Fluroxypyr 3.6% + MCPA 18.2%; Dow AgroSciences, Milan, Italy) was applied just before heading to the conventional management plots, while the organic trials were manually weeded. After machine harvesting, by mid-July (Table S1), the Monlis and ID331 kernels were de-hulled with an Otake FC4S thresher (Satake, Japan), which is a step that was not necessary for the free-threshing bread wheat Blasco. All seeds were stored at 5 °C until further processing.

2.2. Methods

2.2.1. Kernels

Kernel moisture was determined by a GAC2000 moisture analyzer (Dickey–John, Auburn, IL, USA) and corrected to 15% for the einkorns and 16% for Blasco (harder texture) by overnight tempering. Afterwards, the hectoliter weight (kg/hL) was determined with a GAC2000 instrument (Dickey–John, Auburn, IL, USA). The seeds were milled with a Bona–GBR laboratory mill (Bona, Monza, Italy), which separates flour from bran and germ, and the milling yield (% flour/kernels w/w) was computed.

2.2.2. Flour and Dough

Flour particles' size was determined by sifting 100 g of flour for 5 min through a sieve with a 125 μm mesh; the ≥125 vs. <125 ratio was computed. Flour characteristics were then assessed according to the following methods: moisture (AACC 44-15.02 [25]), ash content (AACC 08-03.01 [25]), protein content (N × 5.7; AACC 46-10.01 [25]), yellow pigment (AACC 14-50.01 [25]), dry gluten content (AACC 38-12.02 [25]) with a Glutomatic (Perten, Hägersten, Sweden), Falling number (AACC 56-81.03 [25]) with a 1550 Falling Number (Perten, Hägersten, Sweden), SDS sedimentation test (a breadmaking attitude predictor [26]), starch and amylose contents and α-amylase activity with Megazyme Assay Kits (Megazyme International Ireland Inc., Bray, Ireland), and pentosan and soluble pentosan contents (colorimetric method) [27]. Dough mixing properties were evaluated with a Brabender farinograph (Brabender OHG, Duisburg, Germany) using a 50 g mixer according to the ICC method 115-D [28]. Briefly, after adjusting the maximum consistency of the dough to a fixed value (500 Brabender Units, BU) by altering the quantity of water added, the test was run for 12 min. The parameters recorded were water absorption (amount of water added to set the curve at 500 BU, expressed as percentage of flour at 14% moisture), development time (time in minutes between the origin of the curve and its maximum value i.e., 500 BU), stability time (difference in minutes between the time to the maximum and time when the top of the curve falls below 500 line), and degree of softening (difference in BU between the maximum value and the value at the end of the test).

Dough leavening properties were assessed by rheofermentographic tests and were performed with a Chopin F3 Rheofermentometer (Chopin SA, Villeneuve-La-Garenne, France), according to [29]. The indices recorded from the curves were Hm (mm; maximum development of the dough), h (mm; height of the dough at the end of the test), CO_2 total (mL; total gas production during the test), CO_2 loss (mL; total gas loss during the test), CO_2 retained (mL; gas retained by the dough during the test), and retention coefficient (% CO_2 retained/CO_2 total).

To better understand the fundamental properties of the doughs, rheological analyses were carried out according to [30] with a Physica MCR300 Rheometer, supported by US200 v. 2.5 software (PHYSICA Messtechnic GmbH, Ostfildern, Germany). Measurements were carried out at 25 °C using a corrugated parallel plate system (diametre: 2.5 cm) at a gap of 2 mm and a special humidity cover (H-PTD 150) to prevent moisture losses. Frequency sweep tests were performed over the range 0.1–50 Hz at 0.1% strain on doughs prepared at the same consistency (500 UB) after a resting period of 30 min to equilibrate stresses. The selected 0.1% strain value was obtained from preliminary amplitude sweep tests performed in the range of 0.01–100% strain, at a constant frequency of 1 Hz, to determine the linear viscoelastic region (LVR) of the sample. Each test was performed at least three times. Data were analyzed with US200/32 v. 2.50 rheometer software (PHYSICA Messtechnic GmbH, Ostfildern, Germany) and the value of storage modulus (G′, Pa) and loss modulus (G″, Pa) at 1 Hz and tanδ (ratio between G″ and G′) were computed.

2.2.3. Bread

Two breads per accession were produced according to method 10-10 B [25] with minor modifications [6]. Bread weight (g) was determined with a LP5200P balance (Sartorius AG, Göttingen, Germany), bread volume (cm^3) by rapeseed displacement (method 10-05.01 [25]), and bread height (mm) with a caliper; specific volume (cm^3/g) was computed as volume-to-weight ratio.

Bread quality changes during storage were monitored on eight breads per accession, prepared only from SAL trial flours. Two breads were immediately analyzed, while the others were packaged in food paper bags and stored at 25 °C and 60% relative humidity in an HC 0020 air-conditioned cell (Heraeus, Hanau, Germany). Four times (t_0: 0 h, t_1: 24 h, t_2: 30 h, t_3: 52 h), two bread loaves of each accession were weighed to evaluate weight decrease during storage, then they were transversely cut to obtain three uniform 25-mm-thick slices. Crumb moisture was assessed on two slices while water activity (a_w) analysis was performed on the third central slice with an AQUALAB apparatus (Decagon Devices Inc., Pullman, DC, USA).

Crust color and crumb color of the three slices of each sample at t_0 were assessed in the CIELAB space using a Minolta Chroma Meter CR 210 (Minolta Camera Co., Osaka, Japan) with a standard illuminator C. Additionally, images were captured using a flatbed scanner (Scanjet 6300c; Hewlett Packard, Palo Alto, CA, USA) in 256 grey level at 300 dots per inch and were processed using a dedicated software (Image Pro-Plus 4.5.1.29, Media Cybernetics Inc., Rockville, MD, USA). The parameters evaluated were density red (R), density green (G), density blue (B), and density mean. At the same time, crumb porosity was determined by assessing the number, area, diameter, and shape of the pores, their size distribution (three categories: C1: 0.1–1 mm^2, C2: 1–5 mm^2, C3: >5 mm^2), and % of pore area.

Crumb texture characteristics were assessed by Texture Profile Analysis on the three central slices from each sample at each storage time with an HD.plus Texture Analyzer TA double-column dynamometer (Stable Micro System, Godalming, UK) connected to a Texture Exponent 32 recording system version 4.0.8.0 (Stable Micro System, Godalming, UK). The operating conditions adopted were: load cell 500 N, crossbar speed 2 mm/s, compression plate diameter 36 mm, sample compression up to 40% thickness, and 25 s waiting time between 1st and 2nd compression cycles. The parameters derived from the Force/Time curve were: young modulus (N/mm^2), hardness (N), springiness or elasticity (adimensional), cohesiveness (adimensional), and chewiness (i.e., hardness × cohesiveness × springiness, N).

2.2.4. Statistical Analysis

A two-way analysis of variance (ANOVA) was carried out for most parameters, using environment and genotype, or genotype and storage time, as factors. A one-way ANOVA was performed on the results of the color and porosity parameters, assessed on the breads prepared from the SAL flours. When significant differences ($p \leq 0.05$) were found, Fisher's lowest significant difference (LSD) was computed. Both ANOVA and LSD testing were performed with the software STATGRAPHICS plus v.4 (STATPOINT Technologies Inc., The Plains, VA, USA). Means and standard errors were determined using the software Excel® (Microsoft Corporation, Redmond, DC, USA).

3. Results

3.1. Kernels

Figure 1 reports the values of hectoliter weight, flour yield, and proportion of flour <125 μm, while their behavior in the four locations is depicted in Figure S1. The results in the different locations (Figure S1) suggest that the hectoliter weight of the bread wheat was generally higher than that of the two einkorns, while the flour yield and the percentage of flour particles <125 μm were higher in the einkorns than in the bread wheat Blasco. It is interesting to note that an elevated hectoliter weight, besides indicating healthy caryopses with a compact endosperm typical of "hard" type wheats, such as Blasco, usually forecasts high flour yields [31]. However, despite the smaller seeds and consequent superior teguments proportion [2], as mentioned above, the flour yield of both einkorns seems superior to that of the bread wheat Blasco. This peculiar result was also spotted by Borghi et al. [6] in their survey of 25 einkorns and by Corbellini et al. [7] in their study of 24 einkorns, whose flour yields ranged between 53.0 and 64.4% and were similar to the soft wheat Veronese (59.0%), but were superior to the hard wheat Pandas (50.0–52.6%). The higher extraction rate may be linked to the extra-soft texture of einkorn kernels [1,7], which makes them more easily grindable, and originates a very fine flour [6,32,33]. In this study, indeed, the percentage of einkorn flour particles smaller than 125 μm was, on average, 86.0 g/100 g (in the range of the 84.9–91.7 g/100 g reported by Borghi et al. [6]), while it was 57.8 g/100 g for Blasco. Probably during milling the harder bread wheat kernels broke into larger fragments, thus originating, in comparison to einkorn, a coarser flour. No major differences between ID331 and Monlis were observed.

Figure 1. Bread slices prepared from flours of bread wheat Blasco, einkorn ID331, and einkorn Monlis.

3.2. Flour and Dough

A two-way ANOVA (Table S3) highlighted the existence of significant differences among accessions and among environments, as well as their interactions for moisture, ash, protein, total starch, amylose, soluble pentosans, total pentosans, and yellow pigment content. In all instances, the genotype was the most relevant factor. The results of the four environments, not discussed in this article, are summarized in Table S3.

The ash content of Blasco (Table 1) was significantly lower (0.58 g/100 g DM) than that of the two einkorns (on average, 0.63 g/100 g DM), as evidenced also by other authors [1,32,34]. Additionally, Blasco was less rich in protein and richer in starch than the two T. monococcum accessions.

Table 1. Kernel and flour parameters (mean values ± standard error) of two einkorns (ID331 and Monlis) and one bread wheat (Blasco). n = 8: four environments × two repetitions.

	ID331	MONLIS	BLASCO
Kernel *			
Hectolitre weight (kg/hL)	76.6 ± 1.1	75.8 ± 1.4	85.3 ± 1.4
Flour yield (g/100 g)	61.8 ± 0.7	59.8 ± 1.1	54.0 ± 2.5
Flour <125 µm (g/100 g)	85.3 ±1.6	86.6 ± 1.8	57.8 ± 2.9
Flour			
Moisture (g/100 g)	14.5 ±0.4	14.3 ± 0.4	14.5 ± 0.3
Ash (g/100 g DM)	0.60 [b] ± 0.02	0.67 [a] ± 0.01	0.59 [c] ± 0.01
Protein (g/100 g DM)	16.7 [a] ± 1.5	16.2 [b] ± 1.3	10.5 [c] ± 0.8
Starch (g/100 g DM)	69.3 [b] ± 1.3	68.7 [b] ± 1.8	79.1 [a] ± 1.1
Amylose (g/100 g starch)	25.3 [b] ± 0.1	26.9 [a] ± 0.3	26.8 [a] ± 0.3
Total pentosans (g/100 g DM)	2.49 ± 0.01	2.66 ± 0.08	2.52 ± 0.05
Soluble pentosans (g/100 g DM)	0.95 [b] ± 0.04	1.10 [a] ± 0.03	0.82 [c] ± 0.01
Yellow pigment (mg/kg DM)	8.7 [b] ± 0.6	11.3 [a] ± 0.7	1.0 [c] ± 0.1
Dry gluten (g/100 g DM)	1.58 [a] ±0.13	1.49 [a] ± 0.08	1.15 [b] ± 0.03
Falling number (s)	358 [b] ± 16	365 [b] ± 11	431 [a] ± 13
α-amylase activity (CU/g DM)	0.19 [a] ± 0.02	0.20 [a] ± 0.01	0.15 [b] ± 0.01

Different letters in the same row mean significant differences ($p \leq 0.05$) among samples following LSD test. * The kernel data were not subjected to statistical analysis.

Einkorn is well known for its high protein levels [1,6,7,32,33,35], which in part is due to genetic factors and in part to its smaller kernels having an inferior endosperm-to-external layers ratio and hence, a superior incidence of the protein-rich aleuronic layer [2]. Nevertheless, despite the considerable difference in protein concentration, the amino acid composition of all wheats is similar, with lysine representing the main limiting factor [36–38].

Conversely, the plump Blasco kernels were richer in starch than the smaller, protein-rich einkorn seeds (79.1 vs. 69.0 g/100 g). However, the amylose proportion of total starch was similar between Monlis and Blasco, and was only slightly inferior in ID331 (Table 1). On the other hand, Blasco had similar pentosan and inferior soluble pentosan contents than the einkorns (2.52 vs. 2.57 g/100 g and 0.82 vs. 1.03 g/100 g, respectively). The total pentosan concentration of Blasco was within the variation (2.0–3.0 g/100 g) described for bread wheat [39,40], while the soluble pentosan content fell into the range (0.49–1.23 g/100 g) observed in *T. aestivum* [27,41,42]. To the best of our knowledge, no information about pentosans content in *Triticum monococcum* is available in the literature. Interestingly, when pentosans content in flour increases, the retrogradation of starch decreases, due to their steric interference with the intermolecular associations of starch [43].

A peculiar characteristic of einkorn flour is its yellowness, due to the abundant presence of carotenoid [4,5]. This is confirmed by the yellow pigment content of both ID331 (8.72 mg/kg DM) and Monlis (11.30 mg/kg DM), which are about ten times larger than the value of Blasco (0.98 mg/kg DM) and in the range reported for whole meal flours of a collection of einkorns [1]. Remarkably, the two einkorns tested had concentrations that were twice higher than those commonly reported for yellow durum wheats [44].

The two-way ANOVAs (Table S2) verified the existence of highly significant differences among genotypes, environments, and their interactions for dry gluten content, Falling number, α-amylase activity, and SDS sedimentation value; the genotypic effect was largely the most relevant factor. The results of the four environments, not discussed in this article, are summarized in Table S3 for completeness of information.

As already hinted by the protein content, the dry gluten content was higher in ID331 and Monlis than in Blasco (Table 1). Falling number and α-amylase activity are inversely correlated traits [45]: the higher the α-amylase activity, the shorter the falling time. In this study, Falling Number results were all well above those values suggesting modest pre-germination phenomena (200–300 s) [46]. The significantly higher levels of both FN and α-amylase activities obtained for einkorn flours can be explained with their superior ash content, being that the amylolytic enzymes are mainly localized in the germ and in the external regions of the kernels.

The SDS sedimentation test underlined the good breadmaking propension of the two einkorns tested and of the bread wheat control. Einkorn generally has a poor breadmaking capacity [32,33], but some good accessions are reported [1,6,7,47]. Indeed, ID331 and Monlis are among the most suitable *T. monococcum* to prepare leavened products, and their SDS quality was superior even to the good breadmaking wheat variety Blasco.

The farinographic results (Figure S1) suggested differences between the bread wheat flour and the two einkorn flours only for water absorption (Table 2), as Blasco needed more water (around 6%) than ID331 and Monlis to reach optimum dough consistency. Development time, stability, and degree of softening did not look different. Borghi et al. [6] found that most of the einkorn accessions that they investigated had very low stability (<1 min), but some genotypes reached 2.0–4.5 min. Low development time, poor stability, and a strong degree of softening are indices of a weak flour, with scarce resistance to the mechanical action of kneading and are therefore unsuitable for preparing leavened products. Nevertheless, ID331 and Monlis showed a breadmaking attitude similar to Blasco, a cultivar fit for the manufacturing of high-quality leavened bakery products.

The unreplicated results of the rheofermentographic parameters are summarized in Table 2 and are presented across the four locations in Figure S2. This analysis is helpful to evaluate changes in dough during the leavening phase that can be linked to breadmaking quality [29]. The results in Figure S2 suggest that Monlis and ID331 reached a significantly higher dough development during the test in a shorter time in comparison to Blasco, maybe due to the smaller size of flour (see above) and starch granules [48], whose superior surface-to-volume ratio may favor enzymatic reactions. They also exhibited a limited CO_2 loss, which determined a significantly higher CO_2 retention coefficient, probably attributable to a more compact gluten network due to the higher protein content of the two einkorn flours.

The two-way ANOVA (Table S2) of the viscoelasticity parameters hinted to significant differences between Blasco and the two einkorns, both for the limits of the linear viscoelastic region obtained from the strain sweep test, and the storage and loss moduli obtained from the frequency sweep test. For completeness of information, the results of the four environments, not discussed in this article, are summarized in Table S3.

Table 2. Breadmaking parameters (mean values ± standard error) of two einkorns flours (ID331 and Monlis) and one bread wheat flour (Blasco). $n = 8$: four environments × two repetitions (Brabender, $n = 4$: four environments).

	ID331	MONLIS	BLASCO
SDS sedimentation volume (mL)	91 [a] ± 0.9	88 [a] ± 2	66 [b] ± 7
Brabender farinograph *			
Water absorption (%)	52.7 ± 1.5	52.4 ± 0.8	58.8 ± 0.6
Development time (s)	144 ± 15	168 ± 23	195 ± 107
Stability time (s)	387 ± 226	305 ± 122	429 ± 234
Degree of softening (BU)	65 ± 20	85 ± 15	56 ± 22
Rheofermentograph *			
Dough max height (mm)	72.3 ± 5.3	73.8 ± 4.8	63.0 ± 3.6
Time to max height (min)	121.8 ± 10.8	119.8 ± 20.5	175.0 ± 4.4
CO_2 total (mL)	1136 ± 89	1168 ± 107	1713 ± 79
CO_2 lost (mL)	12 ± 5	15 ± 5	193 ± 26
CO_2 retained (mL)	1124 ± 85	1154 ± 103	1523 ± 85
Retention coefficient (%)	99.0 ± 0.4	99.3 ± 0.1	88.7 ± 1.6
Strain sweep test			
LVR limit for G' (%)	0.70 [a] ± 0.03	0.69 [a] ± 0.02	0.33 [b] ± 0.02
LVR limit for G'' (%)	0.82 [a] ± 0.03	0.79 [a] ± 0.01	0.48 [b] ± 0.02
Frequency sweep test			
G' (Pa) (0.10% strain, 1 Hz)	9216 [b] ± 822	10,748 [a] ± 860	7764 [c] ± 605
G'' (Pa) (0.10% strain, 1 Hz)	4403 [b] ± 270	5131 [a] ± 565	2933 [c] ± 151
Damping factor G''/G'	0.48 [a] ± 0.02	0.48 [a] ± 0.02	0.38 [b] ± 0.02
Bread			
Volume (cm³)	732 [a] ± 128	740 [a] ± 107	671 [b] ± 43
Height (cm)	102 [ab] ± 12	104 [a] ± 10	99 [b] ± 3
Specific volume (cm³/kg)	5.23 [ab] ± 0.97	5.40 [a] ± 0.77	4.55 [b] ± 0.29

Different letters in the same row mean significant differences ($p \leq 0.05$) among samples following LSD test. * The farinograph and rheofermentograph data were not subjected to statistical analysis.

The evaluation of the region of linear viscoelasticity of a sample is an important step: when materials are tested in the linear range, their viscoelastic behaviors do not depend on the magnitude of the stress, the magnitude of the deforming strain, or the rate of application of the strain [49], but on their intrinsic features. The length of the linear viscoelastic region (LVR) can therefore be used as a measurement of dough stability: einkorn doughs remained in the linear viscoelastic region over greater strains than Blasco dough, indicating the presence of a stronger network. Indeed, for Blasco a drop in G' LVR started to occur above the 0.33% strain and became larger at a higher strain, indicating a progressive disorganization of the dough structure beyond this deformation level. A 0.1% strain for the subsequent frequency tests was therefore adopted, as within the LVR of all the samples. Although all the dough samples that were analyzed had the same farinographic consistency (500 BU), significant differences were observed among the samples. Generally, the mechanical spectra of all the samples exhibited a solid-like behavior, with G' always being higher than G''. Einkorn doughs presented significantly higher values of both moduli, with a prevalence of the viscous behavior, as indicated by the higher values of the damping factor (G''/G').

3.3. Bread

3.3.1. Characteristics

The two-way ANOVA for bread volume, height, and specific volume (Table S2) scored significant genetic effects, including especially highly significant environmental effects and their interactions. Interestingly, ID331 and Monlis outcompeted Blasco for loaf volume, height, and specific volume (Table 2), indicating that it is possible to have an einkorn flour with the same technological properties of good quality bread wheat. Cross-sections of the loaves prepared with the flours from the LENObio trial are depicted in Figure 1.

Due to the higher amount of water added following indication of the farinographic test (56.0% vs. 53.4% to reach 500BU), the bread produced from Blasco flour had a significantly higher loaf weight (Table 2), crumb humidity, and a_w (Figure 2, t_0) than the einkorn samples. Monlis exhibited the highest loaf volume (on average, 740 cm^3), followed by ID331 (732 cm^3) and Blasco (671 cm^3). Similar values for loaves obtained from selected einkorn flours are reported [6,7]. These results were reflected in the breads' specific volumes: Monlis originated the breads with the highest value, that is the "lightest" breads (Table 2).

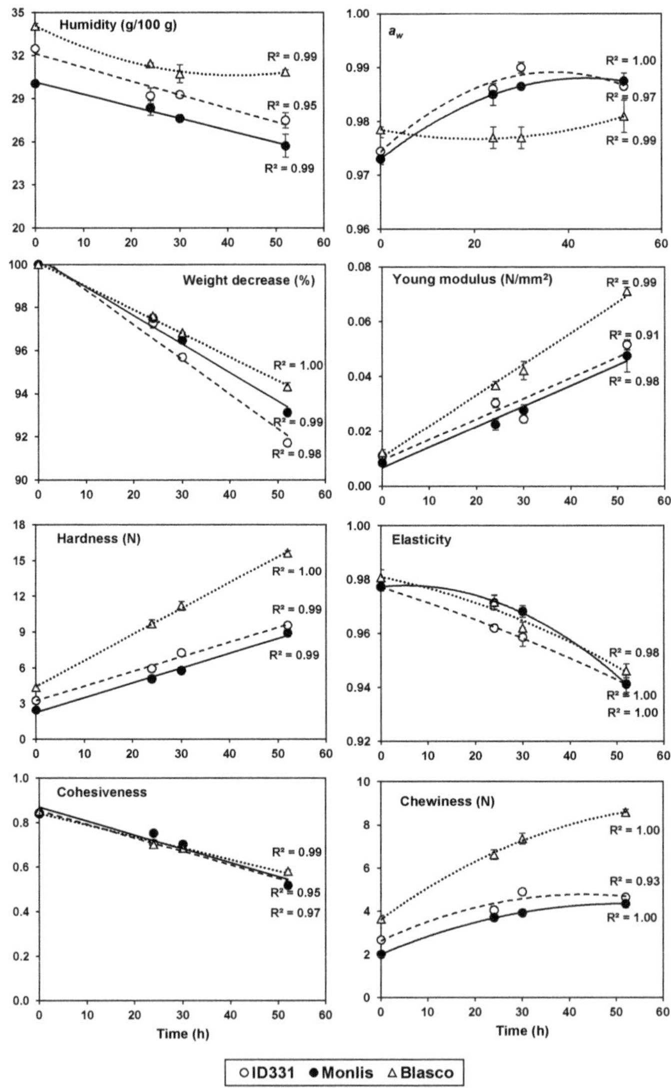

Figure 2. Moisture, water activity (a_w), weight decrease, and texture parameters of breads from two einkorns (ID331 and Monlis) and one bread wheat (Blasco) during storage (h). Error bars represent standard error.

The one-way ANOVA (Table S4) indicated the presence of significant differences among genotypes for color indices in the cases of a^* (crust and crumb) and b^* (crumb). Table 3 reports the average values of L^*, a^*, and b^* evaluated on the crust and crumb of each bread sample. The crust color showed a low variation, with very similar L^*, a^*, and b^* values among samples, as observed also by D'Egidio et al. [31]. A difference was instead evident for crumb color: Blasco had significantly higher a^* values, but a much inferior b^* (yellow index) than both the einkorn genotypes. This reflects the differences in the yellow pigment content of the raw materials (Table 1), which was particularly abundant in the carotenoid-rich einkorns and gives their breads a characteristic and enticing golden yellow color. Table 3 also shows the average values of the color parameters evaluated by Image Analysis; the ANOVA (Table S4) highlighted significant differences among the samples for all parameters: Blasco had significantly lower R (red intensity) and G (green intensity), and had significantly higher B (blue intensity) than the einkorn samples.

Table 3. Crust and crumb colorimetric indices (mean values ± standard error) of breads prepared from flours of two einkorns (ID331 and Monlis) and one bread wheat (Blasco) cropped at Sant'Angelo Lodigiano under conventional management. n = 12: three sections from two slices of two loaves.

		ID331	Monlis	Blasco
Crust	L^*	46.1 ± 3.1	45.7 ± 2.6	46.1 ± 2.7
	a^*	17.3 [ab] ± 0.5	17.0 [b] ± 0.4	17.7 [a] ± 0.7
	b^*	26.7 ± 2.7	26.8 ± 2.2	25.4 ± 1.5
Crumb	L^*	79.6 ± 0.9	79.2 ± 0.5	79.6 ± 1.8
	a^*	−2.6 [b] ± 0.1	−2.9 [c] ± 0.1	0.2 [a] ± 0.08
	b^*	35.2 [a] ± 0.6	35.8 [a] ± 0.2	15.5 [b] ± 0.4
	R	234 [a] ± 5	232 [a] ± 7	219 [b] ± 4
	G	216 [a] ± 5	214 [a] ± 7	207 [b] ± 3
	B	130 [b] ± 3	126 [c] ± 4	172 [a] ± 3

Different letters in the same row mean significant differences ($p \leq 0.05$) among samples following LSD test.

Bread can be considered a solid food foam because the gas developed during fermentation is trapped in its solid matrix. Information about shape, size, and distribution of the pores is very important in order to identify the influence of processing conditions on the quality of the final products and to devise methods to manufacture products with peculiar textural characteristics [50]. Image Analysis was used in this study to assess the porosity of the different breads by determining the number of pores and their geometrical features (area, diameter, shape). All the identified pores were classified according to their size and the results are shown in Table 4, while the ANOVA is presented in Table S5. The Blasco breads presented as many small pores (0.1–1 mm^2) as the einkorn breads, while the number of medium-sized pores (1–5 mm^2) was significantly higher ($p \leq 0.05$). Among einkorns, Monlis bread porosity was characterized by a greater number of large pores (>5 mm^2), with an average area of 9.21 mm^2, significantly higher than the Blasco and ID331 pores of the same class.

Information about the shape and regularity of the pores is obtained by considering that the larger the value, the greater the irregularity. Indeed, the smallest pores had the most regular and uniform shape (range: 1.78–1.89), while the large pores ranged from 2.34 to 2.60, indicating that as their size increased, their form became more irregular and less homogeneous. In our case, ID331 bread had the most regular pores for all the three size classes: its bread presented a porosity similar to Blasco, but was characterized by a more regular shape, thus making the porosity of the crumb more homogeneous.

Table 4. Porosity (mean values ± standard error) of the breads prepared from two einkorn (ID331 and Monlis) and one bread wheat (Blasco) refined flours cropped at Sant'Angelo Lodigiano under conventional management. n = 12: three sections from two slices of two loaves. Porosity classes: C1 = 0.1–1 mm^2, C2 = 1–5 mm^2, C3 = >5 mm^2. Different letters within pores class in the same column indicate significant differences ($p \leq 0.05$) among samples following LSD test.

		N° Pores	N° Pores (%)	Pores Area (mm^2)	Pores Area (%)	Mean Pores Area (mm^2)	Mean Pores Diametre (mm)	Pores Shape
C1	ID331	647 ± 34	75.7 ± 2.8	237.5 [a] ± 13.1	32.6 ± 3.6	0.37 ± 0.01	0.64 [b] ± 0.01	1.78 ± 0.03
	Monlis	611 ± 30	73.9 ± 2.8	219.6 [b] ± 11.4	27.9 ± 4.3	0.36 ± 0.01	0.64 [b] ± 0.01	1.89 ± 0.05
	Blasco	652 ± 32	72.7 ± 1.6	243.5 [a] ± 12.6	30.1 ± 2.2	0.37 ± 0.01	0.65 [a] ± 0.01	1.89 ± 0.04
C2	ID331	193 [b] ± 32	22.5 ± 2.9	377.8 [b] ± 63.2	51.2 [b] ± 4.5	1.96 ± 0.08	1.55 ± 0.04	1.98 [b] ± 0.05
	Monlis	195 [b] ± 23	23.6 ± 2.2	388.0 [b] ± 57.2	48.5 [b] ± 0.8	1.98 ± 0.08	1.57 ± 0.03	2.16 [a] ± 0.08
	Blasco	229 [a] ± 14	25.6 ± 1.2	454.9 [a] ± 31.7	56.2 [a] ± 2.3	1.98 ± 0.05	1.58 ± 0.02	2.12 [a] ± 0.09
C3	ID331	15 [b] ± 3	1.8 [b] ± 0.5	118.1 [b] ± 28.1	16.1 [b] ± 3.6	7.97 [b] ± 0.59	2.93 ± 0.16	2.34 ± 0.34
	Monlis	21 [a] ± 6	2.5 [a] ± 0.7	191.4 [a] ± 51.0	23.6 [a] ± 3.9	9.21 [a] ± 0.94	3.09 ± 0.18	2.39 ± 0.38
	Blasco	15 [b] ± 4	1.7 [b] ± 0.4	112.2 [b] ± 35.5	13.7 [b] ± 3.9	7.37 [b] ± 0.95	2.86 ± 0.13	2.60 ± 0.36

The number of pores and their average area is reflected in the percentage of relative area occupied in the slice and, ultimately, in the total porosity area of the bread. For Blasco, the surface occupied by the small pores was similar to that found in the einkorn samples, while the surface associated with medium-sized pores was significantly higher. Overall, pores area was highest in Blasco bread (810.57 mm^2), mainly because of the greater number of medium-sized pores. In Monlis bread, the large pores contributed considerably to the total porosity, as also confirmed by the less homogeneous internal loaf structure. Nevertheless, the loaves obtained from Monlis were the ones with the largest specific volumes, probably due to the high presence of these large pores. Despite the distribution of pores in the three classes being different among the three samples, the percentage of pores, computed as ratio between total alveolate area and total bread slice area, was similar among the three accessions.

The mechanical and geometrical characteristics of bread influence its behavior during oral processing [50]. To assess these features in a reproducible way, Texture Profile Analysis is used to create controlled stresses and to measure the mechanical characteristics of foods. Figure 2 shows the average values of the texture parameters of the three different breads. The two-way ANOVA (Table S6) stressed the existence of significant differences ($p \leq 0.05$) between accessions only for young modulus (consistency index), hardness, gumminess, and chewiness. Storage times were always significant and so was the genotype x time interaction (except for elasticity). At t_0, Blasco showed the highest consistency, hardness, gumminess, and chewiness, and Monlis exhibited the lowest values, while in general ID331 had results in line with Monlis. These data indicate how the bread loaves obtained from Monlis were the softest ones and were characterized by the best chewability, despite the finer and more regular porosity and the higher moisture content of Blasco breads, which are factors that are conventionally related to a softer structure. Possibly, the superior softness of the einkorn breads may be related to the lower starch content of their flours.

3.3.2. Changes during Storage

Figure 2 also shows the evolution of a_w, weight, young modulus, and texture parameters of the breads' slices during storage. Blasco bread lost most of its moisture in the first 24 h after the production, while the decrease was more progressive for the einkorns. Interestingly, Blasco bread did not present significant water activity changes during storage, while the einkorn breads showed a significant increase after the first 24 h, followed by a plateau. This increase could be due to a reorganization of starch molecules to a different composition of starch with particular reference to the amylose content whose molecules, involved in the phenomenon of retrogradation, have a greater tendency to associate with each other and form hydrogen bonds, releasing water, or even to a reorganization of the protein matrix [51], more abundant in einkorn flours. As expected, the weight of all the breads significantly decreased during storage, but was steeper for the two einkorns after the initial 24 h.

The Young modulus of the Blasco bread at t_0 was higher than that of the einkorn breads. During storage, this index increased for all samples, but was faster in the bread wheat. This behavior is also reflected in the hardness: in fact, the loaves from einkorn flours showed, already at t_0, lower hardness values (ID331: 3.26 N; Monlis: 2.44 N) compared with Blasco (4.35 N), and this difference amplified over time, as the kinetics of the bread from bread wheat were more accelerated than those of einkorn breads. The speed of the process, as measured by the value of the angular coefficient of the regression curves, was 0.24 N/h for Blasco and 0.12–0.14 N/h for the two einkorn samples. Hence, the einkorn breads maintained their softness for a longer time and their retrogradation was slower than that of the bread wheat.

Elasticity and cohesiveness decreased with similar trends in einkorn and wheat breads over the storage time. For that reason, bread chewiness (i.e., the energy required to chew a solid food) was mainly related to the hardness values, which was much higher for wheat bread, particularly during storage. Since the superior softness of the einkorn breads cannot be attributed to a higher moisture content (indeed, Blasco breads were characterized by the highest moisture), such behavior may be related to the lower starch content of einkorn flours and/or to the lower presence of amylose in the samples. Additionally, the slightly higher amylase activity of Monlis and ID331 may have contributed because amylases decrease starch retrogradation, diminish rigidity of the starch gel network, and limit starch–protein interactions [52,53]. Their anti-staling effect is also related to the hydrolyzation of amylopectin and the production of soluble low-molecular-weight branched-chain polymers, which are less prone to retrogradation and influence water movement and accessibility [54].

4. Conclusions

This study demonstrates the possibility of obtaining einkorn breads with technological properties comparable to those of breadmaking-quality bread wheat. However, to achieve this result, high-quality genotypes are necessary. Furthermore, as the quality characteristics of the breads are influenced by processing conditions, einkorn bread manufacturing should utilize short mixing and leavening times to avoid overstressing the doughs. The choice of appropriate einkorn varieties and optimization of the breadmaking process will allow the production of breads with appropriate technological characteristics, endowing the market with an innovative cereal-based food possessing a superior nutritional value and longer shelf life.

Supplementary Materials: The following supporting information can be downloaded at: https://www.mdpi.com/article/10.3390/foods12081610/s1, Figure S1: Hectoliter weight, flour yield, and proportion of flour <125 µm of flours of ID331, Monlis, and Blasco cropped in four locations (SAL, SALbio, LENObio, and ROMA); Figure S2: Reofermentographic results of flours of ID331, Monlis, and Blasco cropped in four locations (SAL, SALbio, LENObio, and ROMA). Table S1: Cropping environments and field management information; Table S2: Two-way ANOVA (* $p \leq 0.05$; ** $p \leq 0.01$; *** $p \leq 0.001$) of composition, rheological analysis, and breadmaking test parameters; Table S3: Environmental values (mean $\pm$ s.e.) of flour and bread parameters of two einkorns (ID331 and Monlis) and one bread wheat (Blasco); Table S4: One-way ANOVA (* $p \leq 0.05$; ** $p \leq 0.01$; *** $p \leq 0.001$) of crust and crumb color parameters; Table S5. One-way ANOVA (* $p \leq 0.05$; ** $p \leq 0.01$; *** $p \leq 0.001$) of porosity; Table S6: Two-way ANOVA (* $p \leq 0.05$; ** $p \leq 0.01$; *** $p \leq 0.001$) of texture profile analysis parameters.

Author Contributions: Conceptualization, A.B. and A.H.; methodology, A.B., M.L., M.M. and A.H.; software, M.M. and A.H.; validation, A.B., M.L., M.M., L.E. and A.H.; formal analysis, A.B., M.L., M.M. and A.H.; investigation, A.B. and L.E.; resources, A.B.; data curation, A.B. and A.H.; writing—original draft preparation, A.B., M.M. and A.H.; writing—review and editing, A.B., M.L., M.M., L.E. and A.H.; supervision, A.B. All authors have read and agreed to the published version of the manuscript.

Funding: This research received no external funding.

53. Amigo, J.M.; Del Olmo Alvarez, A.; Engelsen, M.M.; Lundkvist, H.; Engelsen, S.B. Staling of white wheat bread crumb and effect of maltogenic α-amylases. Part 2: Monitoring the staling process by using near infrared spectroscopy and chemometrics. *Food Chem.* **2019**, *297*, 124946. [CrossRef] [PubMed]

54. Taglieri, I.; Macaluso, M.; Bianchi, A.; Sanmartin, C.; Quartacci, M.F.; Zinnai, A.; Venturi, F. Overcoming bread quality decay concerns: Main issues for bread shelf life as a function of biological leavening agents and different extra ingredients used in formulation. A review. *J. Sci. Food Agric.* **2021**, *101*, 1732–1743. [CrossRef] [PubMed]

Article

Processing and Bread-Making Quality Profile of Spanish Spelt Wheat

Ana Belén Huertas-García [1], Carlos Guzmán [1,*], Maria Itria Ibba [2], Marianna Rakszegi [3], Josefina C. Sillero [4] and Juan B. Alvarez [1]

[1] Departamento de Genética, Escuela Técnica Superior de Ingeniería Agronómica y de Montes, Edificio Gregor Mendel, Campus de Rabanales, Universidad de Córdoba, CeiA3, ES-14071 Córdoba, Spain; b42hugaa@uco.es (A.B.H.-G.); jb.alvarez@uco.es (J.B.A.)

[2] Global Wheat Program, International Maize and Wheat Improvement Center (CIMMYT), Apdo Postal 6-641, Mexico DF, Mexico; m.ibba@cgiar.org

[3] Agricultural Institute, Centre for Agricultural Research, Brunszvik u. 2, 2462 Martonvásár, Hungary; rakszegi.mariann@atk.hu

[4] IFAPA Alameda del Obispo, Avenida Menéndez Pidal s/n, 14004 Cordoba, Spain; josefinac.sillero@juntadeandalucia.es

* Correspondence: carlos.guzman@uco.es

Abstract: Spelt wheat (*Triticum aestivum* L. ssp. *spelta* Thell.) is an ancient wheat that has been widely cultivated for hundreds of years. Recently, this species has been neglected in most of Europe; however, the desire for more natural and traditional foods has driven a revival of the crop. In the current study, eighty-eight traditional spelt genotypes from Spain, together with nine common wheat cultivars and one modern spelt (cv. Anna Maria) were grown during a period of two years in Andalucia (southern Spain). In each, several traits were measured in to evaluate their milling, processing, and end-use quality (bread-making). The comparison between species suggested that, in general, spelt and common wheat showed differences for most of the measured traits; on average, spelt genotypes had softer grains, higher protein content (14.3 vs. 11.9%) and gluten extensibility (alveograph P/L 0.5 vs. 1.8), and lower gluten strength (alveograph W 187 vs. 438×10^{-4} J). In the baking test, both species showed similar values. Nevertheless, the analysis of this set of spelt genotypes showed a wide range for all measured traits, with higher values than common wheat in some spelt genotypes for some traits. This opens up the possibility of using these materials in future breeding programs, to develop either new spelt or common wheat cultivars.

Keywords: wheat quality; genetic resources; ancient wheat; bread-making

Citation: Huertas-García, A.B.; Guzmán, C.; Ibba, M.I.; Rakszegi, M.; Sillero, J.C.; Alvarez, J.B. Processing and Bread-Making Quality Profile of Spanish Spelt Wheat. *Foods* **2023**, *12*, 2996. https://doi.org/10.3390/foods12162996

Academic Editors: Grazia Maria Borrelli and Donatella Bianca Maria Ficco

Received: 7 July 2023
Revised: 4 August 2023
Accepted: 6 August 2023
Published: 9 August 2023

1. Introduction

Since the 1960s, the importance of plant genetic resources has gradually increased, as shown by the development of the "International Treaty on Plant Genetic Resources for Food and Agriculture" [1]. The lack of genetic diversity in crops, globalization, and climate change have shown how easy is for any pathogen or plague to quickly spread around the world [2]. This could be a threat for food security, as modern agriculture is increasingly focused on fewer crops and fewer varieties within each crop [3]. At the same time, a greater awareness of the need to use more sustainable and environmentally friendly agronomic techniques, together with the problems associated with global change, have boosted the search for alternative gene sources, which is one of the strategies used to develop more resilient cultivars under the conditions of global warming. In this context, ancient wheats and wild-wheat relatives, which have adapted to be grown in marginal zones under extreme conditions [4], are considered to host interesting and unexploited genetic variability that could be used in modern wheat-breeding programs to develop more resilient cultivars.

Among these ancient wheats, spelt (*Triticum aestivum* L. ssp. *spelta* Thell., 2n = 6× = 42, A^{u}A^{u}BBDD), originally obtained from the natural hybridization between emmer wheat (*T. turgidum* spp. *dicoccum* Schrank em. Thell., = 4× = 28, A^{u}A^{u}BB) and *Aegilops tauschii* ssp. *strangulata* Coss. (2n = 2× = 14, DD) in the Fertile Crescent (Near East), is, today, the most cultivated species, and several spelt cultivars have been bred in order to improve their productivity [5,6]. For this reason, some ancestral traits like the hulled grain or the semi-branching rachis, have been modified through crosses with common wheat [7–10]. Consequently, two different types of spelt are now present in the farmers' fields: the traditional or pure spelt, and the modern spelt derived from hybridization with common wheat [5]. The variability of these two types of spelt is notably different, with the traditional spelt holding a greater genetic variability than modern spelt. Nevertheless, more studies comparing the variability of modern and traditional spelt are needed.

On the other hand, several studies have suggested the exceptionality of the Iberian spelt (pol. *ibericum* Flaskb.) compared to the rest of the European spelt (Bavarian group, pol. *bavaricum* Vav.), including the old studies of N.I. Vavilov [11–14]. Therefore, while European spelt could derive from a secondary hybridization event between emmer wheat and hexaploid wheat, the Iberian spelt would have originated from the first hybridization event between emmer wheat and *Ae. tauschii* ssp. *strangulata* in Asia [11–14]. Furthermore, the spelt crop in Spain has been scarce until recent times and mainly based in traditional materials. The appearance of modern spelt in Spain is recent, and only two cultivars have been developed since 2018: cvs. Anna Maria and Viso. However, the traditional Spanish spelt stored in germplasm banks is abundant [14]. The current trend with this crop opens up the opportunity to add value to these old materials for their use in the current agricultural context, both as pure spelt and as a source of novel genetic diversity to develop modern spelt [14,15] or common wheat cultivars.

Before this new trend of growing old crops, spelt had already been used in breeding programs as a source of resistance genes for some wheat diseases [16–26]. Now, in the context of the renewed interest in artisan and "more natural" food, spelt is used as raw material for the making of food products (bread, biscuits, pasta, pancakes, etc.) which are present in many bakeries but also in large retailers. For this reason, studies on the processing quality of this crop have increased in importance [27–34].

Wheat processing quality is complex and varies depending on the wheat processor (millers or bakers) and on the final products. Wheat millers for example, value grain size, test weight, and texture, which are associated with the flour yield, and grain protein content, which is a partial indicator of wheat functionality [35,36]. Bakers, on the other hand, value the quantity and quality of the protein in flour, and the rheological properties of the dough made with it. For these reasons, the evaluation of new wheat materials must consider the quality requirements of all wheat processors including both millers and bakers.

Most of the studies conducted on the processing quality of ancient wheat only included a limited number of accessions [37] which reduced their ability to gain a clear understanding of the potential in terms of wheat processing, of such species. In general, the quality characteristics of these ancient wheats have been compared with modern wheat (common wheat–*T. aestivum* L. ssp. *aestivum*). Although this could be right, these data should be evaluated with caution. Many of these ancient wheats have been revived as a modern wheat substitute, and consequently, to establish the quality characteristics of modern wheat as the reference could be inadequate and lead to underestimating all ancient wheats. Obviously, spelt is not common wheat. The development of new cultivars of these ancient wheats should be complementary to modern wheat in the context of the new agri-food industry.

The main objective of this study was the evaluation of a wide collection of Spanish traditional spelt accessions for some grain and technological quality traits, together with their comparison with one modern spelt (cv. Anna Maria) and several common wheat cultivars widely cultivated in Andalucia (southern Spain).

2. Materials and Methods

2.1. Plant Material and Field Trials

Eighty-eight accessions of Spanish traditional spelt, together with ten modern wheat cultivars (nine common wheats and one modern spelt) were used (Table S1). These materials were planted in a randomized complete block design with two replicates during 2019–2020 and 2020–2021 crop seasons in Cordoba (Andalusia, Spain). Due to the high number of materials, the plot size was small (0.13 m^2) and, consequently, the grain yield was limited for some accessions that could be only evaluated from small-scale tests.

The 88 traditional spelt accessions were selected according to their grain protein composition and origin from two wide collections originally provided by the National Small Grains Collection (USDA, Washington, DC, USA) and Centro de Recursos Fitogeneticos (INIA, Madrid, Spain) [38,39]. Of the 10 modern cultivars used as control, 9 were commercial Spanish common wheat cultivars commonly grown in the south of Spain (cvs. Antequera, Arthur Nick, Conil, Galera, Montemayor, Rota, Santaella, Setenil, and Tejada), and fall into different categories within the Spanish quality groups, depending on their performance in each environment. Cvs. Antequera, Conil, Galera, Rota, and Tejada often fall within the Spanish quality group 1 (strong gluten wheat for mechanized bread-making), while cvs. Arthur Nick, Montemayor, Santaella, and Setenil produce grains that are usually classified as quality groups 2–3 (strong–medium gluten wheat for semi-mechanized bread-making).The modern spelt (cv. Anna Maria) is a modern spelt cultivar obtained from hybridization between pure spelt and common wheat.

2.2. Grain and Flour Quality Traits

Thousand kernel weight (TKW, g) and test weight (TW, kg/hL) were obtained using the SeedCount SC5000 digital imaging system (Next Instruments, Australia). The grain (GPC, %) and flour (FPC, %) protein content were determined by near infrared spectroscopy (NIR Systems 6500, Foss, Hillerød, Denmark) based on AACC official methods 39-10.01 and 39-11.01, respectively, which were calibrated based on method 46-11.02 [40]. Grain hardness was measured on samples of 100 kernels with the single-kernel characterization system (SKCS) (Perten Instruments, Springfield, IL, USA) [40]. The polyphenol oxidase (PPO) activity was measured by absorbance at 475 nm according to Anderson and Morris [41], and expressed in $Ug^{-1}min^{-1}$.

For the milling, the two field replicates of each genotype were mixed in order to obtain enough flour. The grain samples were processed applying AACC method 26-95 [40]. All samples were milled into flour using a Brabender Quadrumat Senior mill (CW Brabender, Duisburg, Germany) and flour yield (%) was calculated.

Measurement of sodium dodecyl sulfate (SDS) sedimentation volume (ml) was carried out according to Dick and Quick methodology [42] with the modifications introduced by Peña et al. [43].

2.3. Alveographic and Baking Traits

The dough tenacity (P), extensibility (L), tenacity/extensibility ratio (P/L), tenacity/swelling index ratio (P/G), elasticity index (Ie), and strength (W) were determined by AACC method 54-30.02 using a Chopin alveograph [40]. Due to the limited flour available, dough rheological properties were measured only on 7 common wheat cultivars and 80 spelt accessions.

The bread-making process was conducted on 4 common wheat cultivars and 50 spelt accessions (only of those genotypes for which there was enough flour available to perform the test), using the direct dough method (AACC method 10-10.03), and loaf volume (cc) was determined by rapeseed displacement using a volume meter [40].

2.4. Statistical Methods

The comparison between both species sets was carried out for each trait analyzed by Student's *t*-test. A Pearson correlation analysis was carried out among the grain, flour, and rheological traits within the Spanish spelt set.

For the spelt set, data were analyzed by an analysis of variance (ANOVA) using genotype, year, and genotype × year as variation sources. Because cv. Anna Maria represented the current trend in spelt, the cv. Anna Maria data for each measured traits were used as reference to evaluate and compare the values of each Spanish traditional spelt genotype.

All statistical analyses were performed using Statistix software (version 9).

3. Results

3.1. Comparison among Species

The data obtained for all materials evaluated (Tables S2 and S3) were grouped according to the species (common wheat vs. spelt) in order to compare the two groups. The mean values of each set were analyzed by Student's *t*-test (Table 1).

Table 1. Average values of the common wheat and spelt groups (averaging genotypes and years) and result of the *t*-test done between both values.

Trait	No. Genotypes (Common: Spelt)	Common Wheat (Mean ± s.d.)	Spelt (Mean ± s.d.)	t-Value
		Grain/flour components		
TW (kg/hL)	9:89	77.60 ± 2.77	76.26 ± 1.84	3.94 ***
TKW (g)	9:89	49.73 ± 6.03	51.72 ± 4.81	−2.30 *
GPC (%)	9:89	11.87 ± 1.14	14.27 ± 1.76	−8.01 ***
Hardness (%)	9:89	52.44 ± 15.07	15.78 ± 11.02	18.32 ***
Flour yield (%)	9:89	65.78 ± 6.06	67.53 ± 3.27	−2.78 **
FPC (%)	9:89	10.22 ± 0.72	11.82 ± 0.98	−9.53 ***
SDS-sed (mL)	9:89	15.22 ± 2.65	15.86 ± 3.03	−1.22 ns
PPO activity $(Ug^{-1}min^{-1})$	9:89	4.25 ± 1.75	9.08 ± 2.06	−13.54 ***
		Alveogram parameters		
P (mm)	7:80	141.57 ± 26.18	59.42 ± 16.48	16.93 ***
L (mm)	7:80	84.93 ± 18.13	123.83 ± 23.20	−6.10 ***
P/L (ratio)	7:80	1.79 ± 0.68	0.51 ± 0.24	15.27 ***
P/G (ratio)	7:80	7.08 ± 1.96	2.47 ± 0.89	16.33 ***
W ($\times 10^{-4}$ J)	7:80	437.71 ± 117.33	186.56 ± 57.04	14.16 ***
Ie (%)	7:80	61.33 ± 10.35	46.10 ± 6.45	8.00 ***
		Loaf parameters		
Loaf Volume (cc)	4:50	778.00 ± 28.10	809.50 ± 57.64	−1.04 ns

TW, test weight; TKW, thousand kernel weight; GPC, grain protein content; FPC, flour protein contents; SDS-sed, sodium dodecyl sulfate sedimentation test; PPO activity, polyphenol oxidase activity; P, dough tenacity; L, dough extensibility; G: swelling index; W, dough strength; and Ie, elasticity index. ***, **, *: significant at 99.9, 99, and 95%; ns: not significant.

For grain or flour components, the differences between both species were, in general, small, but significant. The thousand kernel weight (TKW) of spelt was slightly higher than common wheat; however, spelt grains showed lower test weight (TW) values, probably due to the morphology of their grains that have, on average, an elongated shape. This had no influence on the flour yield, although the grain hardness, clearly lower in spelt, could also have played a role on the flour yield.

The protein content was higher in spelt, both in grain and in flour. But this scarcely influenced the gluten strength measured by the SDS-sedimentation test, because the Student's *t*-test analysis indicated that the differences between both species were not significant (Table 1). On the contrary, there were highly significant differences for polyphenol oxidase

(PPO) activity, for which the spelt group exhibited the double mean activity of the common wheat group.

The alveographic parameters showed that while common wheat presented dough with high tenacity (P) and low/moderate extensibility (L), the spelt genotypes showed, in general, low to moderate tenacity (P) and high extensibility (L). In any case, the dough strength (W) was larger in common wheat than in spelt (Table 1). Nevertheless, within both sub-sets there were no significant differences in the bread-making quality of the two groups (loaf volume), although the mean value of spelt was 30 cc higher than that of common wheat.

3.2. Variability for Grain and Flour Quality Traits in Spelt

When the comparison was carried out among the spelt genotypes (Figure 1), both traditional and modern spelt (cv. Anna Maria), the data showed high variation among these genotypes for all measured traits in grain and flour (Table S2). The ANOVA analysis suggested a high influence of the genotype in this variation (Table S4), although the differences between both years were also significant.

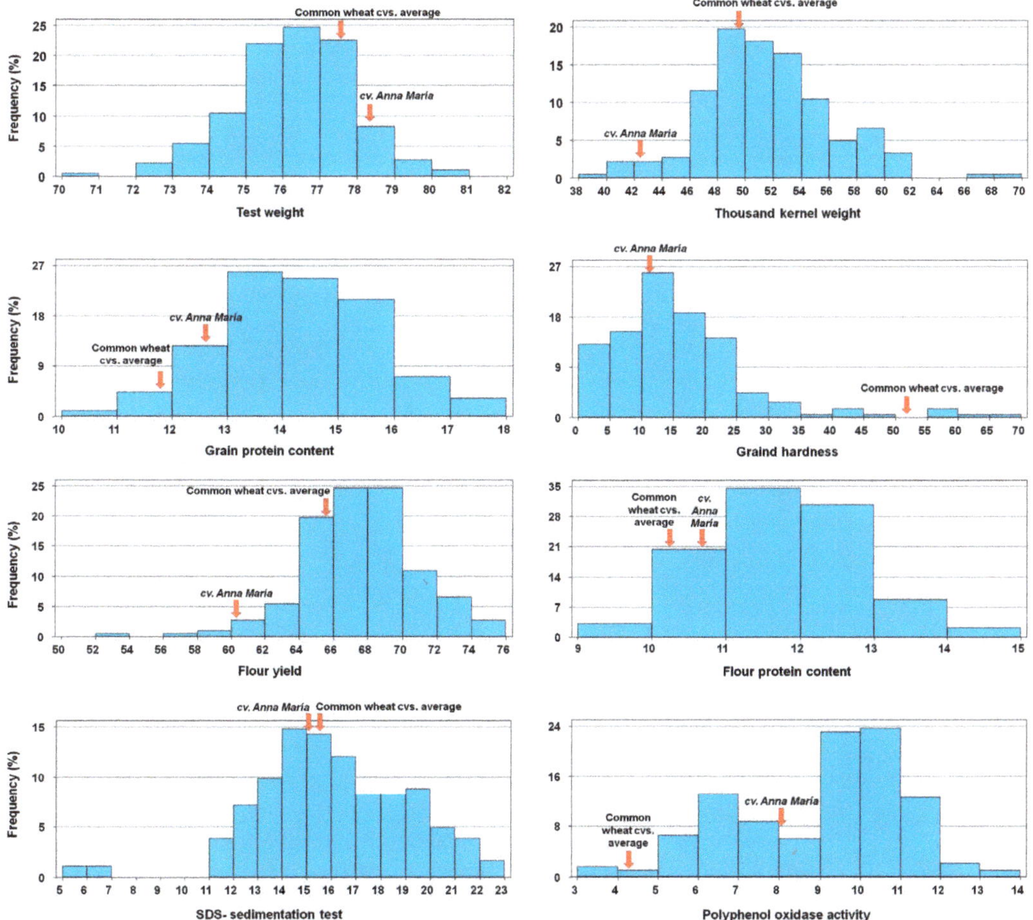

Figure 1. Frequency distribution of the spelt genotypes and average values of the common wheat cultivars for different grain and flour quality traits.

Most of the traditional spelt genotypes showed lower TW values than the cv. Anna Maria; however, the thousand kernel weight (TKW) of the latter was significantly lower than the values of the traditional spelt accessions (Figure 1).

The protein content, both in grain and flour, was highly variable, with cv. Anna Maria being in the low part of the distribution in both cases (Table S2). This high protein content had little effect on the grain hardness, because, in general, the spelt genotypes showed soft or very soft grain, although some accessions showed values of semi-hard grain (Figure 1). The general lower grain hardness associated with the spelt accessions was positively associated with flour yield.

The gluten strength measured as the SDS-sedimentation volume showed values of medium and high, with some exceptions (Figure 1).

For the PPO activity, the range was very wide (3–14 $Ug^{-1}min^{-1}$), and two groups of materials could be distinguished among the spelt genotypes: one with a mean value of 6.5 $Ug^{-1}min^{-1}$, and another with the mean values around 10.5 $Ug^{-1}min^{-1}$.

3.3. Alveogram and Baking Traits in the Spelt Collection

In the previous comparison with common wheat (Table 1), the data showed that spelt doughs had low tenacity, high extensibility, and low to medium gluten strength, as indicated by their W values. The analysis of the 80 genotypes that could be evaluated with the alveograph, showed a wide variation for the traits measured with this equipment (Tables 2 and S3), with different genotypes exhibiting values higher or lower than the average. In this respect, some spelt genotypes could be classified as medium to high gluten strength with W up to 388×10^{-4} J. Spelt genotype BGE 020900 (W = 320×10^{-4} J on average across the two years) could be considered a good donor of this trait for breeding programs. For gluten extensibility, several spelt accessions (such as BGE 001990, PI 348727, and PI 348747) showed very low P/L values (0.3), and could be considered interesting sources of this trait. Accession PI 348465 showed a very interesting combination of both gluten strength and extensibility (W = 283×10^{-4} J and P/L = 0.4) and could be considered the best material found in terms of gluten quality. The modern spelt (cv. Anna Maria) presented values around the average values of the spelt set (Table 2). In this case, the ANOVA analysis also showed the high influence of the genotype in the variation detected (Table S4).

Table 2. Comparison of the alveograph parameters obtained in the traditional spelt accessions and the modern spelt cultivar Anna Maria.

Trait	Traditional Spelt		cv. Anna Maria
	Mean ± s.d.	Range	Mean ± s.d.
P (mm)	59.48 ± 16.58	29.00–138.00	55.00 ± 4.24
L (mm)	123.87 ± 23.31	55.00–186.00	120.50 ± 17.67
P/L (ratio)	0.51 ± 024	0.20–1.70	0.45 ± 0.02
P/G (ratio)	2.47 ± 0.90	1.00–6.70	2.30 ± 0.00
W ($\times 10^{-4}$ J)	186.49 ± 57.18	73.00–388.00	192.00 ± 62.22
Ie (%)	46.04 ± 6.42	28.40–63.20	51.20 ± 10.32

P, dough tenacity; L, dough extensibility; G: swelling index; W, dough strength; and Ie, elasticity index.

As already mentioned, the mean values of spelt and common wheat for loaf volume did not show significant differences (Table 1). However, when the 50 genotypes of the spelt set were independently analysed, these materials exhibited a high variability for this trait, with minimum and maximum values of 600 and 975 cc, respectively (Figure 2). Apart from that, almost 82% of these genotypes had loaf volume between 750 and 875 cc. Genotypes PI 469058 and PI 469051 (885 mL and 848 mL of loaf volume, respectively, on average across the two years) were the best performers for this trait and could be used by breeding programs aimed at the improvement of bread-making quality.

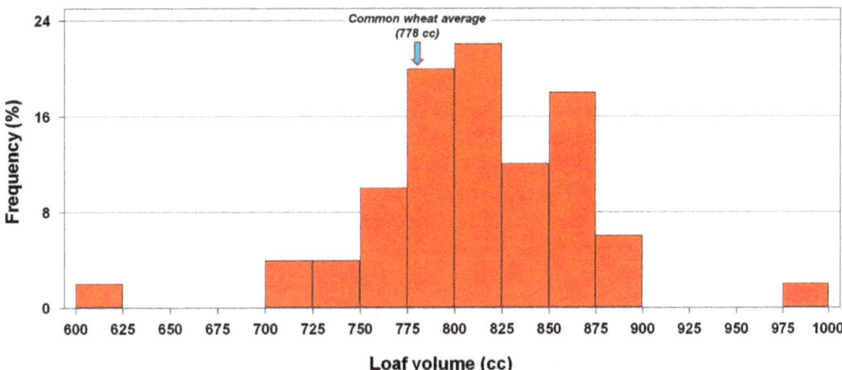

Figure 2. Frequency distribution of the spelt genotypes and average values of the common wheat cultivars for loaf volume.

Finally, a correlation analysis was carried out with the analyzed traits (except loaf volume, due to the lack of this data in many genotypes) within the spelt wheat set (Table S5). A negative correlation was found between test weight and protein content, and grain size (TKW) and alveograph W. Positive correlations were identified between protein content and SDS-sed and gluten extensibility (L), and among the different alveograph parameters.

4. Discussion

Recent changes in the agri-food industry have generated a growing interest in foods and old crops that have been practically lost during the last century [44]. In some cases, this renaissance has been associated with the supposed miraculous properties of these old crops. In general, such statements are not supported by any scientific basis and both the nutritional and nutraceutical properties of ancient wheats have been shown to be very similar to those of modern wheat [45–51]. However, other real benefits, such as the expansion of diversity in food, are little appreciated.

Within the wheat world, both the old varieties that had been replaced by more productive ones, and some of the species that were cultivated in the past have been recovered during recent decades [6,45]. Some of the latter are called "ancient wheats" and consists mainly of three species: einkorn (*T. monococcum* L. ssp. *monococcum*, 2n = 2× = 14, $A^m A^m$), emmer, and spelt. Some of the agronomic characteristics of these ancient wheats were those that led to their disappearance and abandonment when most of the agricultural processes were mechanized. In addition, due to being hulled grain cereals, the need for special dehulling treatment prior to grinding increased costs and affected their profitability. For this reason, their revival is linked to the boom in traditional and gourmet bakeries, where the higher prices of these products can offset production costs. Nowadays larger retailers also offer flour and products made from these types of wheat.

At the same time, this renewed interest has resulted in the development of numerous studies on these species, comparing their characteristics with the ones of modern wheat [50,51]. However, many of these studies have been carried out with a limited number of genotypes [37], which could create bias in the results and undervalue the true role of these old materials. For this reason, the evaluation of large collections of these ancient wheats, with the limitation of the storage materials in germplasm banks, could shed light on these questions, identifying new genotypes with potential utility for breeding programs. In the current study, one collection of 88 Spanish traditional spelt accessions, together with 9 common wheat cultivars and 1 modern spelt cultivar (cv. Anna Maria), were analyzed and compared for several traits related to milling, processing, and end-use quality.

When the analyzed materials are ancient or old wheats, the technological quality must be evaluated with caution. Changes in baking techniques throughout the last century have generated materials adapted to these techniques, which are different from those required for traditional baking, and consequently, the evaluation of ancient wheats according to modern parameters could not be the best strategy. In this regard, our previous studies on the grain composition of the spelt accessions evaluated in the current study showed the presence of rare HMWGs variants in spelt wheat (1, 13 + 16, and 2 + 12) [38,39]. However, it is possible that the high frequency of these variants is an empirical consequence of the way these wheats were and are used in traditional agri-food applications. These characteristics are mainly demanded by bakers, since all the studies carried out suggest the clear influence of glutenins on the viscoelastic properties of wheat dough [52]. However, millers are interested more in other traits more related to the physical and chemical characteristics of the grain such as TW, TKW, protein content, and grain hardness, mainly due to their influence on the flour yield.

Previous studies have shown that the grain size in spelt is larger for the modern material (with common wheat introgression) than for the traditional spelt [9,31,33]. In this study, the traditional spelt genotypes presented a TW and TKW very similar to the common wheat cultivars used as the control, and, compared with the modern spelt (cv. Anna Maria), the latter has a better TW but its TKW was lower. This reinforces the idea that variation in the traditional spelt is high and could be interesting for breeding programs aiming to develop new cultivars with very high grain size [53,54]. In parallel, the spelt genotypes with high grain size showed high protein content (indeed, a significant correlation between TKW and protein content was found) and a soft texture, which positively affected flour yield.

The PPO activity appeared comparatively higher in spelt than in common wheat. This trait is regulated by several enzymes, synthesized by the *Ppo-1* and *Ppo-2* loci at the homoeologous group 2 chromosomes [55–59], and has been associated with the discoloration and darkening of wheat products [60–63], which generates a certain amount of rejection among consumers [41,56,61–64]. Paradoxically, it may happen that today's consumers associate this dark color with the true presence of flour spelt in a food product, while the cream color suggests that the product is made with flour of modern common wheat and not from spelt (more "natural" vs. more "industrial"). Therefore, high PPO activity may not be an undesirable trait for spelt cultivars, although it does deserve attention if spelt is to be used in the breeding of modern common wheat as source of other traits of interest. In this regard, some traditional spelt genotypes showed low PPO activity values ($\leq 5 \; Ug^{-1}min^{-1}$) (Table S2), although this was not the general trend.

Previous studies conducted on spelt revealed that spelt mostly exhibits low to medium gluten strength [28–34]; however, our study revealed that it is also possible to identify genotypes with stronger gluten. In any case, the viscoelastic properties of spelt could be different from those of common wheat. The spelt genotypes showed, in general, more extensible doughs, which was favored by higher protein contents, and, in a few cases, the W values were reasonably high (up to 300×10^{-4} J). This was also due to the strong correlation found between alveograph W and P/L, which is also normal in common wheat sets [65]. However, it was not possible to identify an unambiguous relationship between the W values and loaf volume. As with other measured traits, the variation of these two parameters was high among the spelt genotypes. When the loaf volume was related to the flour protein content, some genotypes with low protein content showed high loaf volume, which suggests the high quality of these gluten proteins.

The current trend in the cereal's world has extended the search for other desirable traits within the grain components, mainly related with nutritional and nutraceutical properties, which would complement the technological properties of the doughs [37,66–68]. Nowadays, the presence of micronutrients such as Fe or Zn in the flour, or dietary fiber in form of soluble arabinoxylans, is highly recommended and this has increased the interest in the ancient wheats, with some studies suggesting that these old materials could be a good source for these traits [69–71]. In this respect, a previous study on these nutritional aspects has revealed that the current spelt genotypes show a notable variation in these traits [72]. These data, together with the data obtained in the current study, highlight the need to increase the evaluation of wide collections of these ancient wheats, in order to detect the true variability in these old materials for different traits, including those associated with processing and nutritional quality. Such analyses will allow the identification of unique germplasms that could be used for the selection and purification of intra-accession variability for the development of traditional and homogeneous spelt varieties, both to be crossed with modern wheat to transfer the trait of interest, to improve modern wheat genetic diversity, and to develop better-adapted spelt cultivars.

5. Conclusions

Ancient wheats can be good sources of interesting agronomic features, mainly rust-resistant genes and quality traits for wheat breeding. The evaluation of the large collections of these old materials would allow for evaluation of the true variability present in these species. In the current study, large variation was found in a set of Spanish spelt landraces, which, in general, showed soft grain, medium–high protein content, low gluten strength, high gluten extensibility, and medium bread-making quality; spelt genotypes showing outstanding values for some of these traits that could be useful for breeding purposes were identified. Additionally, this and similar studies could provide the opportunity to develop new cultivars of spelt with good characteristics for the food industry.

Supplementary Materials: The following supporting information can be downloaded at: https://www.mdpi.com/article/10.3390/foods12162996/s1, Table S1: Plant material used in the study; Table S2: Mean values of the grain and flour measured traits for each season in the materials evaluated (spelt and common wheat); Table S3: Mean values of the alveographic and baking traits for each season in the materials evaluated (spelt and common wheat); Table S4: Effects of genotype, season, and genotype x season (GxS) on quality traits in spelt accessions. Sum of squares, % of the total sum of squares from ANOVA analysis, and coefficient of variation (CV) are indicated; and Table S5: Correlation analysis of quality traits in spelt.

Author Contributions: Conceptualization, C.G. and J.B.A.; formal analysis, A.B.H.-G., M.I.I., M.R. and J.C.S.; data curation, M.I.I., C.G. and M.R.; writing—original draft preparation, A.B.H.-G. and J.B.A.; writing—review and editing, C.G., M.I.I. and M.R.; supervision, C.G. and J.B.A.; project administration, J.B.A.; funding acquisition, C.G. and J.B.A. All authors have read and agreed to the published version of the manuscript.

Funding: This research was supported by grant PID2021-122530OB-I00 from the Spanish State Research Agency (Spanish Ministry of Science and Innovation)—MCIN/AEI/10.13039/501100011033—co-financed with the European Regional Development Fund (FEDER) from the European Union. Carlos Guzman gratefully acknowledges the European Social Fund and the Spanish State Research Agency (Spanish Ministry of Science and Innovation)—MCIN/AEI/10.13039/501100011033—for financial funding through the Ramon y Cajal Program (RYC-2017-21891).

Data Availability Statement: Data are included in the Supplementary Materials.

Acknowledgments: We thank to the National Small Grain Collection (Aberdeen, USA) and Centro de Recursos Fitogeneticos (INIA, Spain) for supplying the analysed materials.

Conflicts of Interest: The authors declare no conflict of interest.

References

1. *International Treaty for Plant Genetic Resources for Food and Agriculture*; Food and Agriculture Organization: Rome, Italy. Available online: https://www.fao.org/3/a-i0510e.pdf (accessed on 8 June 2023).
2. Gautam, H.; Bhardwaj, M.; Kumar, R. Climate change and its impact on plant diseases. *Curr. Sci.* **2013**, *105*, 1685–1691.
3. Esquinas-Alcazar, J. Protecting crop genetic diversity for food security: Political, ethical and technical challenges. *Nat. Rev. Genet.* **2005**, *6*, 946–953. [CrossRef]
4. Srivastava, J.P.; Damania, A.B. Use of collections in cereal improvement in semi-arid areas. In *The Use of Plant Genetic Resources*; Brown, A.H.D., Frankel, O.H., Marshall, D.R., Williams, J.T., Eds.; Cambridge University Press: Cambridge, UK, 1989; pp. 88–104.
5. Campbell, K.G. Spelt: Agronomy, genetics, and breeding. *Plant Breed. Rev.* **1997**, *15*, 187–214.
6. Geisslitz, S.; Scherf, K.A. Rediscovering ancient wheats. *Cereal Foods World* **2020**, *65*, 13.
7. McFadden, E.S.; Sears, E.R. The artificial synthesis of *Triticum spelta*. *Genetics* **1945**, *30*, 14.

8. Schmid, J.; Winzeler, H. Genetic studies of crosses between common wheat (*Triticum aestivum* L.) and spelt (*Triticum spelta* L.). *J. Genet. Breed.* **1990**, *44*, 75–80.

9. Winzeler, H.; Schmid, J.E.; Winzeler, M. Analysis of the yield potential and yield components of F_1 and F_2 hybrids of crosses between wheat (*Triticum aestivum* L.) and spelt (*Triticum spelta* L.). *Euphytica* **1994**, *74*, 211–218. [CrossRef]

10. Suchowilska, E.; Wiwart, M.; Krska, R.; Kandler, W. Do *Triticum aestivum* L. and *Triticum spelta* L. hybrids constitute a promising source material for quality breeding of new wheat varieties? *Agronomy* **2020**, *10*, 43. [CrossRef]

11. Salamini, F.; Ozkan, H.; Brandolini, A.; Schafer-Pregl, R.; Martin, W. Genetics and geography of wild cereal domestication in the Near East. *Nat. Rev. Genet.* **2002**, *3*, 429–441. [CrossRef]

12. Dedkova, O.S.; Badaeva, E.D.; Mitrofanova, O.P.; Zelenin, A.V.; Pukhalskiy, V.A. Analysis of intraspecific divergence of hexaploid wheat *Triticum spelta* L. by C-banding of chromosomes. *Russ. J. Genet.* **2004**, *40*, 1111–1126. [CrossRef]

13. Elía, M.; Moralejo, M.; Rodriguez-Quijano, M.; Molina-Cano, J.L. Spanish spelt: A separate gene pool within the spelt germplasm. *Plant Breed.* **2004**, *123*, 297–299. [CrossRef]

14. Alvarez, J.B. Spanish spelt wheat: From an endangered genetic resource to a trendy crop. *Plants* **2021**, *10*, 2748. [CrossRef]

15. Kulathunga, J.; Reuhs, B.L.; Simsek, S. A review: Novel trends in hulled wheat processing for value addition. *Trends Food Sci. Technol.* **2020**, *106*, 232–241. [CrossRef]

16. Zeven, A.C.; Turkenst, L.J.; Stubbs, R.W. Spelt (*Triticum spelta* L.) as a possible source of race-non-specific resistance to yellow rust (*Puccinia striiformis* West.). *Euphytica* **1968**, *17*, 381–384. [CrossRef]

17. McVey, D.V.; Leonard, K.J. Resistance to wheat stem rust in spring spelts. *Plant Dis.* **1990**, *74*, 966–969. [CrossRef]

18. Zeller, F.J.; Felsenstein, F.G.; Lutz, J.; Holzerland, A.; Katzhammer, M.; Kellermann, A.; Kunzler, J.; Stephan, U.; Oppitz, K. Studies on the resistance of several spelt wheat (*Triticum aestivum* (L.) Thell. ssp. spelta (L.) Thell.) cultivars against wheat powdery mildew (Erysiphe graminis f. sp. tritici) and leaf rust of wheat (Puccinia recondita f. sp. tritici). *Bodenkultur* **1994**, *45*, 147–154.

19. Dyck, P.L.; Sykes, E.E. Genetics of leaf-rust resistance in three spelt wheats. *Can. J. Plant Sci.* **1994**, *74*, 231–233. [CrossRef]

20. Singh, P.K.; Mergoum, M.; Ali, S.; Adhikari, T.B.; Elias, E.M.; Hughes, G.R. Identification of new sources of resistance to tan spot, Stagonospora nodorum blotch, and Septoria tritici blotch of wheat. *Crop Sci.* **2006**, *46*, 2047–2053. [CrossRef]

21. Simon, M.R.; Khlestkina, E.K.; Castillo, N.S.; Borner, A. Mapping quantitative resistance to septoria tritici blotch in spelt wheat. *Eur. J. Plant Pathol.* **2010**, *128*, 317–324. [CrossRef]

22. Mohler, V.; Singh, D.; Singrun, C.; Park, R.F. Characterization and mapping of *Lr65* in spelt wheat 'Altgold Rotkorn'. *Plant Breed.* **2012**, *131*, 252–257. [CrossRef]

23. Singh, D.; Mohler, V.; Park, R.F. Discovery, characterisation and mapping of wheat leaf rust resistance gene *Lr71*. *Euphytica* **2013**, *190*, 131–136. [CrossRef]

24. Peng, F.; Song, N.; Shen, H.; Wu, H.; Dong, H.; Zhang, J.; Li, Y.; Peng, H.; Ni, Z.; Liu, Z.; et al. Molecular mapping of a recessive powdery mildew resistance gene in spelt wheat cultivar Hubel. *Mol. Breed.* **2014**, *34*, 491–500. [CrossRef]

25. Dinkar, V.; Jha, S.K.; Mallick, N.; Niranjana, M.; Agarwal, P.; Sharma, J.B. Vinod, Molecular mapping of a new recessive wheat leaf rust resistance gene originating from *Triticum spelta*. *Sci. Rep.* **2020**, *10*, 22113. [CrossRef] [PubMed]

26. Goriewa-Duba, K.; Duba, A.; Suchowilska, E.; Wiwart, M. An analysis of the genetic diversity of bread wheat × spelt breeding lines in terms of their resistance to powdery mildew and leaf rust. *Agronomy* **2020**, *10*, 658. [CrossRef]

27. Ranhotra, G.S.; Gelroth, J.A.; Glaser, B.K.; Lorenz, K.J. Baking and nutritional qualities of a spelt wheat sample. *Lebensm. Wiss. Technol.* **1995**, *28*, 118–122. [CrossRef]

28. Abdel-Aal, E.S.M.; Hucl, P.; Sosulski, F.W.; Bhirud, P.R. Kernel, milling and baking properties of spring-type spelt and einkorn wheats. *J. Cereal Sci.* **1997**, *26*, 363–370. [CrossRef]

29. Bojnanská, T.; Francáková, H. The use of spelt wheat (*Triticum spelta* L.) for baking applications. *Rostl. Výr.* **2002**, *48*, 141–147. [CrossRef]

30. Wilson, J.D.; Bechtel, D.B.; Wilson, G.W.T.; Seib, P.A. Bread quality of spelt wheat and its starch. *Cereal Chem.* **2008**, *85*, 629–638. [CrossRef]

31. Ratajczak, K.; Sulewska, H.; Grażyna, S.; Matysik, P. Agronomic traits and grain quality of selected spelt wheat varieties versus common wheat. *J. Crop Improv.* **2020**, *34*, 654–675. [CrossRef]

32. Podolska, G.; Aleksandrowicz, E.; Szafrańska, A. Bread making potential of *Triticum aestivum* and *Triticum spelta* species. *Open Life Sci.* **2020**, *15*, 30–40. [CrossRef]

33. Kulathunga, J.; Reuhs, B.L.; Zwinger, S.; Simsek, S. Comparative study on kernel quality and chemical composition of ancient and modern wheat species: Einkorn, emmer, spelt and hard red spring wheat. *Foods* **2021**, *10*, 761. [CrossRef] [PubMed]

34. Hadnađev, M.; Tomić, J.; Škrobot, D.; Dapčević-Hadnađev, T. Rheological behavior of emmer, spelt and khorasan flours. *J. Food Process. Preserv.* **2022**, *46*, e15873. [CrossRef]

35. Matsuo, R.R.; Dexter, J.E. Relationship between some durum wheat physical characteristics and semolina milling properties. *Can. J. Plant Sci.* **1980**, *60*, 49–53. [CrossRef]

36. Breseghello, F.; Sorrells, M.E. Association mapping of kernel size and milling quality in wheat (*Triticum aestivum* L.) cultivars. *Genetics* **2006**, *172*, 1165–1177. [CrossRef]

37. Shewry, P.R.; Hey, S.J. The contribution of wheat to human diet and health. *Food Energy Secur.* **2015**, *4*, 178–202. [CrossRef] [PubMed]

38. Caballero, L.; Martin, L.M.; Alvarez, J.B. Allelic variation of the HMW glutenin subunits in Spanish accessions of spelt wheat (*Triticum aestivum ssp spelta* L. em. Thell.). *Theor. Appl. Genet.* **2001**, *103*, 124–128. [CrossRef]
39. Caballero, L.; Martín, L.M.; Alvarez, J.B. Genetic variability of the low-molecular-weight glutenin subunits in spelt wheat (*Triticum aestivum* ssp *spelta* L. em Thell.). *Theor. Appl. Genet.* **2004**, *108*, 914–919. [CrossRef] [PubMed]
40. AACC. *Approved Methods of the American Association of Cereal Chemists*; American Association of Cereal Chemists: St. Paul, MN, USA, 2010.
41. Anderson, J.V.; Morris, C.F. An improved whole-seed assay for screening wheat germplasm for polyphenol oxidase activity. *Crop Sci.* **2001**, *41*, 1697–1705. [CrossRef]
42. Dick, J.W.; Quick, J.S. A modified screening test for rapid estimation of gluten strength in early-generation durum wheat breeding lines. *Cereal Chem.* **1983**, *60*, 315–318.
43. Peña, R.J.; Amaya, A.; Rajaram, S.; Mujeeb-Kazi, A. Variation in quality characteristics associated with some spring 1B/1R translocation wheats. *J. Cereal Sci.* **1990**, *12*, 105–112. [CrossRef]
44. Padulosi, S.; Hodgkin, T.; Williams, J.; Haq, N.; Engles, J.; Rao, V.; Brown, A.; Jackson, M. 30 underutilized crops: Trends, challenges and opportunities in the 21st century. In *Managing Plant Genetic Diversity*; Engels, J.M.M., Ramanatha, R.V., Brown, A.H.D., Jackson, M.T., Eds.; CABI Publishing: Wallingford, UK, 2002; pp. 323–338.
45. Miedaner, T.; Longin, F. *Neglected Cereals: From Ancient Grains to Superfood*; Agrimedia: Clenze, Germany, 2016.
46. Cooper, R. Re-discovering ancient wheat varieties as functional foods. *J. Tradit. Complement. Med.* **2015**, *5*, 138–143. [CrossRef]
47. Arzani, A.; Ashraf, M. Cultivated ancient wheats (*Triticum* spp.): A potential source of health-beneficial food products. *Compr. Rev. Food Sci. Food Saf.* **2017**, *16*, 477–488. [CrossRef] [PubMed]
48. Bordoni, A.; Danesi, F.; Di Nunzio, M.; Taccari, A.; Valli, V. Ancient wheat and health: A legend or the reality? A review on KAMUT khorasan wheat. *Int. J. Food Sci. Nutr.* **2017**, *68*, 278–286. [CrossRef] [PubMed]
49. Dinu, M.; Whittaker, A.; Pagliai, G.; Benedettelli, S.; Sofi, F. Ancient wheat species and human health: Biochemical and clinical implications. *J. Nutr. Biochem.* **2018**, *52*, 1–9. [CrossRef] [PubMed]
50. Shewry, P.R. Do ancient types of wheat have health benefits compared with modern bread wheat? *J. Cereal Sci.* **2018**, *79*, 469–476. [CrossRef]
51. Brouns, F.; Geisslitz, S.; Guzman, C.; Ikeda, T.M.; Arzani, A.; Latella, G.; Simsek, S.; Colomba, M.; Gregorini, A.; Zevallos, V.; et al. Do ancient wheats contain less gluten than modern bread wheat, in favour of better health? *Nutr. Bull.* **2022**, *47*, 157–167. [CrossRef]
52. Wrigley, C.; Békés, F.; Bushuk, W. *Gliadin and Glutenin: The Unique Balance of Wheat Quality*; AACC International Press: St. Paul, MN, USA, 2006.
53. Packa, D.; Załuski, D.; Graban, Ł.; Lajszner, W. An evaluation of spelt crosses for breeding new varieties of spring spelt. *Agronomy* **2019**, *9*, 167. [CrossRef]
54. Guzmán, C.; Alvarez, J.B. Ancient wheats role in sustainable wheat cultivation. In *Trends in Wheat and Bread Making*; Galanakis, C.M., Ed.; Academic Press: Cambridge, MA, USA, 2021; pp. 29–66.
55. Demeke, T.; Morris, C.F. Molecular characterization of wheat polyphenol oxidase (PPO). *Theor. Appl. Genet.* **2002**, *104*, 813–818. [CrossRef]
56. Jukanti, A.K.; Bruckner, P.L.; Fischer, A.M. Evaluation of wheat polyphenol oxidase genes. *Cereal Chem.* **2004**, *81*, 481–485. [CrossRef]
57. He, X.Y.; He, Z.H.; Zhang, L.P.; Sun, D.J.; Morris, C.F.; Fuerst, E.P.; Xia, X.C. Allelic variation of polyphenol oxidase (PPO) genes located on chromosomes 2A and 2D and development of functional markers for the PPO genes in common wheat. *Theor. Appl. Genet.* **2007**, *115*, 47–58. [CrossRef]
58. Beecher, B.; Skinner, D.Z. Molecular cloning and expression analysis of multiple polyphenol oxidase genes in developing wheat (*Triticum aestivum*) kernels. *J. Cereal Sci.* **2011**, *53*, 371–378. [CrossRef]
59. Taranto, F.; Mangini, G.; Pasqualone, A.; Gadaleta, A.; Blanco, A. Mapping and allelic variations of *Ppo-B1* and *Ppo-B2* gene-related polyphenol oxidase activity in durum wheat. *Mol. Breed.* **2015**, *35*, 80. [CrossRef]
60. Anderson, J.V.; Morris, C.F. Purification and analysis of wheat grain polyphenol oxidase (PPO) protein. *Cereal Chem.* **2003**, *80*, 135–143. [CrossRef]
61. Baik, B.-K.; Czuchajowska, Z.; Pomeranz, Y. Discoloration of dough for oriental noodles. *Cereal Chem.* **1995**, *72*, 198–204.
62. Demeke, T.; Morris, C.F.; Campbell, K.G.; King, G.E.; Anderson, J.A.; Chang, H.-G. Wheat polyphenol oxidase. *Crop Sci.* **2001**, *41*, 1750–1757. [CrossRef]
63. Fuerst, E.P.; Anderson, J.V.; Morris, C.F. Delineating the role of polyphenol oxidase in the darkening of alkaline wheat noodles. *J. Agric. Food Chem.* **2006**, *54*, 2378–2384. [CrossRef]
64. Feillet, P.; Autran, J.-C.; Icard-Vernière, C. Pasta brownness: An assessment. *J. Cereal Sci.* **2000**, *32*, 215–233. [CrossRef]
65. Guzmán, C.; Crossa, J.; Mondal, S.; Govindan, V.; Huerta, J.; Crespo-Herrera, L.; Vargas, M.; Singh, R.P.; Ibba, M.I. Effects of glutenins (Glu-1 and Glu-3) allelic variation on dough properties and bread-making quality of CIMMYT bread wheat breeding lines. *F. Crop. Res.* **2022**, *284*, 108585. [CrossRef]
66. Gomez-Becerra, H.F.; Erdem, H.; Yazici, A.; Tutus, Y.; Torun, B.; Ozturk, L.; Cakmak, I. Grain concentrations of protein and mineral nutrients in a large collection of spelt wheat grown under different environments. *J. Cereal Sci.* **2010**, *52*, 342–349. [CrossRef]

67. Bouis, H.E.; Hotz, C.; McClafferty, B.; Meenakshi, J.V.; Pfeiffer, W.H. Biofortification: A new tool to reduce micronutrient malnutrition. *Food Nutr. Bull.* **2011**, *32*, S31–S40. [CrossRef]
68. Jones, J.M.; García, C.G.; Braun, H.J. Perspective: Whole and refined grains and health—Evidence supporting "make half your grains whole". *Adv. Nutr.* **2020**, *11*, 492–506. [PubMed]
69. Ranhotra, G.S.; Gelroth, J.A.; Glaser, B.K.; Lorenz, K.J. Nutrient composition of spelt wheat. *J. Food Comp. Anal.* **1996**, *9*, 81–84. [CrossRef]
70. Guzmán, C.; Medina-Larqué, A.S.; Velu, G.; González-Santoyo, H.; Singh, R.P.; Huerta-Espino, J.; Ortiz-Monasterio, I.; Peña, R.J. Use of wheat genetic resources to develop biofortified wheat with enhanced grain zinc and iron concentrations and desirable processing quality. *J. Cereal Sci.* **2014**, *60*, 617–622. [CrossRef]
71. Ruibal-Mendieta, N.L.; Delacroix, D.L.; Mignolet, E.; Pycke, J.M.; Marques, C.; Rozenberg, R.; Petitjean, G.; Habib-Jiwan, J.L.; Meurens, M.; Quetin-Leclercq, J.; et al. Spelt (*Triticum aestivum* ssp. spelta) as a source of breadmaking flours and bran naturally enriched in oleic acid and minerals but not phytic acid. *J. Agric. Food Chem.* **2005**, *53*, 2751–2759.
72. Huertas-García, A.B.; Tabbita, F.; Alvarez, J.B.; Sillero, J.C.; Ibba, M.I.; Rakszegi, M.; Guzmán, C. Genetic variability for grain components related to nutritional quality in spelt and common wheat. *J. Agric. Food Chem.* **2023**, *71*, 10598–10606. [CrossRef] [PubMed]

Article

Antioxidants and Phenolic Acid Composition of Wholemeal and Refined-Flour, and Related Biscuits in Old and Modern Cultivars Belonging to Three Cereal Species

Grazia Maria Borrelli, Valeria Menga, Valentina Giovanniello and Donatella Bianca Maria Ficco *

Consiglio per la Ricerca in Agricoltura e l'Analisi dell'Economia Agraria—Centro di Ricerca Cerealicoltura e Colture Industriali, 71122 Foggia, Italy; graziamaria.borrelli@crea.gov.it (G.M.B.); valeria.menga@crea.gov.it (V.M.); v.giovanniello@tiscali.it (V.G.)
* Correspondence: donatellabm.ficco@crea.gov.it

Abstract: Cereals are a good source of phenolics and carotenoids with beneficial effects on human health. In this study, a 2-year evaluation was undertaken on grain, wholemeal and refined-flour of two cultivars, one old and one modern, belonging to three cereal species. Wholemeal of selected cultivars for each species was used for biscuit making. In the grain, some yield-related traits and proteins (PC) were evaluated. In the flours and biscuits, total polyphenols (TPC), flavonoids (TFC), proanthocyanidins (TPAC), carotenoids (TYPC) and antioxidant activities (DPPH and TEAC) were spectrophotometrically determined, whereas HPLC was used for the composition of soluble free and conjugated, and insoluble bound phenolic acids. Species (S), genotype (G) and 'SxG' were highly significant for yield-related and all antioxidant traits, whereas cropping year (Y) significantly affected yield-related traits, PC, TPC, TPAC, TEAC and 'SxGxY' interaction was significant for yield-related traits, TPAC, TYPC, TEAC, DPPH and all phenolic acid fractions. Apart from the TYPC that prevailed in durum wheat together with yield-related traits, barley was found to have significantly higher values for all the other parameters. Generally, the modern cultivars are richest in antioxidant compounds. The free and conjugated fractions were more representative in emmer, while the bound fraction was prevalent in barley and durum wheat. Insoluble bound phenolic acids represented 86.0% of the total, and ferulic acid was the most abundant in all species. A consistent loss of antioxidants was observed in all refined flours. The experimental biscuits were highest in phytochemicals than commercial control. Although barley biscuits were nutritionally superior, their lower consumer acceptance could limit their diffusion. New insights are required to find optimal formulations for better nutritional, sensorial and health biscuits.

Keywords: cereals; wholemeal; biscuits; phenolics; phenolic acid compositions; carotenoids; consumer acceptance

Citation: Borrelli, G.M.; Menga, V.; Giovanniello, V.; Ficco, D.B.M. Antioxidants and Phenolic Acid Composition of Wholemeal and Refined-Flour, and Related Biscuits in Old and Modern Cultivars Belonging to Three Cereal Species. *Foods* **2023**, *12*, 2551. https://doi.org/10.3390/foods12132551

Academic Editor: Grant Campbell

Received: 13 June 2023
Revised: 23 June 2023
Accepted: 26 June 2023
Published: 29 June 2023

1. Introduction

Phenolic compounds and carotenoid pigments, specialized metabolites synthesized during plant development and in response to stress conditions [1], are excellent oxygen radical scavengers. Their intake through the whole-cereal products offers potential health benefits in many chronic diseases [2].

Cereals are a good source of phenolic compounds and carotenoid pigments and, being important components of the human diet, they can contribute to a significant supply of these molecules [3,4].

Phenolic compounds are mostly concentrated in the outer layers of the grain, mainly pericarp and aleurone, and germ [5–7]. Adom et al. [8] showed that the bran/germ fraction of wheat contributes 83% of the total phenolic content of the wholemeal flour. The total phenolic content of bran/germ fractions is 15- to 18-fold higher than that of respective endosperm fractions that contribute only 17% of the total phenolic content. Since external layers are lost during roller-milling, the phenolic compounds are scarce in refined cereal products, and only by consuming wholemeal flour it is possible to benefit from their full levels. Moreover, phenolic compounds may also have an impact on color, flavor and astringency, becoming crucial for the acceptability of the final products by the consumer [9]. Among the phenolic compounds, phenolic acids are the most representative in cereals [10].

As previously observed by other authors [10–12], phenolic acids may mainly occur as insoluble bound linked to cell-wall constituents and as soluble conjugated forms esterified to sugars and to other low molecular weight components. Only 0.5–2% of phenolic acids exist as soluble free forms. Their structural diversity influences the bioavailability: free phenolic acids can cross the intestinal barrier and be found in the blood, while the bound forms of phenolic acids are scarcely digested and recovered in the feces, and only a small part can reach the colon where it exerts its antioxidant activity [13,14]. Unlike phenolics, carotenoids are one of the most important pigments occurring in nature. Several health benefits have been attributed to carotenoids, including the role as provitamin A and antioxidant activity [15]. In cereals, carotenoids are differently distributed in the kernel: α- and β-carotene are mainly located in the germ while lutein, the most abundant pigment, is equally distributed across the kernel [6,7,16–18]. In durum wheat, they are an important quality trait for industry of semolina/couscous and end-products [19,20].

The amount of these specialized metabolites in cereals is highly variable and mostly related to species and variety [8,10,21–23]. Žilić et al. [22], analyzing the antioxidant content over one year in a cereal collection, showed that total phenolic and total flavonoid content was higher in hull-less barley, followed by hull-less oat, rye, durum wheat and bread wheat. Interestingly, the highest antioxidant activity observed in hull-less barley was ascribed to a specific subclass of flavonoids, being more effective as antioxidants than vitamin C, E and carotenoids [24,25]. Comparing old and modern durum wheat genotypes, some authors suggested that breeding has qualitatively influenced the profiles of phenolic compounds [26]. Little information is available on this matter in old and modern genotypes, characterized by different year of release and yield potential, in multi-cereal species for more crop years.

Furthermore, environmental factors (E, including year, location, as well as agronomic practices), genetic (G) effects and 'GxE' all contribute to determining phenotypic variation for phenolics, with environmental effects larger than genotypic differences [10,27,28]. Contrarily, for carotenoid pigments a strong G effect was evidenced, particularly in durum wheat [29].

Among cereal species, durum wheat, emmer wheat and barley are an important source of carbohydrates for human consumption. Durum wheat is the preferred raw material for pasta making, couscous and some types of bread, mainly cultivated in Southern regions of Italy. According to the year of release, durum wheat cultivars were grouped into modern developed after the introduction in dwarfing genes in the 1950s, and into old those developed before that time [30]. Old cultivars are characterized by greater rusticity and lower yield while the modern ones differ in term of better yield and quality [31]. Previous research findings showed that the total polyphenol content in both old and modern durum wheat cultivars was similar, but the old cultivars had a higher number of unique compounds not observed in modern varieties [32].

In the last decades, farmers and consumers addressed much attention to emmer wheat, which is phylogenetically related to durum wheat [33]. This renewed interest for emmer is mostly due to the grain bioactive substances and to the possibility of using conventional or organic farming practices with low chemical inputs [34]. Especially, the old cultivars, although lower yielding than modern ones, are suitable for the development of more sustainable crop systems [34].

Barley has been recognized for its adaptability to both highly productive agricultural systems and marginal area. It is also high in dietary fiber (mainly β-glucans), minerals and other phytochemicals such as phenolic acids and flavonoids [24,35]. In particular, proanthocyanidins, the major types of flavonoids in barley grain, are oligomeric and polymeric flavan-3-ols that exert strong antioxidant activity and known also for their ability to bind proteins affecting sensory acceptability [36]. Barley cultivars may have yellow, blue or purple color caused by accumulation of flavonoids compounds in distinct layers of grain [37].

Naked (hull-less) barley is a form of domesticated barley that have an easier-to-remove hull, thanks to which it could have multiple food applications for human consumption in bread preparation, breakfast snacks and beverages (alcoholic and nonalcoholic) [38].

The aim of this research is: To explore the differences, in a 2-year evaluation, of phytochemicals as phenolics and carotenoids, and antioxidant activity in wholemeal and refined-flour, of old and modern cultivars belonging to three cereal species, durum wheat, emmer wheat and barley; to evaluate the effect of species, genotype, environment and their interaction on these traits; to study the antioxidants in the biscuits obtained by selected cultivars for each species; and to provide information on consumer acceptability of monovarietal biscuits in comparison with a commercial product.

2. Materials and Methods

2.1. Plant Materials

Three species, durum wheat (*Triticum turgidum*, ssp. *durum* Desf.), hulled emmer wheat (*Triticum turgidum* L. spp. *dicoccum Shrank*) and barley (*Hordeum vulgare* L. spp. *vulgare*), available at Research Centre for Cereal and Industrial Crops (CREA-CI), were cropped in Foggia (southern Italy) at the experimental fields of the CREA-CI (41°28′ N, 15°34′ E; 76 m a.s.l.), over two crop years (2015–2016 and 2016–2017). For each species, old and modern cultivars were chosen (Table 1). The seeds were planted in 10 m^2 plots according to a Randomized Complete Block Design with three replications. Standard cultural practices for each species were applied. Meteorological data on two crop years were obtained from an on-site weather station (Table S1). The plants were harvested mechanically after physiological maturity. All seeds were stored at 4 °C until further processing.

Table 1. Area of origin and pedigree of the genotypes used in this study.

Taxonomic Classification	Accession	Cultivar/Landrace-Origin	Year of Release	Genotype	
Durum wheat	Cappelli	Cultivar-Selection from Tunisian population 'Jean Retifah'	1915	Old	
	Fortore	Cultivar–Capeiti-8/Valforte	1995	Modern	
Emmer wheat	Molisano	Landrace-Molise region, Central Italy	//	Old	
	PadrePio	Cultivar–Simeto/Molise	2016	Modern	

Table 1. *Cont.*

Taxonomic Classification	Accession	Cultivar/Landrace-Origin	Year of Release	Genotype	
Barley	L94	Ethiopian landrace line (black and naked grains)	//	Old	
	Priora	Cultivar–Arda/Mondo (white and naked grains)	2000	Modern	

2.2. Wholegrain Analysis

Protein content (PC) corrected for dry matter (DM) and Test weight (TW) were determined by NIR (Infratec Nova Analyzer, Foss Italia, Padova, Italy). Thousand Kernel Weight (TKW) was calculated from the mean weight of three sets of 100 grains per plot for each sample.

2.3. Processing

2.3.1. Flours Production

To produce the wholemeal flour, the kernels of each cereal species were ground in a sample mill (Tecator Cyclotec 1093; Foss Italia, Padova, Italy) using a 0.5 mm sieves. To obtain semolina in durum wheat or refined flours in other cereals, the seeds were conditioned at 16.5% (wet basis) moisture and were milled at experimental mill (Labormill 4RB, Bona, Monza, Italy) with four rolls and 42 and 54 GG sieves (sieve 180 µm), which separates flour from bran and germ. Henceforth, the semolina and the other flours were named refined flours.

2.3.2. Biscuit Production

Fortore (durum wheat), Molisano (emmer) and L94 (barley) were selected for interesting levels of phytochemicals and used for biscuit production. Grains were milled to wholemeal flour by means of a granite stone mill (diameter 300 mm model Getreidemühle, Colombini Sergio s.a.s, Abbiategrasso, Milan, Italy). Commercial control (CTRL, 100% wheat) was used in the experimentation. Sucrose, eggs, sunflower oil, salt and vanilla essence were purchased at local retailers. Three independent biscuit-making production trials were performed by Frasca Bakery (Foggia, Italy), involved in the present experiment. The biscuit-making process consisted of: (i) kneading for 3 min sucrose (400 g), sunflower oil (320 mL), eggs (4), salt (4 g), vanilla essence and baking (20 g) by an electric mixer with flat beater (PL16 5B, Conti s.r.l, Bussolengo (Verona, Italy), then adding 1 kg wholemeal flour and kneading for 3 min, and finally adding water (250 mL) and kneading for 3 min; (ii) the dough was rolled out on a tray using a rolling pin and cut into desired shapes using a biscuit cutter; (iii) baking in a steam tube deck oven (Mondial 43, Mondial Forni spa, Verona, Italy) for 15 min at 180 °C. Biscuits were finely crushed in a mortar for subsequent analyses.

2.4. Chemical Compounds

2.4.1. Carotenoids

Total carotenoids pigments, referred to as yellow pigments (TYPC), were analyzed according to method 14–50 of AACC International, as modified by in Beleggia et al. [39] for microsamples. The data were expressed as micrograms per gram on dry matter ($\mu g\ g^{-1}$ DM). All assays were conducted in triplicate.

2.4.2. Phenolics

Phenolic compounds were extracted according to Suriano et al. [40], with minor modifications. The samples (0.5 g) were extracted using 10 mL methanol (80:20) acidified with 1% 12 N HCl, for 30 min in an ultrasonic bath. After centrifugation, the supernatants were used for the determination of phenolics and antioxidant activity. Total polyphenol content (TPC) was determined using Folin–Ciocalteu reagent, according to the modified method of Suriano et al. [40], and expressed as μg gallic acid equivalents (GAE) g^{-1} DM. Total flavonoid content (TFC) was determined according to the method of Kim et al. [41], and expressed as μg catechin equivalents (CE) g^{-1} DM. The total proanthocyanidins (TPAC) were determined according to the modified vanillin assay of Sun et al. [42], and expressed as μg catechin equivalents (CE) g^{-1} DM. All assays were conducted in triplicate.

2.4.3. Phenolic Acid and Flavonoid Composition

Soluble free and conjugated, and insoluble bound phenolic acids and flavonoids were extracted, separated and quantified according to the method described in Suriano et al. [40], with some modifications, using an Agilent 1200 Series HPLC system (Agilent Technologies, Waldbronn, Germany) equipped with a diode array detector. Separation of phenolic acids was achieved using a reversed phase C18 column (InfinityLAB Poroshell 120 RC-C18, 100×2.1 mm; particle size = 2.7 μm) from Agilent (Santa Clara, CA, USA). The column temperature was 35 °C, and the mobile phase consisted of (A) water with phosphoric acid 10^{-3} M and (B) acetonitrile at a flow rate of 0.5 mL/min, using the following linear gradient program: 5% B for 2 min, from 5% to 30% B for 10 min, from 30% B to 55% B for 1 min, from 55% to 70% for 2 min, isocratic at 70% for 1 min, linear gradient from 70% to 5% B for 6 min. Two microliters of sample were injected using an autosampler. The wavelengths used for quantification of the phenolic acids were 280 and 320 nm. The quantification was based on the peak area of the following standards: p-Hydroxybenzoic acid, Vanillic acid, Caffeic acid, Syringic acid, Vanillin, Ferulic acid, Sinapic acid, p-Coumaric acid, Protocatechuic acid, Trans-cinnamic acid and Cis-cinnamic acid and Syringaldeide. Moreover, some standards of flavonoids in cereals were used: Quercitin, Catechin and Naringenin. An example of phenolic and flavonoid chromatograms during the whole cereal food supply chain was reported in Figure S1. All used reagents were obtained from Merk Life Science S.r.l, Milano, Italy. All assays were made in triplicate.

2.4.4. Antioxidant Activities

The antioxidant activity was determined using two different assays: the DPPH and TEAC methods. DPPH radical scavenging capacity was determined according to Suriano et al. [40], using a Trolox calibration curve, and measuring the absorbance at 517 nm. Data were expressed as μmol Trolox equivalents (TE) g^{-1} DM. The TEAC Trolox equivalent antioxidant capacity was determined according to the method of Fares et al. [43], by using a Trolox standard curve, on the basis of the percentage inhibition of absorbance at 734 nm of the radical cation $ABTS^{\bullet+}$ and expressed as μmol Trolox equivalents (TE). All assays were conducted in triplicate.

2.5. Consumer Acceptance

A sensory evaluation questionnaire was used in this study to assess the degree of liking of the different biscuits based on their sensory appeal, with respect to the CTRL. Thirty untrained participants performed the test, aged between 25 and 65 years, 80% females and 20% males. The sensory attributes evaluated were odor, sweetness, flavor, crumbliness, crispness, color and could also include overall acceptability. A five-point hedonic scale was used to evaluate the attributes for consumer acceptance, varying from disliked extremely (1) to liked extremely (5) [44]. Biscuits were coded with 4 random letters and water was served to participants for mouth cleaning between samples evaluation.

2.6. Statistical Analysis

For all the datasets, one-way analysis of variance (ANOVA) was performed to estimate differences ascribable to the species (S), genotype (G) or year (Y) effect, while two-factor ANOVA was applied to study the effect of the 'SxG' and 'SxGxY' interactions. Whenever a significant F value was obtained for single factors or their interaction, Tukey HSD test was performed at $p < 0.05$ level. Pearson correlations (r) of the means among phenolics and antioxidant activities were calculated. Statistical analyses were performed using the STATISTICA program (StatSoft Italia srl, vers. 8.0, 2007). A Principal Component Analysis (PCA) was performed using a correlation matrix to visualize differences and similarities of PC, yield-related traits and antioxidants in the three species for two years by using the JMP software (SAS Institute Inc., Cary, NC, USA version 8).

3. Results and Discussion

3.1. Whole Grain Quality and Yield-Related Traits

The characterization of the grains of all samples, with regard to PC and yield-related parameters, TW and TKW, was performed and the results were shown in Table 2.

Table 2. Qualitative and yield-related traits in wholegrain samples.

Species	Genotype	Type	Year	PC (g kg^{-1}, DM)	TW (Kg hL^{-1})	TKW (g)
Interaction of Species x Genotype x Year effects (SxGxY)						
Durum wheat	Cappelli	Old	2015/16	15.90	81.43 a,b	54.60 a,b
Durum wheat	Cappelli	Old	2016/17	16.50	81.18 a,b	52.43 b,c
Durum wheat	Fortore	Modern	2015/16	14.30	78.16 b,c	48.33 c,d
Durum wheat	Fortore	Modern	2016/17	13.83	82.51 a	57.73 a
Emmer wheat	Molisano	Old	2015/16	14.83	73.06 e,f	42.30 e,f
Emmer wheat	Molisano	Old	2016/17	16.70	69.37 f	40.13 f
Emmer wheat	PadrePio	Modern	2015/16	16.63	80.98 a,b	45.60 d,e
Emmer wheat	PadrePio	Modern	2016/17	18.40	77.96 b,c	46.87 d
Barley	L94	Old	2015/16	13.87	74.60 c–e	40.67 f
Barley	L94	Old	2016/17	14.87	75.21 c–e	40.07 f
Barley	Priora	Modern	2015/16	14.93	80.15 a,b	47.07 d
Barley	Priora	Modern	2016/17	16.23	78.93 a–c	49.03 c,d
	$F_{(2.24)}$			0.99	4.0	8.56
	p value			n.s.	*	**
Interaction of Species x Genotype effects (SxG)						
Durum wheat	Cappelli	Old		16.20 a,b	81.31 a	53.52 a
Durum wheat	Fortore	Modern		14.07 d	80.34 a	53.03 a
Emmer wheat	Molisano	Old		15.77 b,c	71.21 c	41.22 c
Emmer wheat	PadrePio	Modern		17.52 a	79.47 a	46.23 b
Barley	L94	Old		14.37 c,d	74.90 b	40.37 c
Barley	Priora	Modern		15.58 b–d	79.54 a	48.05 b
	$F_{(2.30)}$			16.67	17.24	7.53
	p value			***	***	**
Single effect (Species) (S)						
	Durum wheat			15.13 b	80.82 a	53.28 a
	Emmer wheat			16.64 a	75.34 b	43.73 b
	Barley			14.98 b	77.22 b	44.21 b
	$F_{(2.33)}$			6.56	7.98	24.89
	p value			**	***	***

Table 2. *Cont.*

Species	Genotype	Type	Year	PC (g kg^{-1}, DM)	TW (Kg hL^{-1})	TKW (g)
	Single effect (Genotype) (G)					
	Old genotypes			15.44	75.81 b	45.03 b
	Modern genotypes			15.72	79.78 a	49.11 a
	$F_{(1.34)}$			0.33	11.28	5.05
	p value			n.s.	**	*
	Single effect (Year) (Y)					
	2015/16			15.08 b	78.06	46.43
	2016/17			16.09 a	77.53	47.71
	$F_{(1.34)}$			5.03	0.15	0.44
	p value			*	n.s.	n.s.

PC = protein content; TW = test weight; TKW = thousand kernel weight. For each parameter, different letters indicate significant differences according to the Tukey's test ($p \leq 0.05$). *, **, ***, significant at 0.05, 0.01 and 0.001 probability level, respectively; n.s., not significant.

ANOVA showed significant interactions between species and genotype 'SxG' for PC and yield-related traits. Considering the PC, a pronounced effect due to the species (S) was found with a minor significant effect of the year (Y) ($p < 0.05$). Among durum wheats, the old cultivar was characterized by higher protein percentage compared with modern one, as a consequence of breeding programs for higher yields at the expense of grain quality [31,45,46]. The yield increase, essentially due to a greater carbon availability to the grains, is accompanied by the decrease in protein content, by dilution effects [47]. The opposite was observed in emmer wheat and barley. According to Geisslitz et al. [48], higher proteins were observed for ancient wheats, einkorn, emmer and spelt, compared to modern wheat species, common and durum wheat. In fact, in our study, the emmer cultivars showed highest protein content (16.65 g kg^{-1}, DM, on average) compared to modern durum wheat cultivars (15.14 g kg^{-1}, DM, on average).

The yield-related parameters which resulted were significantly affected by species, genotype and 'SxG' and 'SxGxY' interactions whereas crop years have no effect on them. The higher values were observed in emmer wheat and in barley modern cultivars in both crop years, as a result of genetic gains in yield in both species [49–52]. In durum wheat, a different response was observed over the two crop years: Fortore showed the highest values of TW and TKW in 2016–2017 crop years and the lowest in the previous one, while the yield-related response of Cappelli was more stable in the two crop years. This confirms the behavior of old durum wheat cultivars which, although having a lower yield potential, are characterized by a lower sensitivity to environmental conditions and a greater stability of their productions [46,53].

3.2. Effects of Species, Genotype and Crop Year on the Content of Phenolic Compounds and Carotenoids, and Antioxidant Activities in Wholemeal

On wholemeal of all cultivars of the three species grown in two crop years, phenolics (TPC, TFC and TPAC) and TYPC, and DPPH- and ABTS-radical scavenging activities were determined and the effects of species (S), genotype (G), year (Y) and their combined interactions were measured by ANOVA (Table 3). S, G and 'SxG' were highly significant for all parameters ($p < 0.001$), whereas Y significantly affected only TPC, TPAC and TEAC and 'SxGxY' interactions were significant only for TPAC, TYPC, DPPH and TEAC.

Table 3. Mean values of Total polyphenols (TPC), flavonoids (TFC), proanthocyanidins (TPAC), carotenoids (TYPC) and antioxidant activities (DPPH and TEAC), for wholemeal of all genotypes of the three species analyzed in two crop years.

Species	Genotype	Type	Year	TPC (μg GAE g^{-1})	TFC (μg CE g^{-1})	TPAC (μg CE g^{-1})	TYPC (μg g^{-1})	DPPH (μmol TE g^{-1})	TEAC (μmol TE g^{-1})
Interaction of Species x Genotype x Year effects (SxGxY)									
Durum wheat	Cappelli	Old	2015/16	1065.09	288.82	118.52 d	5.66 c	1.98 c,d	2.52 e–g
Durum wheat	Cappelli	Old	2016/17	1173.80	276.07	115.25 d	5.75 c	2.09 c,d	2.77 d
Durum wheat	Fortore	Modern	2015/16	896.52	313.82	145.16 d	7.58 a	1.74 d	2.71 d,e
Durum wheat	Fortore	Modern	2016/17	925.88	301.71	128.32 d	6.79 b	1.81 c,d	2.57 e,f
Emmer wheat	Molisano	Old	2015/16	1067.93	289.47	113.09 d	4.49 e	1.99 c,d	2.34 f,g
Emmer wheat	Molisano	Old	2016/17	1167.83	314.84	125.53 d	4.99 d	2.20 c,d	2.28 g
Emmer wheat	PadrePio	Modern	2015/16	1037.73	295.60	132.68 d	5.71 c	2.16 c,d	2.31 g
Emmer wheat	PadrePio	Modern	2016/17	1066.12	325.12	134.86 d	6.08 c	2.27 c	2.37 f,g
Barley	L94	Old	2015/16	2565.04	717.66	824.41 c	3.78 f	8.85 b	8.10 c
Barley	L94	Old	2016/17	2668.70	731.91	822.78 c	3.87 f	8.38 b	8.16 c
Barley	Priora	Modern	2015/16	2704.27	871.41	1639.18 a	2.92 g	11.43 a	10.95 a
Barley	Priora	Modern	2016/17	2823.60	863.26	1314.95 b	3.10 g	11.06 a	10.06 b
	$F_{(2.24)}$			*1.79*	*1.2*	*73.39*	*7.32*	*0.23*	*10.89*
	p value			n.s.	n.s.	***	***	**	***
Interaction of Species x Genotype (SxG)									n.s.
Durum wheat	Cappelli	Old		1119.44 c	282.45 d	116.88 c	5.70 b	2.04 c,d	2.64 c
Durum wheat	Fortore	Modern		911.20 e	307.76 c	136.74 c	7.19 a	1.78 d	2.64 c
Emmer wheat	Molisano	Old		1117.88 c	302.15 c	119.31 c	4.74 c	2.09 c,d	2.31 d
Emmer wheat	PadrePio	Modern		1051.93 d	310.36 c	133.77 c	5.89 b	2.22 c	2.34 d
Barley	L94	Old		2616.87 b	724.78 b	823.59 b	3.83 d	8.62 b	8.13 b
Barley	Priora	Modern		2763.93 a	867.34 a	1477.06 a	3.01 e	11.24 a	10.51 a
	$F_{(2.30)}$			*82.11*	*122.99*	*1231.45*	*176.87*	*210.67*	*283.97*
	p value			***	***	***	***	***	***
Single effect (Species) (S)									
	Durum wheat			1015.32 c	295.10 c	126.81 b	6.45 a	1.91 c	2.46 b
	Emmer wheat			1084.90 b	306.26 b	126.54 b	5.32 b	2.16 b	2.33 b
	Barley			2690.40 a	796.06 a	1150.33 a	3.42 c	9.93 a	9.32 a
	$F_{(2.33)}$			*9224.26*	*7527.75*	*12,747.46*	*1068.10*	*7155.16*	*9508.79*
	p value			***	***	***	***	***	***
Single effect (Genotype) (G)									
	Old genotypes			1618.06 a	436.46 b	353.26 b	4.76 b	4.25 b	4.36 b
	Modern genotypes			1575.69 b	495.15 a	582.52 a	5.36 a	5.08 a	5.16 a
	$F_{(1.34)}$			*13.83*	*237.67*	*1438.67*	*126.27*	*177.7*	*294.06*
	p value			***	***	***	***	***	***
Single effect (Year) (Y)									
	2015/16			1556.10 b	462.80	495.50 a	5.02	6.49	4.82 a
	2016/17			1637.66 a	468.82	440.28 b	5.10	4.64	4.70 b
	$F_{(1.34)}$			*51.24*	*2.5*	*83.47*	*1.90*	*0.81*	*6.55*
	p value			***	n.s.	***	n.s.	n.s.	*

For each parameter, different letters indicate significant differences according to the Tukey's test ($p \leq 0.05$). *, **, ***, significant at 0.05, 0.01 and 0.001 probability level, respectively; n.s., not significant.

Apart from the TYPC, barley was found to have significantly higher values for all the other parameters with respect to both wheats, of 38.5% on average for TPC and TFC, of 11% TPAC and of 24% on average for the two antioxidant capacities. The highest TPC, TFC and TPAC and antioxidant activities in barley are in agreement with previous results [36,54,55] confirming it to be an excellent dietary source of natural antioxidants with good health potential [56]. Although both ABTS and DPPH methods measured the antioxidant activity, the different levels can be explained by their different mechanisms [56,57]. The advantage of the ABTS radical is its high reactivity, and thus the likely ability to react with a broader range of antioxidants [58]. On the contrary, the DPPH method provides lower values related to Trolox than the ABTS method due to higher stability (and thus lower reactivity) of the DPPH radical. This agreed with durum and emmer wheat response, but not with barley. It is known that DPPH radical reacts with polyphenols (catechins, proanthocyanidins), but not with the phenolic acids and sugars [59]. That could explain the higher levels of DPPH compared to TEAC observed in barley, richest in polyphenols.

Durum wheat distinguished itself from the other species for the highest TYPC, with major levels in the modern cultivars as a result of breeding for this quality trait related to consumer preference for bright yellow color of pasta [19,20].

Among the cultivars, generally the highest values of all determinations were observed in the modern ones, with the only exception of the TPC for which the highest content was in the old wheat cultivars, Molisano and Cappelli [12,23]. A particular trend was observed for barley in which the modern cultivar Priora distinguished itself for the highest TPC values within and among species. Farther, Priora showed the highest values for all traits except for TYPC, higher in L94. This is consistent with what has been observed by other authors, as the L94, a black-colored cultivar, has more carotenoids compared to Priora, which has white seeds [18,40].

The crop years do not have a clear trend, and where the effect is significant, the response is variable, with higher values in the first year for TPAC and TEAC activity and in the second year for TPC. The lowest rainfall observed in the 2016/2017 crop season, in particular during the grain filling, from April to June (106.1 vs. 163.5 mm; Table S1) could have determined a drought stress condition, resulting in a greater stress-induced synthesis of antioxidants, particularly polyphenols, to serve as free radical scavengers, mitigating oxidative and dehydration stress [60].

3.3. Effects of Species, Genotype and Crop Year on the Content of Some Phenolic Compounds and Carotenoids and Antioxidant Activities in Refined-Flours

TPC, TFC, TPAC and TYPC and DPPH- and ABTS-radical scavenging activities were analyzed on refined-flours of all genotypes of the three species grown in two crop years and the relative loss of each parameter compared to wholemeal was measured. The results are shown in Figure 1.

While the proanthocyanidins disappear completely in the refined flours of durum and emmer wheats, a consistent loss in their content was observed in barley, higher in the modern cultivar Priora than in the old L94 (73% vs. 56%). As observed by Irakli et al. [61], the flavanols were more concentrated in the bran, with a content three times higher than pearled flour. Furthermore, both barley cultivars showed the higher losses of TPC and TFC (55% and 57% on average, respectively), resulting in the highest reductions in DPPH and TEAC activities (63% and 52% on average, respectively), according to Van Hung [3], without consistent differences between old and modern cultivars. Conversely, the lowest decreases in TYPC were observed in this species (15%, on average) (Figure 1). Compared to barley, lower losses were generally observed in emmer and durum wheats. In these species, the responses in old and modern cultivars were different, except for DPPH activity and TPC in emmer (55% and 18%, on average, respectively). In durum wheat, the highest TPC and TFC losses were observed in the modern cultivar Fortore that, contrarily, showed the lower TYPC and both antioxidant activity losses. In wheat emmer, a different variation rate was observed for the other traits. In particular, the old cultivar Molisano showed the lowest decrease in TEAC scavenging activity compared to the modern PadrePio (15% vs. 39%, respectively). This agreed with Skendi et al. [62], who found in emmer landrace flours higher antioxidant activity than their commercial counterparts. As the TPC was lost to a minor extent in both wheats, their maintenance in refined-flour means it might be interesting to use this raw material to produce improved end-products. In particular, Cappelli was confirmed as a cultivar able to preserve useful compounds for health-promoting purposes [12,30].

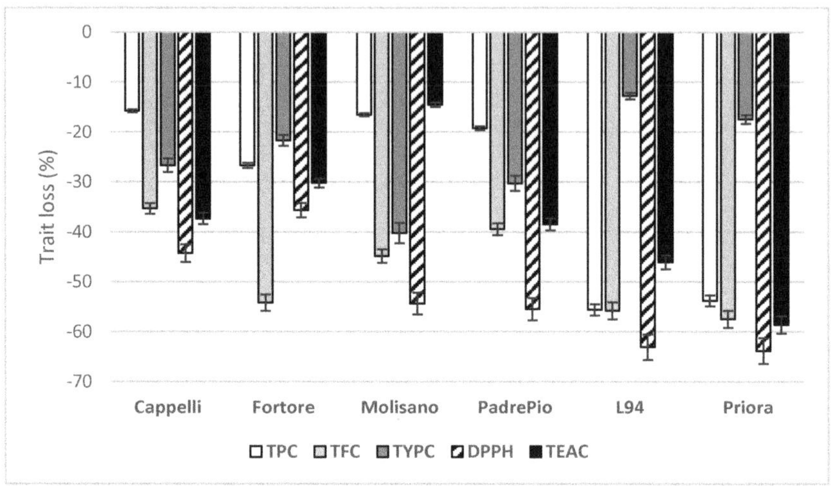

Figure 1. Variation in total polyphenols (TPC), flavonoids (TFC), carotenoids (TYPC) and antioxidant activities (DPPH and TEAC) in refined flours, for all cultivars of the three species analyzed in two crop years. Durum wheats: Cappelli and Fortore; emmer wheat: Molisano and PadrePio; barley: L94 and Priora.

3.4. Phenolic Acid Composition in Wholemeals and Refined-Flours

Soluble free and conjugated, and insoluble bound phenolic acids were investigated in wholemeals of the old and modern genotypes of the three species, and the results are shown in Table 4.

Table 4. Mean value of total content of soluble free and conjugated, and insoluble bound phenolic acids and flavonoids in the wholemeal of old and modern genotypes of the three species analyzed in two crop years. Data are expressed as $\mu g\ g^{-1}$, DM.

Species	Genotype	Crop Years	TSF Phenolic Acids	TSF Flavonoids	TSC Phenolic Acids	TSC Flavonoids	TIB Phenolic Acids	TIB Flavonoids
Durum wheat	Cappelli	2015/16	20.11 c,d	n.d.	26.83 de	n.d.	465.87 a	n.d.
		2016/17	20.58 c,d	n.d.	24.72 e	n.d.	389.53 b	n.d.
	Fortore	2015/16	13.01 g	n.d.	29.06 d,e	2.19 a	384.25 b	n.d.
		2016/17	14.12 f,g	n.d.	38.11 b	1.71 b	314.66 b	n.d.
Emmer wheat	Molisano	2015/16	23.17 c	5.37 d	36.18 b,c	0.44 d	238.06 c	n.d.
		2016/17	19.11 c–f	5.57 d	31.44 d	0.59 c	266.97 c	n.d.
	PadrePio	2015/16	72.73 a	2.47 e	52.24 a	n.d.	242.67 c	0.74 c,d
		2016/17	43.32 b	2.37 e	42.07 b	n.d.	276.30 c	1.12 c
Barley	L94	2015/16	19.20 c–e	18.90 c	27.26 d,e	0.55 c,d	502.04 a	5.19 a,b
		2016/17	43.03 b	25.80 b	30.06 d,e	0.51 c,d	423.87 a,b	4.49 b
	Priora	2015/16	14.70 e–g	52.76 a	27.72 d,e	n.d.	482.55 a	5.17 a,b
		2016/17	15.78 d–g	53.89 a	25.45 e	n.d.	375.15 b	5.65 a
$F_{(2.24)}$ *(SxGxY)*			*48.96*	*11.96*	*14.85*	*4.49*	*1.00*	*3.84*
p value			***	***	***	*	*	*

TSF = Total Soluble Free; TSC = Total Soluble Conjugated; TIB = Total Insoluble Bound. For each parameter, different letters indicate significant differences according to the Tukey's test ($p \leq 0.05$). * and *** significant at 0.05 and 0.001 probability level, respectively.

The ANOVA results were significant for all phenolics. Large variability was seen in the total content of these compounds across the species and the genotypes investigated. Overall, free and conjugated fractions resulted as more representative in emmer wheat (12%, on average) compared to durum wheat and barley (5%, on average), in line with the results of Andersson et al. [63], while the bound fraction was prevalent in barley and among wheats, in durum wheat, in according to Brandolini et al. [64]. The common phenolic acids were mainly ferulic acid, vanillin, coumaric acid and cis- and trans-cinnamic acids (Table S2).

Insoluble bound phenolic acids represented 86.0%, on average, of the total phenolic acids, and ferulic acid was the most abundant in all species, with a variation range from 334.45 $\mu g\ g^{-1}$ DW in barley to 293.80 $\mu g\ g^{-1}$ DW in emmer, on average (Table S2), according to previous studies [10,11,40,64]. Besides ferulic acid, sinapic acid was the second-most abundant phenolic acid, followed by coumaric acid and cis-cinnamic acid, and other minor components in common to all cultivars (i.e., vanillin, syringic acid, syringaldeide and vanillic acid) or genotype-dependent (protocatechuic acid, p-hydroxybenzoic acid, caffeic acid and trans-cinnamic acid). In barley and durum wheat, the highest-bound phenolic acids were observed in the old cultivars with the lowest levels in the second crop season (Tables 4 and S2). The highest values observed in the old cultivar Cappelli agreed with Menga et al. [12]. The opposite trend was observed for emmer wheat. Although phenolic acids have been involved in biotic and abiotic stress tolerance [1], the modern cultivars were mainly developed for yield performance and nutritional and qualitative traits, and not specifically for phenolic acid accumulation, and this could explain the major levels of these compounds in the old cultivars compared to modern ones [60]. As evidenced by other authors, a variability exists in the profile and quantity of phenolic acids among species. For instance, unlike Li et al. [11], who found sinapic acid in its free form only in durum wheat, in this study, sinapic acid was present in the free form only in emmer and in the bound form in all species, with prevalence in emmer wheat. In both fractions, the old cultivar Molisano had the highest content of this compound. The different phenolic acid profile of our results compared with those reported in the literature could be due to the different cultivars as well as the condition of extraction and the chromatography system.

The flavonoids, although mainly present in the stem and leaves of plants [65], were also found in bran and germ section of kernels [8]. Similarly to phenolic acids, the flavonoids are found in free, conjugated and bound form (Table S2). In durum wheats, results showed they were absent. In free fraction, little quercetin amounts were observed in emmer Molisano in both crop years, while catechin was found in free and bound flavonoids in barley, with Priora having two-fold higher content compared to L94 in the free fraction. Naringenin is the common flavonoid to all fractions of emmer wheat and barley. Naringenin is a flavonoid belonging to flavanones subclass, widely spread in beans, citrus fruits, bergamot, tomatoes and other fruits, and in little amounts in cereals, with a possible role on plant growth, and stress responses in plants [66]. In barley, some authors [67,68] found naringenin, quercetin and catechin as potential biomarkers, involved in a significant in vitro reduction in the Fusarium graminearum, a devastating disease of Triticeae, causing yield losses, and also indirectly affecting the quality of grains.

A Pearson correlation was calculated among phenolic compounds and antioxidant activities (Table S3), confirming that TPC, TFC, TPAC and catechin significantly contribute to both radical scavenging activities ($p < 0.001$). Other phenolic acids such as caffeic, syringic, coumaric and trans-cinnamic acids may perform a minor role in both antioxidant activities, with the exception of ferulic acid that positively correlated only with ABTS and sinapic acid that negatively affected the antioxidant potential. This agrees with Menga et al. [12] and Horvat et al. [69].

The refined flours have much lower phenolic acid content than the wholemeal. A total of 28.2% of total phenolic acids, on average, was found in refined flours in the three species (Table S4) as also observed by Guan et al. [70]. The different losses compared to Menga et al. [12] could be explained by the inclusion in this study of other two species, besides durum wheat, showing different endogenous levels of phenolic acids.

Ferulic acid is the most abundant in all samples, accounting for up to 90% of total phenolic acids [62]. Our data are in according to Giordano et al. [71] and Skendi et al. [62], who observed an amount of free ferulic acid in refined-flours that does not surpass the value of 1.4 μg g^{-1} on average, whereas the amount of bound ferulic acid varied between 80.23 μg g^{-1} (durum wheat) and 128.94 μg g^{-1} (barley), suggesting that not only the kernel tissue but also the species and the genotype affect the amount of ferulic acid content.

In free fraction, other phenolic acids were identified as vanillin in durum wheat and barley, and syringic and p-coumaric acids in barley, with prevalence of old genotype. In the conjugated fraction, vanillin and p-coumaric acid are in common, while vanillic acid was found in durum wheat and barley and syringic acid in barley. The other two compounds in common for all species in insoluble bound phenolic acids were sinapic acid and p-coumaric acid. Interestingly, major phenolic acids were found in barley refined flour, particularly in L94, with the order ferulic acid > trans-cinnamic acid > sinapic acid > p-coumaric acid > caffeic acid > p-hydroxybenzoic acid > protocatechuic acid > vanillic acid. The highest levels of the insoluble bound form of phenolic acids were in the modern genotype for emmer wheat while; for the other species, results for the old were prevalent, reflecting the wholemeal trend. Similar to the phenolic acids, the flavonoids were found in free, conjugated and bound form. Among the studied species, only barley contained catechin in the bound flavonoid fraction, without differences between old and modern cultivars (Table S4), in the range observed by Idehen et al. [24]. With wholemeal being higher in antioxidants, we will concentrate our attention on this type of flour becoming the raw material for biscuit making.

3.5. Principal Component Analysis of Grain and Wholemeal between Phenolics, Phenolic Acids, Quality and Yield-Related Traits

In order to analyze multiple variables in the grain and wholemeal of the three species and responses in two crop years, a principal component analysis (PCA) was undertaken, and results were reported by translating multiple data into a score plot and loading plot (Figure 2a,b). The principal component 1 (PC1) explained 44.4% of the total variance, whilst the principal component 2 (PC2) explained 18.5% of the variance (Table S5). PC1 discriminated the species, barley genotypes being on the positive and durum wheat and emmer wheat on the negative axes. In turn, durum wheat and emmer wheat genotypes were separated along the PC2. Major antioxidant traits were mainly influenced by PC1, whereas TW, TKW and some other phenolics by PC2. In particular, the first factor was highly and positively associated with the TPC, TFC, TPAC, DPPH, TEAC, caffeic acid, syringic acid, p-coumaric acid, ferulic acid, trans-cinnamic acid, catechin and negatively with the TYPC, protocatechuic acid and sinapic acid. The second factor showed a positive association with yield-related traits (TW, TKW), and some phenolic acids (vanillic acid, vanillin, cis-cinnamic acid) and negatively with p-hydroxybenzoic acid and the flavonoid naringenin. Instead, both factors were negatively associated to PC.

The general inverse relation between the yield-related traits (TW and TKW) and p-hydroxybenzoic acid and naringenin observed along the Factor 2 indicated that yield-related traits are at odds with some phenolics, as previously observed in Menga et al. [12]. Regarding the cultivars, a discrimination was observed between the old and modern ones for emmer wheat along Factor 1. Instead, the crop years had a different trend only for durum wheat cultivars. On the basis of the PCA, the barley cultivars were closely related to most of the phytochemicals and to the antioxidant activities. These data are in agreement with previous studies [37,40,72]. For the emmer, the old cultivar Molisano is associated to naringenin and p-hydroxybenzoic acid, while the modern PadrePio by PC, protocatechuic acid and sinapic acid in the two years. Finally, durum wheat cultivars Fortore and Cappelli were distributed along the negative left quadrant and were associated with yield-related traits and with TYPC. It is noteworthy that TYPC in durum wheat is a criterion for the marketing and the nutritional quality of end-products, such as pasta [19,20].

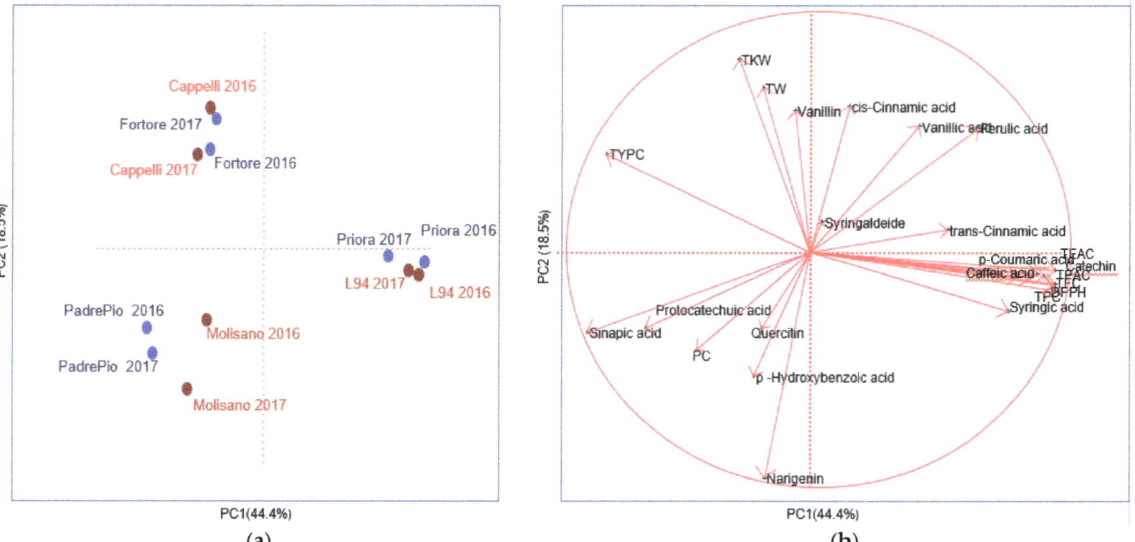

Figure 2. Principal component analysis (PCA) score plot (**a**) and loading plot (**b**) of the trait analyzed in the three species. In red, the old, and in blue, the modern cultivars belonging to the three species. TPC = total polyphenols; TFC = total flavonoids; TPAC = total proanthocyanidins; TYPC = total carotenoids; DPPH = antioxidant activity; TEAC = antioxidant activity; PC = protein content; TW = test weight; TWK = thousand kernel weight.

3.6. Effect of Biscuit Processing in Phenolic Compound Level and Composition

For each species, a cultivar was selected for interesting levels of phytochemicals and used for biscuit making: in particular, Fortore (durum wheat) was high in TFC and TYPC, Molisano (emmer wheat) was high in TPC and free and conjugated flavonoids and L94 (barley) was high in conjugated and bounded flavonoids.

The experimental biscuits belonging to the three species were nutritionally superior to commercial control as they have highest TPC, TFC, TPAC and TYPC, as well as both antioxidant activities (Figure 3). Among the three species, durum wheat prevailed for the TYPC, while barley was confirmed to have the best performance for all the other analyzed parameters.

With respect to the corresponding wholemeal, flavonoids decreased in all samples, while in Fortore and Molisano an increase in total polyphenols and proanthocyanidins was observed (24% and 55% on average, respectively). A greater increase was found for these traits in the CTRL. On the contrary, they decreased in L94 (18% and 31%, respectively) (Table S6a; Figure 3). The antioxidant activities reflect this trend, with results partially overlapped with the results of Li et al. [73] on muffins. In all biscuits, a positive effect of baking on the free, conjugated and bound phenolic acids was observed, confirming the general response reported by Abdel-Aal and Rabalski [21] for cookies and muffins.

As product-making processes were the same for all our biscuits, the unique difference in the recipe was the flour belonging to the different cultivars that may have contributed to changes in phenolic contents among the end-products.

Ferulic acid was the principal phenolic acid in the free, conjugated or bound extracts of the end-products, showing the highest values in bound form (about thirty-fold higher when compared to control biscuit). With respect to corresponding wholemeal, an increase in free ferulic acid content was observed in Fortore and L94 (41% and 56%, respectively), the conjugated form was negatively affected, particularly in barley and emmer wheat (45% of loss, on average) while bound fraction decreased in the same extent in all samples (47% on average) (Tables 5 and S6b). The increment observed in free ferulic acid could be due to the release of bound forms from the food matrix during the baking process [21,43].

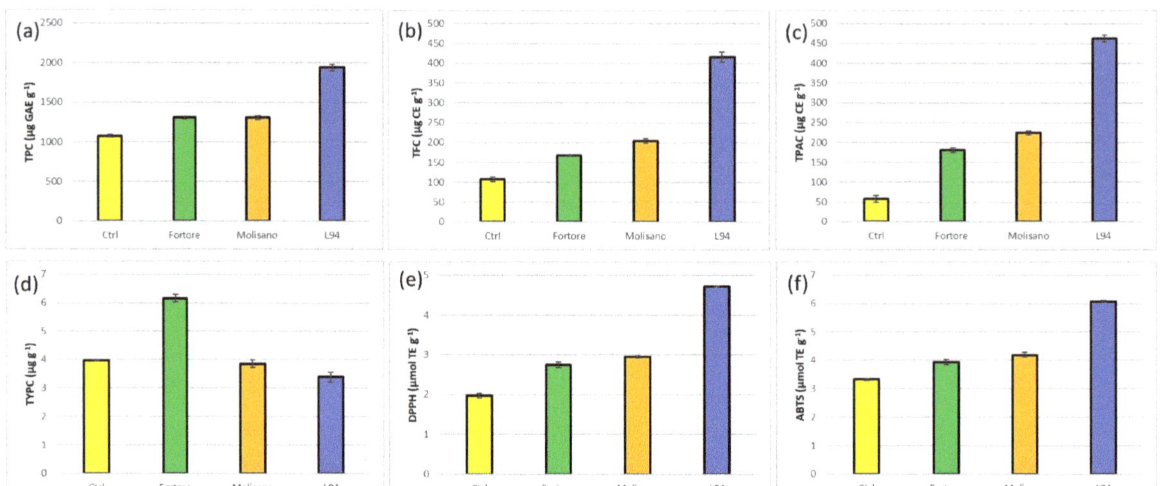

Figure 3. Total polyphenols (TPC) (**a**), total flavonoids (TFC) (**b**), total proanthocyanidins (TPAC) (**c**), total carotenoids (TYPC) (**d**), antioxidant activity (DPPH) (**e**), antioxidant activity (ABTS) (**f**) of biscuits from commercial CTRL (yellow), Fortore (Durum wheat, green), Molisano (Emmer wheat, orange) and L94 (Barley, blue). (Means and standard deviations; values expressed on dry matter).

Ferulic acid, vanillic acid and vanillin were the common phenolic acids in all extracts of the three products. Vanillin being a component of the biscuit recipe, its high levels are not dependent on baking conditions or the flour, and they have no relevance. Major phenolic acids were found in the bound fraction (p-coumaric, sinapic, p-hydroxybenzoic, syringic, caffeic acids and syringaldeide), while p-coumaric and sinapic acids were found in the conjugated fraction. In the commercial control, only p-coumaric acid in the bound fraction was detected.

In general, other than the release of bound phenolics from the food matrix, different mechanisms could be involved in changing phenolic acids during baking, such as polymerization and oxidation of phenolics, thermal degradation and production of Maillard reaction products, as supposed by other authors [74,75].

Table 5. Comparison of phenolic acid composition, free- and conjugated-soluble and insoluble bound, in biscuits derived from the best genotype choosen within each of the three species and from commercial control. (Means and standard deviations; values expressed as µg g^{-1}, DM).

| | Durum Wheat | | | Emmer Wheat | | | Barley | | | Commercial | | |
| | Fortore | | | Molisano | | | L94 | | | Control | | |
	Soluble Free	Soluble Conjugated	Insoluble Bound	Soluble Free	Soluble Conjugated	Insoluble Bound	Soluble Free	Soluble Conjugated	Insoluble Bound	Soluble Free	Soluble Conjugated	Insoluble Bound
Vanillic acid	38.24 ± 0.94	37.87 ± 1.43	9.75 ± 0.47	12.96 ± 0.01	19.17 ± 1.24	5.16 ± 0.43	15.20 ± 1.02	19.17 ± 1.24	6.4 ± 0.49	4.20 ± 0.4	52.91 ± 0.15	1.23 ± 0.11
Vanillin	1127.44 ± 8.49	661.12 ± 21.87	40.13 ± 1.06	1237.92 ± 2.26	876.22 ± 28.42	29.89 ± 1.77	1149.56 ± 5.83	871.23 ± 35.49	29.61 ± 1.19	1400.84 ± 12.16	868.12 ± 29.77	38.59 ± 0.41
Ferulic acid	6.52 ± 0.17	11.23 ± 0.49	327.04 ± 22.02	5.36 ± 0.57	9.17 ± 0.38	265.41 ± 5.2	5.72 ± 0.41	9.20 ± 0.34	347.09 ± 0.07	1.12 ± 0.11	0.88 ± 0.04	10.91 ± 1.13
Sinapic acid	1.40 ± 0.06	5.84 ± 0.34	10.03 ± 0.49	4.08 ± 0.82	5.47 ± 0.64	9.25 ± 0.15	n.d.	12.16 ± 0.30	11.29 ± 0.09	n.d.	n.d.	n.d.
Catechin	n.d.	n.d.	n.d.	n.d.	n.d.	n.d.	n.d.	n.d.	n.d.	n.d.	n.d.	n.d.
p-Hydroxybenzoic acid	n.d.	n.d.	1 ± 0.05	n.d.	n.d.	0.91 ± 0.03	n.d.	n.d.	0.75 ± 0.07	n.d.	n.d.	n.d.
Syringic acid	n.d.	n.d.	1.04 ± 0.15	n.d.	n.d.	0.8 ± 0.07	n.d.	n.d.	1.16 ± 0.39	n.d.	n.d.	n.d.
p-Coumaric acid	n.d.	1.39 ± 0.01	9.36 ± 0.72	n.d.	0.83 ± 0.04	16.61 ± 0.19	n.d.	8.27 ± 0.15	12.49 ± 1.22	n.d.	n.d.	9.27 ± 0.04
Syringaldeide	n.d.	n.d.	2.89 ± 0.32	n.d.	n.d.	3.59 ± 0.05	n.d.	n.d.	3.87 ± 0.07	n.d.	n.d.	n.d.
Caffeic acid	n.d.	n.d.	0.87 ± 0.13	n.d.	n.d.	1.87 ± 0.07	n.d.	n.d.	0.87 ± 0.13	n.d.	n.d.	n.d.

3.7. Sensory Biscuits Profile

The degree of liking of the biscuits obtained from different species in comparison to commercial control was assessed by the consumer test evaluation based on its sensory appeal (color, odor, sweetness, crumbliness, crispness, flavor and overall acceptability). A radar graph represents the sensory data (Figure 4). Sweetness, crumbliness, flavor and overall acceptability in Molisano were rated by the consumers with even higher scores than the control sample. Then, Fortore biscuits emerged for crispness and odor and finally, L94 was not appreciated, except for crumbliness. In all examined samples, the crust color was from bright yellow in biscuit control to brownish yellow. Color and external aspect of biscuits could be affected by reducing sugars, which caramelize during the baking process producing brown color [76]. Considering sweetness, the opposite trend was evidenced for Molisano and L94 showing the highest and lowest values, respectively. This diverse response could be due to different sugar content and composition of the flours influencing other than the sensory characteristics, also the structure and texture of dough and subsequent cooking performance [77].

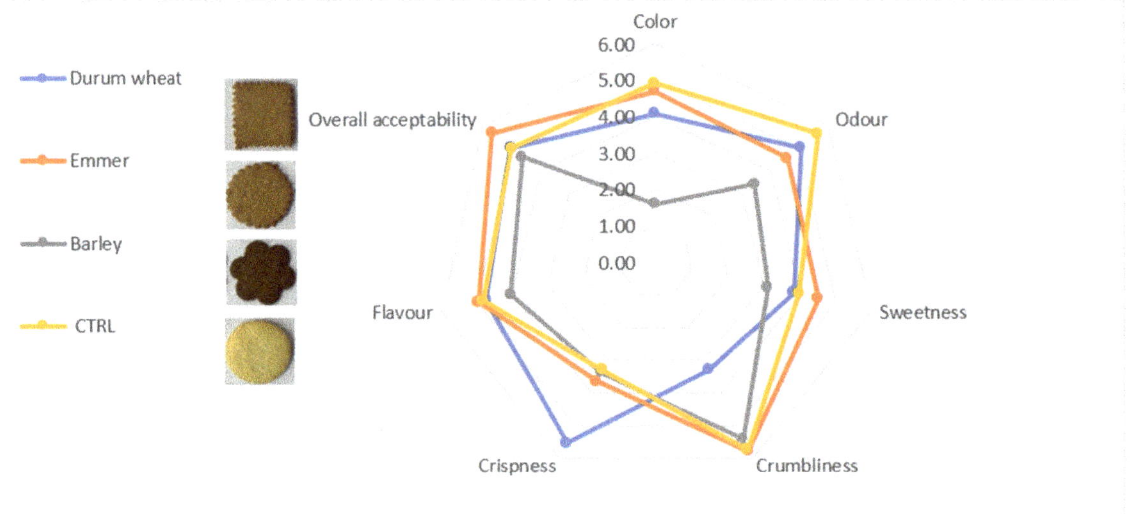

Figure 4. Radar plot obtained from consumer test evaluation of different biscuits.

The aromatic attributes scores were quite different referred to odor (smell) and flavor (taste). The odor was better in control biscuits while the L94 was the worst. Instead, for flavor considered the whole package as combination of taste, odor and chemical sensations, the emmer, the durum and the control biscuit samples clearly differed from the barley one. Although barley was one of the best sources of phytochemicals, it was not appreciated by consumers and not marketable.

4. Conclusions

This work provides new insights into the quali-quantitative composition of some cereal bioactive metabolites in relation to their potential antioxidant activity in wholemeal and cereal-based products. Wholemeal of cereals represents a rich source of phenolics, mostly phenolic acids and flavonoids. Among the studied species, the barley resulted to show the best performance for all traits, except for total carotenoids. The phenolic acids were more representative in emmer in free and conjugated fractions, while the bound fraction, representing the 86% of the total, was prevalent in barley and durum wheat. Considering the old and the modern cultivars, generally the modern ones contained higher levels of antioxidants, except for TPC and TPAC that, instead, were more affected by growing season.

As the baking processes resulted in a loss of phenolics in biscuits as compared to wholemeal flour, the choice of raw materials richest in these compounds becomes crucial in obtaining better final products. In this optic, although barley biscuits were richer in antioxidants, the less consumer acceptance could limit its diffusion. Considering the popularity of this product, the optimization of ingredients or the blending of wheat flour with selected fractions of barley to obtain enriched biscuits should always go hand-in-hand with sensory evaluation, to reach health benefits and to be easily marketable.

Supplementary Materials: The following supporting information can be downloaded at: https://www.mdpi.com/article/10.3390/foods12132551/s1, Table S1: Monthly total rainfall and temperature (minimum and maximum) at the experimental field of Foggia during the two growing seasons, 2016 and 2017; Table S2: Soluble free, soluble conjugated and insoluble bound phenolic acids in the wholemeal for the two cultivars of three species analyzed in two crop years. Values are means± standard deviation (SD) of three independent evaluation; Table S3: Pairwaise significant correlations between antioxidant activities and phenolic compounds; Table S4: Soluble free, soluble conjugated and insoluble bound phenolic acids in the refined-flours of the two cultivars of three species analyzed in two crop years; Table S5: Eigenvalue and percentage of variation explained by the first eleven factors for all traits analyzed; Table S6a: Total polyphenols (TPC), flavonoids (TFC), proanthocyanidins (TPAC), carotenoids (TYPC) and antioxidant activities (DPPH and TEAC) in the wholemeal of three species and a commercial CTRL. Values are means± standard deviation (SD) of three independent evaluation; Table S6b: Soluble free, soluble conjugated and insoluble bound phenolic acids in the wholemeal of three species and a commercial CTRL. Values are means ± standard deviation (SD) of three independent evaluation; Figure S1: Example of phenolic and flavonoid chromatogram during the whole cereal food supply chain.

Author Contributions: Conceptualization, D.B.M.F.; methodology, V.G.; validation, D.B.M.F. and G.M.B.; formal analysis, V.M.; investigation, D.B.M.F. and G.M.B.; resources, D.B.M.F.; data curation, D.B.M.F., G.M.B. and V.M.; writing—original draft preparation, G.M.B.; writing—review and editing, G.M.B. and V.M.; supervision, D.B.M.F. All authors have read and agreed to the published version of the manuscript.

Funding: This study includes results obtained within the project entitled "Nobili Cereali" funded by the PSR 2014–2020 Regione Campania—Misura 16.1.1–Azione 2 "Sostegno ai POI", coordinated by Roberto Rubino.

Data Availability Statement: The data presented in this study are available on request from the corresponding author. The data are not publicly available.

Acknowledgments: The authors acknowledge Silvana Paone for the agronomical support, Antonio Troccoli for meteorological data and Michele Savino for his technical support.

Conflicts of Interest: The authors declare no conflict of interest.

References

1. Marone, D.; Mastrangelo, A.M.; Borrelli, G.M.; Mores, A.; Laidò, G.; Russo, M.A.; Ficco, D.B.M. Specialized metabolites: Physiological and biochemical role in stress resistance, strategies to improve their accumulation, and new applications in crop breeding and management. *Plant Physiol. Biochem.* **2022**, *172*, 48–55. [CrossRef] [PubMed]
2. Sreenivasulu, N.; Fernie, A.R. Diversity: Current and prospective secondary metabolites for nutrition and medicine. *Curr. Opin. Biotechnol.* **2022**, *74*, 164–170. [CrossRef] [PubMed]
3. Van Hung, P. Phenolic Compounds of Cereals and Their Antioxidant Capacity. *Crit. Rev. Food Sci. Nutr.* **2016**, *56*, 25–35. [CrossRef]
4. Francavilla, A.; Joye, I.J. Anthocyanins in Whole Grain Cereals and Their Potential Effect on Health. *Nutrients* **2020**, *12*, 2922. [CrossRef] [PubMed]
5. Barron, C.; Surget, A.; Rouau, X. Relative amounts of tissues in mature wheat grain and their carbohydrate and phenolic acid composition. *J. Cereal Sci.* **2007**, *45*, 88–96. [CrossRef]
6. Fu, B.X.; Chiremba, C.; Pozniak, C.J.; Wang, K.; Nam, S. Total Phenolic and Yellow Pigment Contents and Antioxidant Activities of Durum Wheat Milling Fractions. *Antioxidants* **2017**, *6*, 78. [CrossRef]

7. Ficco, D.B.M.; Borrelli, G.M.; Miedico, O.; Giovanniello, V.; Tarallo, M.; Pompa, C.; De Vita, P.; Chiaravalle, A.E. Effects of grain debranning on bioactive compounds, antioxidant capacity and essential and toxic trace elements in purple durum wheats. *LWT—Food Sci. Technol.* **2020**, *118*, 108734. [CrossRef]

8. Adom, K.K.; Sorrells, M.E.; Liu, R.H. Phytochemicals and Antioxidant Activity of Milled Fractions of Different Wheat Varieties. *J. Agric. Food Chem.* **2005**, *53*, 2297–2306. [CrossRef]

9. Bhasin, T. Sensory characteristics of wholegrain. *Adv. Obesity Weight Manag. Control* **2020**, *10*, 191–194. [CrossRef]

10. Laddomada, B.; Durante, M.; Mangini, G.; D'amico, L.; Lenucci, M.S.; Simeone, R.; Piarulli, L.; Mita, G.; Blanco, A. Genetic variation for phenolic acids concentration and composition in a tetraploid wheat (*Triticum turgidum* L.) collection. *Genet. Resour. Crop Evol.* **2017**, *64*, 587–597. [CrossRef]

11. Li, L.; Shewry, R.; Ward, J.L. Phenolic Acids in Wheat Varieties in the Healthgrain Diversity Screen. *J. Agric. Food Chem.* **2008**, *56*, 9732–9739. [CrossRef]

12. Menga, V.; Giovanniello, V.; Savino, M.; Gallo, A.; Colecchia, S.A.; De Simone, V.; Zingale, S.; Ficco, D.B.M. Comparative Analysis of Qualitative and Bioactive Compounds of Whole and Refined Flours in Durum Wheat Grains with Different Year of Release and Yield Potential. *Plants* **2023**, *12*, 1350. [CrossRef] [PubMed]

13. Scalbert, A.; Morand, C.; Manach, C.; Rémésy, C. Absorption and metabolism of polyphenols in the gut and impact on health. *Biomed. Pharmacother.* **2002**, *56*, 276–282. [CrossRef] [PubMed]

14. Vitaglione, P.; Napolitano, A.; Fogliano, V. Cereal dietary fibre: A natural functional ingredient to deliver phenolic compounds into the gut. *Trends Food Sci. Technol.* **2008**, *19*, 451–463. [CrossRef]

15. Xavier, A.A.; Pérez-Gálvez, A. Carotenoids as a Source of Antioxidants in the Diet. *Subcell. Biochem.* **2016**, *79*, 359–375. [PubMed]

16. Panfili, G.; Fratianni, A.; Irano, M. Improved Normal-Phase High-Performance Liquid Chromatography Procedure for the Determination of Carotenoids in Cereals. *J. Agric. Food Chem.* **2004**, *52*, 6373–6377. [CrossRef] [PubMed]

17. Borrelli, G.; De Leonardis, A.; Platani, C.; Troccoli, A. Distribution along durum wheat kernel of the components involved in semolina colour. *J. Cereal Sci.* **2008**, *48*, 494–502. [CrossRef]

18. Ndolo, V.U.; Beta, T. Distribution of carotenoids in endosperm, germ, and aleurone fractions of cereal grain kernels. *Food Chem.* **2013**, *139*, 663–671. [CrossRef]

19. Troccoli, A.; Borrelli, G.M.; De Vita, P.; Fares, C.; Di Fonzo, N. Durum Wheat Quality: A Multidisciplinary Concept (mini review). *J. Cereal Sci.* **2000**, *32*, 99–113. [CrossRef]

20. Ficco, D.B.M.; Mastrangelo, A.M.; Trono, D.; Borrelli, G.M.; De Vita, P.; Fares, C.; Beleggia, R.; Platani, C.; Papa, R. The colours of durum wheat: A review. *Crop Pasture Sci.* **2014**, *65*, 1–15. [CrossRef]

21. Abdel-Aal, E.-S.M.; Rabalski, I. Bioactive Compounds and their Antioxidant Capacity in Selected Primitive and Modern Wheat Species. *Open Agric. J.* **2008**, *2*, 7–14. [CrossRef]

22. Žilić, S.; Šukalović, V.; Dodig, D.; Maksimović, V.; Maksimović, M.; Basić, Z. Antioxidant activity of small grain cereals caused by phenolics and lipid soluble antioxidants. *J. Cereal Sci.* **2011**, *54*, 417–424. [CrossRef]

23. Zrcková, M.; Capouchová, I.; Paznocht, L.; Eliášová, M.; Dvořák, P.; Konvalina, P.; Janovská, D.; Orsák, M.; Bečková, L. Variation of the total content of polyphenols and phenolic acids in einkorn, emmer, spelt and common wheat grain as a function of genotype, wheat species and crop year. *Plant Soil Environ.* **2019**, *65*, 260–266. [CrossRef]

24. Idehen, E.; Tang, Y.; Sang, S. Bioactive phytochemicals in barley. *J. Food Drug Anal.* **2017**, *25*, 148–161. [CrossRef] [PubMed]

25. Li, M.; Qian, M.; Jiang, Q.; Tan, B.; Yin, Y.; Han, X. Evidence of Flavonoids on Disease Prevention. *Antioxidants* **2023**, *12*, 527. [CrossRef]

26. Di Loreto, A.; Bosi, S.; Montero, L.; Bregola, V.; Marotti, I.; Sferrazza, R.E.; Dinelli, G.; Herrero, M.; Cifuentes, A. Determination of phenolic compounds in ancient and modern durum wheat genotypes. *Electrophoresis* **2018**, *39*, 2001–2010. [CrossRef]

27. Fernandez-Orozco, R.; Li, L.; Harflett, C.; Shewry, P.R.; Ward, J.L. Effects of Environment and Genotype on Phenolic Acids in Wheat in the HEALTHGRAIN Diversity Screen. *J. Agric. Food Chem.* **2010**, *58*, 9341–9352. [CrossRef]

28. Pasqualone, A.; DelVecchio, L.N.; Mangini, G.; Taranto, F.; Blanco, A. Variability of total soluble phenolic compounds and antioxidant activity in a collection of tetraploid wheat. *Agric. Food Sci.* **2014**, *23*, 307–316. [CrossRef]

29. Blanco, A.; Colasuonno, P.; Gadaleta, A.; Mangini, G.; Schiavulli, A.; Simeone, R.; Digesù, A.M.; De Vita, P.; Mastrangelo, A.M.; Cattivelli, L. Quantitative trait loci for yellow pigment concentration and individual carotenoid compounds in durum wheat. *J. Cereal Sci.* **2011**, *54*, 255–264. [CrossRef]

30. Giacosa, A.; Peroni, G.; Rondanelli, M. Phytochemical Components and Human Health Effects of Old versus Modern Italian Wheat Varieties: The Case of Durum Wheat Senatore Cappelli. *Nutrients* **2022**, *14*, 2779. [CrossRef]

31. Giunta, F.; Bassu, S.; Mefleh, M.; Motzo, R. Is the Technological Quality of Old Durum Wheat Cultivars Superior to That of Modern Ones When Exposed to Moderately High Temperatures during Grain Filling? *Foods* **2020**, *9*, 778. [CrossRef] [PubMed]

32. Dinelli, G.; Segura Carretero, A.; Di Silvestro, R.; Marotti, I.; Fu, S.; Benedettelli, S.; Ghiselli, L.; Fernandez Gutierrez, A. Determination of phenolic compounds in modern and old varieties of durum wheat using liquid chromatography coupled with time-of-flight mass spectrometry. *J. Chromatogr. A* **2009**, *1216*, 7229–7240. [CrossRef] [PubMed]

33. Mac Key, J. Wheat: Its concept, evolution and taxonomy. In *Durum Wheat Breeding: Current Approaches and Future Strategies*; Royo, C., Nachit, M.M., Di Fonzo, N., Araus, J.L., Pfeiffer, W.H., Slafer, G.A., Eds.; Food Products Press: Binghamton, NY, USA, 2005; Chapter 1; pp. 3–61, ISBN 1-56022-966-7.

34. Dhanavath, S.; Rao, U.P. Nutritional and Nutraceutical Properties of Triticum dicoccum Wheat and Its Health Benefits: An Overview. *J. Food Sci.* **2017**, *82*, 2243–2250. [CrossRef] [PubMed]
35. Lahouar, L.; El Arem, A.; Ghrairi, F.; Chahdoura, H.; Ben Salem, H.; El Felah, M.; Achour, L. Phytochemical content and antioxidant properties of diverse varieties of whole barley (*Hordeum vulgare* L.) grown in Tunisia. *Food Chem.* **2014**, *145*, 578–583. [CrossRef]
36. Zhu, F. Proanthocyanidins in cereals and pseudocereals. *Crit. Rev. Food Sci. Nutr.* **2019**, *59*, 1521–1533. [CrossRef]
37. Iannucci, A.; Suriano, S.; Codianni, P. Genetic Diversity for Agronomic Traits and Phytochemical Compounds in Coloured Naked Barley Lines. *Plants* **2021**, *10*, 1575. [CrossRef]
38. Dickin, E.; Steele, K.; Edwards-Jones, G.; Wright, D. Agronomic diversity of naked barley (*Hordeum vulgare* L.): A potential resource for breeding new food barley for Europe. *Euphytica* **2012**, *184*, 85–99. [CrossRef]
39. Beleggia, R.; Platani, C.; Nigro, F.; Cattivelli, L. A micro-method for the determination of Yellow Pigment Content in durum wheat. *J. Cereal Sci.* **2010**, *52*, 106–110. [CrossRef]
40. Suriano, S.; Iannucci, A.; Codianni, P.; Fares, C.; Russo, M.; Pecchioni, N.; Marciello, U.; Savino, M. Phenolic acids profile, nutritional and phytochemical compounds, antioxidant properties in colored barley grown in southern Italy. *Food Res. Int.* **2018**, *113*, 221–233. [CrossRef]
41. Kim, D.-O.; Jeong, S.W.; Lee, C.Y. Antioxidant capacity of phenolic phytochemicals from various cultivars of plums. *Food Chem.* **2003**, *81*, 321–326. [CrossRef]
42. Sun, B.; Ricardo-Da-Silva, J.M.; Spranger, I. Critical Factors of Vanillin Assay for Catechins and Proanthocyanidins. *J. Agric. Food Chem.* **1998**, *46*, 4267–4274. [CrossRef]
43. Fares, C.; Platani, C.; Baiano, A.; Menga, V. Effect of processing and cooking on phenolic acid profile and antioxidant capacity of durum wheat pasta enriched with debranning fractions of wheat. *Food Chem.* **2010**, *119*, 1023–1029. [CrossRef]
44. Pereira, D.; Correia, P.M.R.; Guiné, R.P.F. Analysis of the physical-chemical and sensorial properties of Maria type cookies. *Acta Chim. Slovaca* **2013**, *6*, 269–280. [CrossRef]
45. De Santis, M.A.; Giuliani, M.M.; Giuzio, L.; De Vita, P.; Lovegrove, A.; Shewry, P.R.; Flagella, Z. Differences in gluten protein composition between old and modern durum wheat genotypes in relation to 20th century breeding in Italy. *Eur. J. Agron.* **2017**, *87*, 19–29. [CrossRef]
46. De Vita, P.; Nicosia, O.L.D.; Nigro, F.; Platani, C.; Riefolo, C.; Di Fonzo, N.; Cattivelli, L. Breeding progress in morpho-physiological, agronomical and qualitative traits of durum wheat cultivars released in Italy during the 20th century. *Eur. J. Agron.* **2007**, *26*, 39–53. [CrossRef]
47. Lawlor, D.W. Carbon and nitrogen assimilation in relation to yield: Mechanisms are the key to understanding production systems. *J. Exp. Bot.* **2002**, *53*, 773–787. [CrossRef]
48. Geisslitz, S.; Longin, C.F.H.; Scherf, K.A.; Koehler, P. Comparative Study on Gluten Protein Composition of Ancient (Einkorn, Emmer and Spelt) and Modern Wheat Species (Durum and Common Wheat). *Foods* **2019**, *8*, 409. [CrossRef]
49. Abeledo, L.G.; Calderini, D.; Slafer, G. Genetic improvement of barley yield potential and its physiological determinants in Argentina (1944–1998). *Euphytica* **2003**, *130*, 325–334. [CrossRef]
50. De Vita, P.; Riefolo, C.; Codianni, P.; Cattivelli, L.; Fares, C. Agronomic and qualitative traits of *T. turgidum* ssp. dicoccum genotypes cultivated in Italy. *Euphytica* **2006**, *150*, 195–205. [CrossRef]
51. Pagnotta, M.A.; Mondini, L.; Codianni, P.; Fares, C. Agronomical, quality, and molecular characterization of twenty Italian emmer wheat (Triticum dicoccon) accessions. *Genet. Resour. Crop Evol.* **2009**, *56*, 299–310. [CrossRef]
52. Rodrigues, O.; Minella, E.; Costenaro, E.R. Genetic Improvement of Barley (*Hordeum vulgare* L.) in Brazil: Yield Increase and Associated Traits. *Agric. Sci.* **2020**, *11*, 417–430. [CrossRef]
53. De Vita, P.; Mastrangelo, A.; Matteu, L.; Mazzucotelli, E.; Virzì, N.; Palumbo, M.; Storto, M.L.; Rizza, F.; Cattivelli, L. Genetic improvement effects on yield stability in durum wheat genotypes grown in Italy. *Field Crops Res.* **2010**, *119*, 68–77. [CrossRef]
54. Zieliński, H.; Kozłowska, H. Antioxidant Activity and Total Phenolics in Selected Cereal Grains and Their Different Morphological Fractions. *J. Agric. Food Chem.* **2000**, *48*, 2008–2016. [CrossRef] [PubMed]
55. Menga, V.; Fares, C.; Troccoli, A.; Cattivelli, L.; Baiano, A. Effects of genotype, location and baking on the phenolic content and some antioxidant properties of cereal species. *Int. J. Food Sci. Technol.* **2010**, *45*, 7–16. [CrossRef]
56. Zhao, H.; Fan, W.; Dong, J.; Lu, J.; Chen, J.; Shan, L.; Lin, Y.; Kong, W. Evaluation of antioxidant activities and total phenolic contents of typical malting barley varieties. *Food Chem.* **2008**, *107*, 296–304. [CrossRef]
57. Mareček, V.; Mikyška, A.; Hampel, D.; Čejka, P.; Neuwirthová, J.; Malachová, A.; Cerkal, R. ABTS and DPPH methods as a tool for studying antioxidant capacity of spring barley and malt. *J. Cereal Sci.* **2017**, *73*, 40–45. [CrossRef]
58. Stratil, P.; Klejdus, B.; Kubáň, V. Determination of phenolic compounds and their antioxidant activity in fruits and cereals. *Talanta* **2007**, *71*, 1741–1751. [CrossRef]
59. Kaneda, H.; Kobayashi, N.; Furusho, S.; Sahara, H.; Koshino, S. Reducing activity and flavor stability of beer. *MBAA TQ* **1995**, *32*, 90–94.
60. Laddomada, B.; Blanco, A.; Mita, G.; D'amico, L.; Singh, R.P.; Ammar, K.; Crossa, J.; Guzmán, C. Drought and Heat Stress Impacts on Phenolic Acids Accumulation in Durum Wheat Cultivars. *Foods* **2021**, *10*, 2142. [CrossRef]
61. Irakli, M.; Lazaridou, A.; Mylonas, I.; Biliaderis, C.G. Bioactive Components and Antioxidant Activity Distribution in Pearling Fractions of Different Greek Barley Cultivars. *Foods* **2020**, *9*, 783. [CrossRef]

62. Skendi, A. Alternatives to increase the antioxidant capacity of bread with phenolics. In *Trends in Wheat and Bread Making*; Galanakis, C.M., Ed.; Academic Press: Cambridge, MA, USA, 2021; Chapter 11; pp. 311–341. [CrossRef]

63. Andersson, A.A.; Dimberg, L.; Åman, P.; Landberg, R. Recent findings on certain bioactive components in whole grain wheat and rye. *J. Cereal Sci.* **2014**, *59*, 294–311. [CrossRef]

64. Brandolini, A.; Castoldi, P.; Plizzari, L.; Hidalgo, A. Phenolic acids composition, total polyphenols content and antioxidant activity of Triticum monococcum, Triticum turgidum and Triticum aestivum: A two-years evaluation. *J. Cereal Sci.* **2013**, *58*, 123–131. [CrossRef]

65. Wojakowska, A.; Perkowski, J.; Góral, T.; Stobiecki, M. Structural characterization of flavonoid glycosides from leaves of wheat (Triticum aestivum L.) using LC/MS/MS profiling of the target compounds. *J. Mass Spectrom.* **2013**, *48*, 329–339. [CrossRef] [PubMed]

66. Yildiztugay, E.; Ozfidan-Konakci, C.; Kucukoduk, M.; Turkan, I. Flavonoid Naringenin Alleviates Short-Term Osmotic and Salinity Stresses Through Regulating Photosynthetic Machinery and Chloroplastic Antioxidant Metabolism in Phaseolus vulgaris. *Front. Plant Sci.* **2020**, *11*, 682. [CrossRef] [PubMed]

67. Bollina, V.; Kumaraswamy, G.K.; Kushalappa, A.C.; Choo, T.M.; Dion, Y.; Rioux, S.; Faubert, D.; Hamzehzarghani, H. Mass spectrometry-based metabolomics application to identify quantitative resistance-related metabolites in barley against Fusarium head blight. *Mol. Plant Pathol.* **2010**, *11*, 769–782. [CrossRef] [PubMed]

68. Kumaraswamy, G.K.; Bollina, V.; Kushalappa, A.C.; Choo, T.M.; Dion, Y.; Rioux, S.; Mamer, O.; Faubert, D. Metabolomics technology to phenotype resistance in barley against Gibberella zeae. *Eur. J. Plant Pathol.* **2011**, *130*, 29–43. [CrossRef]

69. Horvat, D.; Šimić, G.; Drezner, G.; Lalić, A.; Ledenčan, T.; Tucak, M.; Plavšić, H.; Andrić, L.; Zdunić, Z. Phenolic Acid Profiles and Antioxidant Activity of Major Cereal Crops. *Antioxidants* **2020**, *9*, 527. [CrossRef]

70. Guan, W.; Zhang, D.; Tan, B. Effect of Layered Debranning Processing on the Proximate Composition, Polyphenol Content, and Antioxidant Activity of Whole Grain Wheat. *J. Food Process. Preserv.* **2023**, *2023*, 1083867. [CrossRef]

71. Giordano, D.; Locatelli, M.; Travaglia, F.; Bordiga, M.; Reyneri, A.; Coïsson, J.D.; Blandino, M. Bioactive compound and antioxidant activity distribution in roller-milled and pearled fractions of conventional and pigmented wheat varieties. *Food Chem.* **2017**, *233*, 483–491. [CrossRef]

72. Suriano, S.; Savino, M.; Codianni, P.; Iannucci, A.; Caternolo, G.; Russo, M.; Pecchioni, N.; Troccoli, A. Anthocyanin profile and antioxidant capacity in coloured barley. *Int. J. Food Sci. Technol.* **2019**, *54*, 2478–2486. [CrossRef]

73. Li, W.; Pickard, M.D.; Beta, T. Effect of thermal processing on antioxidant properties of purple wheat bran. *Food Chem.* **2007**, *104*, 1080–1086. [CrossRef]

74. Duodu, K.G. Effects of processing on antioxidant phenolics of cereal and legume grains. In *Advances in Cereal Science: Implications to Food Processing and Health Promotion*; Awika, J.M., Piironen, V., Bean, S., Eds.; American Chemical Society: Washington, DC, USA, 2011; pp. 31–54.

75. Shahidi, F.; Yeo, J.-D. Insoluble-Bound Phenolics in Food. *Molecules* **2016**, *21*, 1216. [CrossRef] [PubMed]

76. Imeneo, V.; Romeo, R.; Gattuso, A.; De Bruno, A.; Piscopo, A. Functionalized Biscuits with Bioactive Ingredients Obtained by Citrus Lemon Pomace. *Foods* **2021**, *10*, 2460. [CrossRef] [PubMed]

77. Biguzzi, C.; Schlich, P.; Lange, C. The impact of sugar and fat reduction on perception and liking of biscuits. *Food Qual. Prefer.* **2014**, *35*, 41–47. [CrossRef]

Article

Cornelian Cherry (*Cornus mas*) Powder as a Functional Ingredient for the Formulation of Bread Loaves: Physical Properties, Nutritional Value, Phytochemical Composition, and Sensory Attributes

Veronika Šimora [1,2], Hana Ďúranová [1], Ján Brindza [3], Marvin Moncada [4], Eva Ivanišová [5], Patrícia Joanidis [1], Dušan Straka [1], Lucia Gabríny [1] and Miroslava Kačániová [2,6,*]

[1] AgroBioTech Research Centre, Slovak University of Agriculture, Trieda Andreja Hlinku 2, 94976 Nitra, Slovakia

[2] Institute of Horticulture, Faculty of Horticulture and Landscape Engineering, Slovak University of Agriculture, Trieda Andreja Hlinku 2, 94976 Nitra, Slovakia

[3] Institute of Plant and Environmental Sciences, Slovak University of Agriculture, Trieda Andreja Hlinku 2, 94976 Nitra, Slovakia

[4] Department of Food, Bioprocessing, and Nutrition Science, North Carolina State University, Raleigh, NC 27606, USA

[5] Institute of Food Sciences, Slovak University of Agriculture, Trieda Andreja Hlinku 2, 94976 Nitra, Slovakia

[6] Department of Bioenergy, Food Technology and Microbiology, Institute of Food Technology and Nutrition, University of Rzeszow, 4 Zelwerowicza Str., 35-601 Rzeszow, Poland

* Correspondence: miroslava.kacaniova@gmail.com

Citation: Šimora, V.; Ďúranová, H.; Brindza, J.; Moncada, M.; Ivanišová, E.; Joanidis, P.; Straka, D.; Gabríny, L.; Kačániová, M. Cornelian Cherry (*Cornus mas*) Powder as a Functional Ingredient for the Formulation of Bread Loaves: Physical Properties, Nutritional Value, Phytochemical Composition, and Sensory Attributes. *Foods* **2023**, *12*, 593. https://doi.org/10.3390/foods12030593

Academic Editors: Donatella Bianca Maria Ficco and Grazia Maria Borrelli

Received: 17 December 2022
Revised: 25 January 2023
Accepted: 28 January 2023
Published: 31 January 2023

Abstract: In the current study, Cornelian cherry powder (CCP, *Cornus mas*) was investigated as a functional ingredient for bread production. Experimental bread loaves were prepared using five levels of CCP (0, 1, 2, 5, and 10% w/w) to replace wheat flour in bread formulation. The final products were analyzed regarding their proximate composition, content of selected biologically active substances, antioxidant activity (AA), volume, and sensory attributes. Increasing the incorporation of CCP led to significantly ($p < 0.05$) higher concentrations of carbohydrate, ash, energetic value, total polyphenols, phenolic acids and AA, and reduced fat and protein contents ($p < 0.05$). Moreover, up to 5% addition of CCP positively affected the volume (642.63 ± 7.24 mL) and specific volume (2.83 ± 0.02 cm^3/g) of bread loaves, which were significantly ($p < 0.05$) higher compared to the control (no addition of CCP; 576.99 ± 2.97 mL; 2.55 ± 0.002 cm^3/g). The sensory attributes chewiness, crumb springiness, bitterness, and sourness had lower scores ($p < 0.05$) in bread formulated with 10% CCP compared to the control. Overall, results show that the bread loaves produced with up to 5% CCP addition were considered the preferred formulation among the experimental samples tested, taking into consideration their composition, bioactive content, sensory, and physical properties.

Keywords: bakery; biologically active substances; DPPH assay; nutritional value; sensory properties

1. Introduction

At a global level, bakery goods constitute an essential part of human nutrition [1]. The development of innovative products using value-added ingredients has become an important trend in the bread manufacturing industry in an effort to meet the demand of a new generation of consumers seeking healthier lifestyles [2]. Fruits are a remarkable source of natural bioactive compounds with great potential of incorporation in the formulation of multiple products [3–5]. However, fresh fruits are highly perishable due to their extremely high moisture content (MC; 75–95%) [6], which leads to estimated post-harvest losses of around 30% [7]. Therefore, efficient preservation techniques are necessary to extend their shelf-life and marketability [8]. In this regard, drying technology has proven to be a feasible, convenient, and well-accepted processing strategy for improving food storage stability [9,10]. Several types of dehydration technologies are available, e.g., solar drying, hot air drying, vacuum drying, spray drying, osmotic drying, microwave drying, and freeze drying [11]. Among them, hot air drying is currently the most widely used method in the agri-food sphere. Indeed, a more homogeneous, sanitary, and coloured dehydrated product can be quickly obtained using this process [12]. Ultimately, powdered dried fruits provide the producers with enhanced nutritional and health properties for the many types of cereal products [13]. In effect, the addition of powdered fruits to the recipe of the bakeware offers an important protective capacity against a number of civilization disorders due to their antioxidant, anti-inflammatory, anti-mutagenic, and anti-carcinogenic characteristics [14]. For example, enriched foods showed enhanced antioxidant power, which can help prevent some chronic diseases [15].

In this regard, many fruit-derived materials have been examined, including mango peel [4], banana [3], pomegranate peel [16], orange peel [17], and apple pomace [18]. From a technologically advanced point of view, fruit-derived ingredients are naturally gluten-free, and this might modify important attributes of the flour blend and consequently the properties of the final bread [19]. For instance, decreased volume and cohesiveness of bread loaves formulated with mango peel powder (more than 5% *w/w*) addition has been noted by Chen et al. [4]. Moreover, decreased volume, specific volume, moisture, elasticity, and cohesiveness of bread loaves with increasing concentrations of pomegranate peel powder (0%, 1%, 3%, 5%, and 7%) have been found by Zhang et al. [16]. On the other hand, sensory ratings shown by Baba et al. [3] revealed no significant differences in taste, aroma, and appearance of wheat bread containing banana powder (up to 30%) as compared to the control sample, but the bread with the 30% addition had lower overall acceptability. Hence, the careful selection of ingredients and the determination of their adequate concentration [20] are of great importance for a successful food practice.

Cornelian cherries (CC; *Cornus mas* L.) are currently gaining increasing attention from the scientific research community [21]. These oval or pear-shaped edible fruits with color ranging from red to purple constitute an important source of vitamin C and polyphenols, mainly flavonoids, anthocyanins, and iridoids [21–23]. Such substances are linked to a wide range of biological effects and pharmacological properties, including antimicrobial, anti-inflammatory, anti-cancer, anti-diabetic, and anti-atherosclerotic activities [24–26]. Most often, the cherries are consumed fresh or as a dried delicacy. Due to their beneficial properties, they can serve as purposeful ingredients for the commercial food sector [27]. In recent years, CC has also been applied as a flavoring ingredient in ice creams, desserts, and cakes [28]. Furthermore, CC and their juices or extracts have inspired the production of novel foods, such as beer [29], soup [30], vinegar [31], or burgers [32]. Regarding baked goods, İlyasoğlu et al. [33] replaced wheat flour (WF) with CC (10 g 100 g^{-1} composite flour) in cookie formulations and reported enhanced contents of omega-3 fatty acids, total phenolic content, and antioxidant capacity in the final product.

To the best of our knowledge, the addition of Cornelian cherry powder (CCP) to bread formulation has not yet been investigated. Therefore, this study is the first report showing the production and quality assessment of bread loaves produced with CCP. For this, the incorporation of different CCP concentrations (1, 2, 5, and 10% *w/w*) for the partial substitution of WF in the formulation of bread loaves was performed, and the quality characteristics of the final products were determined. To achieve our goal, we assessed the compositional profile, antioxidant activity (AA), phenolic compounds, key physical attributes, and sensory properties of bread loaves. As a result, recommendations about the most appropriate level of CCP for bread production are discussed. This report unveils the potentiality of Cornelian cherry powder as a functional ingredient for bakery products.

2. Materials and Methods

2.1. Materials

Wheat flour (WF; T-650 type) was purchased from a grinding mill (Pohronský Ruskov JSC, Pohronský Ruskov, Slovakia). Other raw materials for bread making, such as salt

(Solivary Trade Ltd., Trenčín, Slovakia), saccharose (Slovenské Cukrovary Ltd., Sereď, Slovakia), and compressed yeast (Thymos Ltd., Veľká Lomnica, Slovakia), were obtained from a local grocery store. The chemical reagents were of analytical quality, and all of them were purchased from Reachem (Bratislava, Slovakia) or Sigma Aldrich (Saint Louis, MO, USA).

2.2. Preparation of Cornelian Cherry Powder (CCP)

Fresh, fully ripe Cornelian cherry (CC; *Cornus mas*) fruits were collected from the SUA Botanical Garden (Slovak University of Agriculture, SUA, Nitra, Slovakia) and subsequently selected, cleaned, and pitted. For the preparation of Cornelian cherry powder (CCP), the fruits were dried at 45 °C until complete dehydration using a cabinet dryer (Universal oven UF 160, Memmert GmbH + Co.KG, Büchenbach, Germany). After this, the dried CC was homogenized (ETA Gratus 0028 90030, ETA-Slovakia Ltd., Bratislava, Slovakia) and sieved to obtain powder particles with a diamter of 0.5 mm. The produced CCP was packed in polyethylene (PE) bags and stored at room temperature in a dark place until analysis and/or use.

2.3. Composite Flour Preparation

Five composite flours were prepared by partially replacing WF with CCP according to the following ratios: WF:CCP, 100:0 (control sample), 99:1 (1% addition of CCP), 98:2 (2% addition of CCP), 95:5 (5% addition of CCP), and 90:10 (10% addition of CCP), respectively. The WF/CCP blends were individually packaged in PE bags and kept at room temperature until the bread making process.

2.4. Bread Loaf Preparation

Bread loaf experimental treatments (Figure 1) were prepared following the methodology described by Valková et al. [18]. The bread formulation consisted of 500 g of WF or WF/CCP blends, saccharose (1% of flour), salt (2% of flour), water (60% of flour), and yeast (2% of flour). Initially, compressed yeasts were reactivated in a saccharose solution at 32 °C for 5 min. All ingredients were blended for 10 min in two steps (first step: 4 min at 1500 rpm; second step: 6 min at 3000 rpm) in a mixer (DIOSNA SP 12; DIOSNA Dierks and Söhne GmbH, Osnabrück, Germany) using a dough hook accessory to ensure proper hydration of flour. Then, the dough was carefully portioned into 250 ± 5 g pieces and placed into oiled and floured tins. The tins were transferred to a fermentation cabinet (MIWE cube, Pekass Ltd., Pilsen, Czech Republic) set at 32 °C and 85% relative humidity, and allowed to proof for 40 min. The bread loaves were baked in two phases (Phase I: 180 °C with the addition of 160 mL steam at the same temperature; Phase II: 210 °C for 7 min, no steam) in a steamer oven (Laboratory oven MIWE cube, Pekass Ltd., Pilsen, Czech Republic). The baked loaves were removed from the tins and left to cool at room temperature for 2 h until cutting. In total, three batches of each type of bread were produced (15 bread loaves in total).

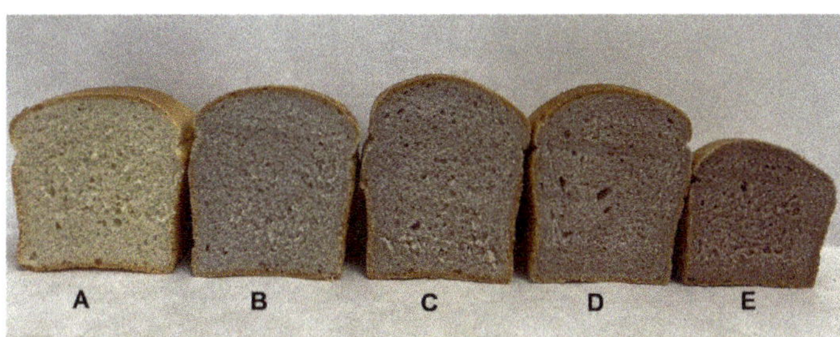

Figure 1. Experimental bread loaves. (**A**) control sample, (**B**) 1% Cornelian cherry powder addition, (**C**) 2% Cornelian cherry powder addition, (**D**) 5% Cornelian cherry powder addition, and (E) 10% Cornelian cherry powder addition.

2.5. Determination of Compositional Profile and Energetic Value

The compositional value of CCP and experimental bread loaves was assessed. For this purpose, moisture, ash, crude protein, fat, and total carbohydrate contents, as well as the energetic values, were determined.

The moisture content (MC) was measured with an automatic moisture analyzer, DBS 60-3 (Kern and Sohn GmbH, Altstadt, Germany), according to the manufacturer's instructions and the ASTM D 6980 method. Briefly, 1 g of sample was placed on the sample plate and tested at 120 °C for the required time (10–15 min). The total ash and crude protein contents were determined in accordance with AACC standard 08-01 using a muffle furnace (Neberterm, Germany) and the semi-micro Kjeldahl method (factor of converted nitrogen to protein was 6.25). The total fat content was analyzed using the Ankom XT15 Fat Extractor (Ankom Technology, Fairport, NY, USA) in accordance with the manufacturer's instructions. The total carbohydrate content (TCC) and energy were calculated by Equations (1) and (2), according to Valková et al. [18] and Arraibi et al. [34]:

$$\text{TCC (\%)} = 100 - \text{moisture (\%)} - \text{protein (\%)} - \text{lipids (\%)} - \text{ash (\%)} \tag{1}$$

$$\text{Energy (kcal/100 g)} = 4 \times (\% \text{ proteins} + \% \text{ carbohydrates}) + 9 \times (\% \text{ fat}) \tag{2}$$

2.6. Determination of Radical Scavenging Activity and Polyphenolic Compounds

Firstly, samples of ethanolic extracts (CCP, and bread treatments) were prepared. For each extraction, 0.2 g (CCP) and 0.5 g (bread treatments) of sample received 20 mL or 40 mL of 80% ethanol, respectively, and were extracted for 2 h, followed by centrifugation at $4000\times g$ for 10 min in a Rotofix 32A (Hettich, Spenge, Germany). The supernatant was used for the determination of antioxidant activity (AA), total polyphenols content (TPC), total phenolic acids content (TPAC), and flavonoid content (FC).

The AA, TPC, TPAC, and FC of the samples were analyzed using the 2,2-diphenyl-1-picrylhydrazyl (DPPH) assay [18], the colorimetric assay utilizing the Folin-Ciocalteu (F-C) reagent [18], and according to the procedures of Valková et al. [18] and Ivanišová et al. [35], respectively. The AA was expressed as Trolox equivalent antioxidant capacity (TEAC) in milligrams per gram of dry weight (dw). Gallic acid (for TPC analysis), caffeic acid (for TPAC analysis), and quercetin (for FC analysis) standards were used, and the results were expressed as gallic acid equivalents (GAE), caffeic acid equivalents (CAE), and quercetin equivalents (QE) in milligrams per gram of dw, respectively.

2.7. Volume Analysis

The volume (mL) and specific volume (cm^3/g) of bread samples were assessed using an automatically laser-based scanning device, the VolScan Profiler VSP 300 (Stable Micro Systems, Godalming, UK), according to the manufacturer's recommendation (AACC approved method 10.16.01).

2.8. Sensory Assessment

Sensory analysis was performed by 10 panelists (three men and seven women, aged 26–47), trained according to the standard STN EN ISO 8586. The evaluation of bread samples took place at the Sensory Laboratory of the Research Centre AgroBioTech (Slovak University of Agriculture, SUA, Nitra, Slovakia) during the late morning. The experimental samples (control and four treatments, 1, 2, 5, and 10% (w/w)) were coded with 3-digit numbers and presented to the panelists at the same time. Between evaluations of individual samples, the panelists were instructed to drink water.

Eleven sensory descriptors were evaluated (on a 15-point unstructured scale) and divided into the following categories: color (crust color, crumb color), texture (pore uniformity, crumb springiness, chewiness), aroma, taste (sweet taste, bitter taste, sour taste, aftertaste), and overall impression. The descriptors for sensory rating were selected according to García-Gómez et al. [36], and based on preliminary training sessions with selected panelists.

2.9. Statistical Analysis

All the analyses were conducted at least in triplicate, and the data was reported as mean value $\pm$ standard deviation. One-way analysis of variance (ANOVA) and the Tukey test (Prism 8.0.1 program, GraphPad Software, San Diego, CA, USA) were applied to establish statistically significant differences between the samples at the level of $p < 0.05$.

3. Results

3.1. Characterization of Cornelian Cherry Powder

The compositional profile, energetic value, polyphenolic compounds, and AA of CCP are shown in Table 1. CCP had a low energetic value, and low fat and protein contents, but a high total carbohydrate and ash contents. Regarding DPPH free radical scavenging activity and polyphenolic compounds, the CCP exhibited high AA, and high total polyphenols, total phenolic acids, and flavonoid contents.

Table 1. Composition, energetic value, antioxidant activity, and polyphenolic compounds of Cornelian cherry powder.

Parameters	CCP
Fat (%)	0.20 ± 0.03
Carbohydrate (%)	86.88 ± 0.93
Protein (%)	0.78 ± 0.05
Ash (%)	5.47 ± 0.10
Energetic value (kcal/100 g)	352.44 ± 3.67
Moisture (%)	6.67 ± 0.85
AA (mg TEAC/g)	8.75 ± 0.01
TPC (mg GAE/g)	9.08 ± 0.54
TPAC (mg CAE/g)	2.62 ± 0.15
FC (mg QE/g)	3.62 ± 0.30

Values are expressed as the mean $\pm$ standard deviation ($n = 3$). CCP—Cornelian cherry powder; AA—antioxidant activity expressed as mg of Trolox equivalents per gram dry weight; TPC—total polyphenols content expressed as mg of gallic acid equivalents per gram dry weight; TPAC—total phenolic acids content expressed as mg of caffeic acid equivalents per gram dry weight; FC—flavonoid content expressed as mg of quercetin equivalents per gram dry weight.

3.2. Compositional Profile and Energetic Value of Experimental Bread Loaves

Increasing the percentage of CCP in the bread formulations led to a progressive and significant ($p < 0.05$) increase in total carbohydrate and ash contents, and energetic value, as shown in Table 2. Interestingly, no differences in total carbohydrates and energetic value were noted between bread loaves with 5% and 10% of CCP. In addition, a significantly ($p < 0.05$) linear reduction in fat and protein contents was observed when more CCP was added to the formulation of bread loaves (Table 2), which agrees with the low fat and protein contents found for CCP (Table 1). No significant differences were observed for the MC of bread treatments ($p > 0.05$).

Table 2. Compositional profiles of experimental bread loaves.

Parameters	Incorporation Ratio of CCP (% *w/w*) [1]				
	0	1	2	5	10
Fat (%)	6.57 ± 0.05 [a]	6.26 ± 0.06 [b]	6.00 ± 0.04 [c]	5.52 ± 0.06 [c]	5.05 ± 0.08 [d]
Carbohydrate (%)	69.58 ± 0.19 [d]	70.05 ± 0.12 [c]	70.68 ± 0.15 [b]	71.65 ± 0.44 [a]	72.45 ± 0.40 [a]
Protein (%)	13.90 ± 0.08 [a]	13.61 ± 0.06 [b]	13.12 ± 0.03 [c]	12.60 ± 0.08 [d]	12.06 ± 0.06 [e]
Ash (%)	0.62 ± 0.03 [e]	0.70 ± 0.02 [d]	0.75 ± 0.02 [c]	0.87 ± 0.04 [b]	0.99 ± 0.03 [a]

Table 2. *Cont.*

Parameters	Incorporation Ratio of CCP (% *w/w*) [1]				
	0	1	2	5	10
Moisture (%)	9.32 ± 0.20	9.38 ± 0.16	9.46 ± 0.09	9.35 ± 0.46	9.45 ± 0.41
Energetic value (kcal/100 g)	393.08 ± 0.94 [a]	391.01 ± 0.93 [b]	389.16 ± 0.34 [c]	386.71 ± 1.75 [d]	383.47 ± 1.53 [d]

[1] Incorporation ratio of partial substitution of wheat flour; please see item 2.3 for further details. Values are expressed as the mean $\pm$ standard deviation (n = 3). Data in the same line with different superscript letters are significantly different (Tukey's test, $p < 0.05$). CCP—Cornelian cherry powder.

3.3. Antioxidant Activity and Polyphenolic Compounds of Experimental Bread Loaves

The AA and content of selected polyphenolic compounds in experimental bread loaves enriched with CCP are summarized in Table 3. For all analyzed parameters, a significant and linear increase in bioactivity was demonstrated as higher ratios of CCP were used for the production of bread loaves ($p < 0.05$), with the exception of the 1% CCP treatment, which had similar results compared to the control sample ($p > 0.05$). Moreover, flavonoid compounds were not detected in any of the bread treatments (Table 3).

Table 3. Antioxidant activity and selected polyphenolic compounds of experimental bread loaves.

Parameters	Incorporation Ratio of CCP (% *w/w*) [1]				
	0	1	2	5	10
AA (mg TEAC/g)	0.60 ± 0.01 [d]	0.61 ± 0.04 [d]	0.69 ± 0.03 [c]	0.76 ± 0.02 [b]	1.22 ± 0.02 [a]
TPC (mg GAE/g)	3.82 ± 0.11 [d]	3.98 ± 0.09 [d]	4.59 ± 0.19 [c]	4.90 ± 0.03 [b]	5.12 ± 0.02 [a]
TPAC (mg CAE/g)	0.49 ± 0.09 [d]	0.55 ± 0.03 [d]	0.68 ± 0.05 [c]	0.77 ± 0.02 [b]	1.18 ± 0.05 [a]
FC (mg QE/g)	ND	ND	ND	ND	ND

[1] Incorporation ratio of partial substitution of wheat flour; please see item 2.3 for further details. Values are expressed as the mean $\pm$ standard deviation (n = 3). Data in the same line with different superscript letters are significantly different (Tukey's test, $p < 0.05$). CCP—Cornelian cherry powder; AA—antioxidant activity expressed as mg of Trolox equivalents per gram dry weight; TPC—total polyphenols content expressed as mg of gallic acid equivalents per gram dry weight; TPAC—total phenolic acids content expressed as mg of caffeic acid equivalents per gram dry weight; FC—flavonoid content expressed as mg of quercetin equivalents per gram dry weight. ND—not detected.

3.4. Volume of Experimental Bread Loaves

Breads produced with WF/CCP blends of 1%, 2%, and 5% CCP showed significantly different ($p < 0.05$) volume and specific volume (Table 4). The addition of CCP led to a significant increase in both parameters, volume and specific volume, which is a desirable attribute for bread loaves. However, bread loaves prepared with 10% CCP had significantly ($p < 0.05$) lower volume and specific volume compared to all treatments and the control sample. The highest results were observed for 2% CCP, which resulted in an increase of 15.8% and 16.4% for volume and specific volume, respectively.

Table 4. Volume of the experimental bread loaves.

Incorporation Ratio of CCP (% *w/w*) [1]	Volume (mL)	Specific Volume (cm³/g)
0	576.99 ± 2.97 [d]	2.55 ± 0.002 [d]
1	604.38 ± 8.48 [c]	2.69 ± 0.05 [c]
2	668.17 ± 6.56 [a]	2.97 ± 0.04 [a]
5	642.63 ± 7.24 [b]	2.83 ± 0.02 [b]
10	443.63 ± 1.22 [e]	1.94 ± 0.02 [e]

[1] Incorporation ratio of partial substitution of wheat flour; please see item 2.3 for further details. Values are expressed as the mean $\pm$ standard deviation (n = 3). Data in the same column with different superscript letters are significantly different (Tukey's test, $p < 0.05$). CCP—Cornelian cherry powder.

3.5. Sensory Properties of Experimental Bread Loaves

The crust and crumb colors of bread loaves formulated with $\geq$2% and $\geq$5% CCP, respectively, were reported as significantly ($p < 0.05$) darker compared to the control (Table 5). Significant differences in the bread aroma were identified between the control sample and the bread enriched with 2% and 10% CCP, respectively. Further, the evaluators perceived an aftertaste ($p < 0.05$) in samples produced with $\geq$1% CCP, while the highest sour and bitter taste scores ($p < 0.05$) were reported for bread produced with the highest (10%) CCP ratio. Additionally, this sample was reported as the least chewable and having poor crumb springiness. Regarding the pore uniformity and sweet taste parameters, there were no significant differences between experimental samples. When evaluating the overall impression, the sample with the 2% addition was perceived as the tastiest and had the highest score, which is significantly superior to the control ($p < 0.05$).

Table 5. Sensory analysis of the experimental bread loaves.

Parameters	Incorporation Ratio of CCP (% *w/w*) [1]				
	0	1	2	5	10
Crust color	4.00 ± 0.00 [b]	5.48 ± 1.00 [cd]	7.03 ± 2.10 [ad]	9.16 ± 2.34 [a]	8.77 ± 2.73 [ac]
Crumb color	2.50 ± 0.00 [d]	5.98 ± 1.73 [ce]	7.44 ± 1.67 [be]	9.02 ± 1.76 [ae]	11.52 ± 1.47 [a]
Pore uniformity	8.00 ± 0.00 [a]	8.48 ± 1.59 [a]	7.85 ± 2.09 [a]	8.38 ± 1.54 [a]	9.79 ± 2.33 [a]
Aroma	4.00 ± 0.00 [b]	4.50 ± 0.65 [ab]	5.69 ± 1.48 [a]	6.14 ± 2.45 [ab]	7.89 ± 3.44 [a]
Crumb springiness	13.00 ± 0.00 [a]	12.50 ± 1.36 [a]	13.17 ± 2.02 [a]	10.70 ± 3.09 [ac]	7.92 ± 2.63 [bc]
Chewiness	6.00 ± 0.00 [bc]	6.17 ± 0.80 [bc]	5.97 ± 1.09 [bc]	6.64 ± 1.87 [ac]	8.45 ± 1.37 [a]
Sweet taste	3.00 ± 0.00 [a]	3.50 ± 0.86 [a]	3.68 ± 0.83 [a]	3.65 ± 1.43 [a]	4.04 ± 2.07 [a]
Bitter taste	1.00 ± 0.00 [bc]	1.21 ± 0.30 [ac]	1.43 ± 0.80 [ac]	2.47 ± 1.49 [ac]	2.72 ± 1.70 [a]
Sour taste	1.00 ± 0.00 [c]	1.65 ± 0.85 [bc]	1.44 ± 0.41 [b]	2.56 ± 1.07 [b]	6.62 ± 2.93 [a]
Aftertaste	3.00 ± 0.00 [b]	4.11 ± 0.96 [a]	4.63 ± 1.36 [a]	5.99 ± 1.81 [a]	7.92 ± 2.88 [a]
Overall impression	11.00 ± 0.00 [bcde]	11.58 ± 1.37 [ae]	12.12 ± 1.04 [a]	10.38 ± 2.40 [ac]	8.54 ± 2.69 [ad]

[1] Incorporation ratio of partial substitution of wheat flour; please see item 2.3 for further details. Values are expressed as the mean ± standard deviation ($n = 3$). Data in the same line with different superscript letters are significantly different (Tukey's test, $p < 0.05$). CCP—Cornelian cherry powder.

4. Discussion

In general, the functional characteristics of a raw material affect its interaction with other food components and strongly determine its final application [37]. Wheat flour is used as a major staple raw material in bread production [38], and it is the most abundant source of calories and protein in the human diet [39]. Although it is also a great source of nutrients, its content of bioactive compounds and AA is poor as a consequence of the refining during processing [40]. To improve the nutritional profile and the biological activity of bakery products, the partial replacement of WF with phytochemical-rich, functional plant-based flours or powders is an interesting strategy. In this sense, horticultural crops are known to provide a rich source of diverse nutritional molecules, many of them possessing antioxidant activity, which has been reported as capable of protecting the human body against oxidative cellular reactions [41]. Cornelian cherry-derived ingredients have been tested as part of several formulations to produce enhanced food products [28,32,33,42–52], but not in bread formulations yet. Therefore, this is the first report of such research activities with solid potential for practical applications to bread manufacturing.

Regarding the characterization of CCP, Tontul et al. [53] recorded similar values for MC (8.03 ± 0.14%) in CCP dried at 50 °C. Since moisture in sugar-rich powders acts as a plasticizer [54], the lower MC of CCP used in our study may have a positive effect on its cohesive properties. Importantly, foods with reduced MC are considered safe due to the growth mitigation of undesirable microorganisms (especially molds), thus improving the shelf-life of the product [55]. We found CCP to have a higher ash content compared to dried CC fruits (2.83 ± 0.35%), as demonstrated by Petkova and Ognyanov [56]. Further, total carbohydrates represent the most essential source of energetic value in CC fruits [57], which is also in line with our findings. Likewise, relatively low crude protein levels (ranging from 1.43 to 2.71%) have also been demonstrated in CC fruits by Serbia by Bijelić et al. [58]. In fact, raw CC fruits typically have low fat content (1.49 ± 0.02%) [59], which was also confirmed by our results. Differences between our results and those of cited research studies can be explained by different genotypes used, as well as the influence of environmental growth conditions [60].

It is important to characterize not only the overall AA but also the individual antioxidant components responsible for such activity, which are present in diverse fruits [61–64]. Our findings demonstrate that oven-dried CCP produced in this study has high contents of total polyphenols, phenolic acids, and flavonoids along with a strong AA and reiterate results found in fresh CC fruits determined by Dupak et al. [65] and Szczepaniak et al. [66]. In contrast, Popović et al. [67] have identified a lower content of total polyphenols in dried samples of 10 CC genotypes. In addition, AA and the contents of total phenolic acids and flavonoids were lower in CC pulp analyzed by Dupak et al. [65] in comparison with our CCP. In fact, these discrepancies are expected since different processing protocols and parameters, fruit genotypes and varieties, and maturity stages affect the aforementioned results [25,68]. Considering our results, it can be hypothetically assumed that eating both our CCP and/or products enriched with the CCP could be beneficial for human health in the sense of their ability to eliminate harmful oxidative stress in the organism, thus reducing the risk of chronic disease incidence.

The effect of CCP addition on the nutritional composition and key quality attributes of bread loaves enriched with four concentrations (1%, 2%, 5%, and 10%) of CCP was further evaluated. Carbohydrates are the prime macronutrients in bread. The content of carbohydrates in the bread formulations increased linearly with the incorporation of CCP, because CCP is also a major source of this macronutrient (>85%, Table 1). In addition, increasing additions of CCP led to progressively lower protein and fat contents and higher ash contents in the enriched bread loaves, also as a reflex of the original CCP composition (Table 1). The same trend was noted by Topdaş et al. [28], who analyzed the impact of different CC fruit paste (5%, 10%, and 15%) additions to the composition of ice cream. Similarly, a lower fat content in CCP-enriched biscuits compared to control samples was reported by İlyasoğlu et al. [33]. Given these findings, CCP may be a promising ingredient in the preparation of low-fat goods for the food sector.

Moreover, our results pointed to markedly higher concentrations of phenolic compounds (TPC and TPAC) and AA in the CCP-containing bread loaves in comparison to the control. This finding agrees with the research conducted by İlyasoğlu et al. [33], Topdaş et al. [28], and Haghani et al. [51], which revealed increased levels of TPC and higher AA, as well as higher concentrations of CC in products used to prepare biscuits, conventional ice cream, and probiotic ice cream, respectively. Enhanced TPC and antioxidant capacity of white chocolate and dairy desserts after addition of CCP and CC juice, respectively, were also observed in the studies performed by Cerit et al. [45] and Ivanova et al. [47]. On the other hand, the total absence of FC identified in all our bread samples may be related to the thermolability of these biologically active substances [69] and possible complete degradation during the baking process. Indeed, flavonoids are major phenolic compounds with natural antioxidant capacity (mediated via their functional hydroxyl groups in their structure) [70], reported in CC fruits [63] and also here in our developed CCP (Table 1). The microencapsulation technique using appropriate wall materials has been reported in the literature [71] as an efficient stabilization approach for the preservation of polyphenolic extracts from Cornelian cherries [47]. Therefore, the spray-drying and microencapsulation of plant extracts to produce powdered ingredients with preserved biologically active compounds destined for bakery products will be considered in future studies by our research team.

The addition of non-traditional ingredients to the bakery goods not only affects their nutritional composition and bioactivity properties but may also influence important physical attributes. One of the most crucial physical properties of bread is its volume, which strongly determines the consumers' preferences and predicts its quality, as well [72]. In our study, bread samples enriched with up to 5% CCP displayed significantly higher values for volume and specific volume compared to control wheat bread. We hypothesize that this increase may be related to the presence of pectin and other hydrocolloids (estimated to be about 5.7% in dried CC fruits) in the CCP composition [73]. Confirming our hypothesis, Rosell et al. [74] reported that mixing WF with hydrocolloids increases dough stability and loaf volume due to their ability to absorb water and gelling properties, as well. In view of this, during heating a gel network may be formed, which can consequently strengthen the expanding dough cells, thereby improving gas retention and bread volume [75]. In addition, Das et al. [76] found that hydrocolloids prevent the small cells found in bread dough from clumping together to generate larger cells. Having a larger number of small cells can form a more uniform matrix that acts as a CO_2-trapping network. Hydrocolloids improving the volume of bread loaves were also recognized in the research conducted by Kang et al. [77] and Zhao et al. [78]. Conversely, the reduction in volume and specific volume observed in our bread samples supplemented with 10% CCP may be attributed to an undesirably higher content of hydrophilic compounds (including carbohydrates) in these bread loaves, which could theoretically cause excessively higher viscosity. Furthermore, the reduced volume and increased stiffness of bread loaves enriched with higher CCP addition may also be due to a relevant reduction in the amount of gluten in the doughs [79], caused by the significant depletion of WF content in that formulation. Overall, our findings suggest that the addition of CCP up to 5% is a promising strategy to produce bread loaves and other related products with enhanced bioactivity and preserved bread quality parameters.

The sensory properties of a food product play a major role in its consumer acceptance and marketability [80]. When dealing with the incorporation of alternative ingredients into food formulations, the goal is to enhance the nutritional, bioactive, and physical attributes of the product without compromising its sensory acceptability [81]. In effect, higher levels of non-traditional bakery ingredients in bread formulations can greatly affect their taste and aroma [15], which are considered important sensory characteristics along with the texture [82]. Fresh, mature CC fruits have an intrinsic cherry-like, tart-sweet, and sour flavor with characteristic aroma [27,63] which might interfere with the sensory perception of a final product. Indeed, a gradual increase of aroma and taste (sweet, bitter, sour taste, aftertaste) scores of bread samples with the increment of CCP incorporation ratio was observed. Lowest scores for bitter, sour, and aftertaste were observed for bread formulated with 1% and 2% CCP addition. We hypothesize that bioactive compounds such as phenolics and iridoids found in CCP, which display many biological activities, such as anti-inflammatory, antioxidative, anti-cancer, anti-atherogenic, antidiabetic, and neuroprotective attributes [83], may play an important role in the observed sensory findings [84,85]. Further, as the CCP ratio increased, the crumb and crust color of CCP-enriched samples became darker (Figure 1). We suppose that this color modification could affect consumers' acceptability in a positive manner. In effect, previously it was shown that the red color of bread (caused by red beetroot addition) was preferred by consumers as compared to that supplemented only with white beetroot [86]. In addition to these observations, the texture of our bread loaves was also modified by CCP addition. The increased chewiness of bread was positively correlated with the amount of CCP (10% addition), which is consistent with a previous study [87] documenting the superior chewiness of bread samples enriched with artichoke fiber. Similarly, the significant difference in springiness identified only between the bread with 10% CCP addition and the control reflects a dose-dependent effect of CCP addition on this specific parameter. Finally, the best overall impression score was reported for samples with 2% CCP addition, and it was significantly higher compared to the control, whereas the other treatments showed similar results ($p > 0.05$). Altogether, bread loaves prepared with 2% or 5% replacement ratios of WF with CCP proceed to be promising bakery products with desirable overall characteristics. Our results present a novel, functional approach for the development of wheat bread with enhanced attributes for the current health-oriented bakery market.

5. Conclusions

Our study investigated the production and application of Cornelian cherry powder as a potential functional food ingredient for the partial replacement of WF in bread formulations. Our results show that the incorporation of CCP (replacement ratios between 1–10% w/w) in wheat bread formulations produces final products with significantly different composition, bioactivity, volume, and sensory attributes compared to the control. Indeed, increasing ratios of CCP lead to bread loaves with significantly higher ($p < 0.05$) carbohydrate and ash contents, energetic value, TPC, TPAC, and AA, but lower fat and protein contents. Further, our findings showed that CCP added at 1% to 5% ratios significantly ($p < 0.05$) improved the volume and specific volume of experimental bread loaves, and the highest overall impression score, significantly higher compared to the control, was reported for samples with 2% CCP addition. Overall, here we demonstrate that CCP can partially replace WF when used up to 5% (w/w) for bread formulations without negatively impacting key physical properties and sensory attributes, while enhancing the concentration of phenolic antioxidants. Moreover, we believe that the incorporation strategy shown here can be successfully applied to multiple bakery products for the production of healthier and more functional food products for the emerging health-oriented market.

Author Contributions: Conceptualization, V.Š. and J.B.; methodology, V.Š., E.I., P.J., and D.S.; software, V.Š., H.Ď., and M.M.; validation, V.Š., L.G., and M.K.; formal analysis, V.Š., H.Ď., J.B., E.I., and M.M.; investigation, V.Š., P.J., and D.S.; resources, L.G. and M.K.; data curation, L.G. and M.K.; writing—original draft preparation, V.Š. and H.Ď.; writing—review and editing, V.Š. and H.Ď.; visualization, V.Š., H.Ď., L.G., and M.K.; supervision, L.G. and M.K.; project administration, L.G.; funding acquisition, M.K. All authors have read and agreed to the published version of the manuscript.

Funding: This publication was supported by the Operational Program Integrated Infrastructure within the project: Demand-driven research for the sustainableland innovative food, Drive4SIFood 313011V336, co-financed by the European Regional Development Fund.

Data Availability Statement: Not applicable.

Acknowledgments: This work has been supported by grants from the KEGA, grant number 010SPU-4/2021. The authors are thankful to Roberta Hoskins for language corrections.

Conflicts of Interest: The authors declare no conflict of interest.

References

1. Cappelli, A.; Cini, E. Challenges and opportunities in wheat flour, pasta, bread, and bakery product production chains: A systematic review of innovations and improvement strategies to increase sustainability, productivity, and product quality. *Sustainability* **2021**, *13*, 2608. [CrossRef]
2. Mitelut, A.C.; Popa, E.E.; Popescu, P.A.; Popa, M.E. Trends of innovation in bread and bakery production. In *Trends in Wheat and Bread Making*, 1st ed.; Galanakis, C.M., Ed.; Elsevier: Amsterdam, The Netherlands, 2021; pp. 199–226. [CrossRef]
3. Baba, M.D.; Manga, T.A.; Daniel, C.; Danrangi, J. Sensory evaluation of toasted bread fortified with banana flour: A preliminary study. *AJFSN* **2015**, *2*, 9–12.
4. Chen, Y.; Zhao, L.; He, T.; Ou, Z.; Hu, Z.; Wang, K. Effects of mango peel powder on starch digestion and quality characteristics of bread. *Int. J. Biol. Macromol.* **2019**, *140*, 647–652. [CrossRef] [PubMed]
5. Irigoytia, M.B.; Irigoytia, K.; Sosa, N.; de Escalada Pla, M.; Genevois, C. Blueberry by-product as a novel food ingredient: Physicochemical characterization and study of its application in a bakery product. *JSFA* **2022**, *102*, 4551–4560. [CrossRef]
6. Otoni, C.G.; Avena-Bustillos, R.J.; Azeredo, H.M.; Lorevice, M.V.; Moura, M.R.; Mattoso, L.H.; McHugh, T.H. Recent advances on edible films based on fruits and vegetables—A review. *CRFSFS* **2017**, *16*, 1151–1169. [CrossRef]
7. Ahmad, M.S.; Siddiqui, M.W. Factors affecting postharvest quality of fresh fruits. In *Postharvest Quality Assurance of Fruits*, 1st ed.; Springer International Publishing: Cham, Switzerland, 2015; pp. 7–32.
8. Sridhar, A.; Ponnuchamy, M.; Kumar, P.S.; Kapoor, A. Food preservation techniques and nanotechnology for increased shelf life of fruits, vegetables, beverages and spices: A review. *Environ. Chem. Lett.* **2021**, *19*, 1715–1735. [CrossRef] [PubMed]
9. Castro, A.M.; Mayorga, E.Y.; Moreno, F.L. Mathematical modelling of convective drying of fruits: A review. *J. Food Eng.* **2018**, *223*, 152–167. [CrossRef]
10. Calín-Sánchez, Á.; Lipan, L.; Cano-Lamadrid, M.; Kharaghani, A.; Masztalerz, K.; Carbonell-Barrachina, Á.A.; Figiel, A. Comparison of traditional and novel drying techniques and its effect on quality of fruits, vegetables and aromatic herbs. *Foods* **2020**, *9*, 1261. [CrossRef]

11. Abbasi, S.; Azari, S. Novel microwave–freeze drying of onion slices. *Int. J. Food Sci.* **2009**, *44*, 974–979. [CrossRef]
12. Wankhade, P.K.; Sapkal, R.S.; Sapkal, V.S. Drying characteristics of okra slices using different drying methods by comparative evaluation. In Proceedings of the World Congress on Engineering and Computer Science, San Francisco, CA, USA, 24–26 October 2012; Volume 2, pp. 24–26.
13. Martins, Z.E.; Pinho, O.; Ferreira, I.M.P.L.V.O. Food industry by-products used as functional ingredients of bakery products. *Trends Food Sci. Technol.* **2017**, *67*, 106–128. [CrossRef]
14. Maqsood, S.; Adiamo, O.; Ahmad, M.; Mudgil, P. Bioactive compounds from date fruit and seed as potential nutraceutical and functional food ingredients. *Food Chem.* **2020**, *308*, 125522. [CrossRef] [PubMed]
15. Mildner-Szkudlarz, S.; Bajerska, J.; Zawirska-Wojtasiak.; Górecka, D. White grape pomace as a source of dietary fibre and polyphenols and its effect on physical and nutraceutical characteristics of wheat biscuits. *JSFA* **2013**, *93*, 389–395. [CrossRef] [PubMed]
16. Zhang, Y.; Song, K.Y.; Joung, K.Y.; Shin, S.Y.; Kim, Y.S. Quality characteristics of steamed bread containing pomegranate (*Punica granatum* L.) peel powder. *KJFCS* **2017**, *33*, 54–64. [CrossRef]
17. Han, L.; Zhang, J.; Cao, X. Effects of orange peel powder on rheological properties of wheat dough and bread aging. *Nutr. Food Sci.* **2021**, *9*, 1061–1069. [CrossRef] [PubMed]
18. Valková, V.; Ďúranová, H.; Havrlentová, M.; Ivanišová, E.; Mezey, J.; Tóthová, Z.; Gabríny, L.; Kačániová, M. Selected physico-chemical, nutritional, antioxidant and sensory properties of wheat bread supplemented with apple pomace powder as a by-product from juice production. *Plants* **2022**, *11*, 1256. [CrossRef] [PubMed]
19. Švec, I.; Kapačinskaité, R.; Hrušková, M. Wheat dough fermentation and bread trial results under the effect of quinoa and canahua wholemeal additions. *Czech J. Food Sci.* **2020**, *38*, 49–56. [CrossRef]
20. Mondal, A.; Datta, A.K. Bread baking—A review. *J. Food Eng.* **2008**, *86*, 465–474. [CrossRef]
21. De Biaggi, M.; Donno, D.; Mellano, M.G.; Riondato, I.; Rakotoniaina, E.N.; Beccaro, G.L. *Cornus mas* (L.) fruit as a potential source of natural health-promoting compounds: Physico-chemical characterisation of bioactive components. *Plant Foods Hum. Nutr.* **2018**, *73*, 89–94. [CrossRef]
22. Milenković-Anđelković, A.S.; Anđelković, M.Z.; Radovanović, A.N.; Radovanović, B.C.; Nikolić, V. Phenol composition, DPPH radical scavenging and antimicrobial activity of Cornelian cherry (*Cornus mas*) fruit and leaf extracts. *Hem. Ind.* **2015**, *69*, 331–337. [CrossRef]
23. Sozański, T.; Kucharska, A.Z.; Rapak, A.; Szumny, D.; Trocha, M.; Merwid-Ląd, A.; Szeląg, A. Iridoid–loganic acid versus anthocyanins from the *Cornus mas* fruits (cornelian cherry): Common and different effects on diet-induced atherosclerosis, PPARs expression and inflammation. *Atherosclerosis* **2016**, *254*, 151–160. [CrossRef]
24. Dinda, B.; Kyriakopoulos, A.M.; Dinda, S.; Zoumpourlis, V.; Thomaidis, N.S.; Velegraki, A.; Dinda, M. *Cornus mas* L (cornelian cherry), an important European and Asian traditional food and medicine: Ethnomedicine, phytochemistry and pharmacology for its commercial utilization in drug industry. *J. Ethnopharmacol.* **2016**, *193*, 670–690. [CrossRef]
25. Hassanpour, H.; Yousef, H.; Jafar, H.; Mohammad, A. Antioxidant capacity and phytochemical properties of cornelian cherry (*Cornus mas* L.) genotypes in Iran. *Sci. Hortic.* **2011**, *129*, 459–463. [CrossRef]
26. Pawlowska, A.M.; Camangi, F.; Braca, A. Quali-quantitative analysis of flavonoids of *Cornus mas* L. (*Cornaceae) fruits*. *Food Chem.* **2010**, *119*, 1257–1261. [CrossRef]
27. Kazimierski, M.; Reguła, J.; Molska, M. Cornelian cherry (*Cornus mas* L.)–characteristics, nutritional and pro-health properties. *Acta Sci. Pol. Technol. Aliment.* **2019**, *18*, 5–12. [CrossRef]
28. Topdaş, E.F.; Çakmakçi, S.; Akiroğlu, K. The antioxidant activity, vitamin c contents, physical, chemical and sensory properties of ice cream supplemented with cornelian cherry (*Cornus mas* L.) paste. *Kafkas Univ. Vet. Fak. Derg.* **2017**, *23*, 691–697. [CrossRef]
29. Adamenko, K.; Kawa-Rygielska, J.; Kucharska, A.Z. Characteristics of Cornelian cherry sour non-alcoholic beers brewed with the special yeast *Saccharomycodes ludwigii*. *Food Chem.* **2020**, *312*, 125968. [CrossRef]
30. Karademir, E.; Yalçın, E. Effect of fermentation on some quality properties of cornelian cherry tarhana produced from different cereal/pseudocereal flours. *Qual. Assur. Saf.* **2019**, *11*, 127–135. [CrossRef]
31. Kawa-Rygielska, J.; Adamenko, K.; Kucharska, A.Z.; Piórecki, N. Bioactive compounds in cornelian cherry vinegars. *Molecules* **2018**, *23*, 379. [CrossRef] [PubMed]
32. Salejda, A.M.; Kucharska, A.Z.; Krasnowska, G. Effect of Cornelian cherry (*Cornus mas* L.) juice on selected quality properties of beef burgers. *J. Food Qual.* **2018**, *2018*, 1563651. [CrossRef]
33. İlyasoğlu, H.; Arslan Burnaz, N.; Arpa Zemzemoğlu, T.E. Flaxseed and Cornelian cherry: Development of a functional cookie using response surface methodology. *J. Food Process. Preserv.* **2022**, *46*, e16954. [CrossRef]
34. Arraibi, A.A.; Liberal, Â.; Dias, M.I.; Alves, M.J.; Ferreira, I.C.; Barros, L.; Barreira, J.C. Chemical and bioactive characterization of Spanish and Belgian apple pomace for its potential use as a novel dermocosmetic formulation. *Foods* **2021**, *10*, 1949. [CrossRef] [PubMed]
35. Ivanišová, E.; Grygorieva, O.; Abrahamova, V.; Schubertova, Z.; Terentjeva, M.; Brindza, J. Characterization of morphological parameters and biological activity of jujube fruit (*Ziziphus jujuba* Mill.). *J. Berry Res.* **2017**, *7*, 249–260. [CrossRef]
36. García-Gómez, B.; Fernández-Canto, N.; Vázquez-Odériz, M.L.; Quiroga-García, M.; Muñoz-Ferreiro, N.; Romero-Rodríguez, M.Á. Sensory descriptive analysis and hedonic consumer test for Galician type breads. *Food Control* **2022**, *134*, 108765. [CrossRef]

37. Ocheme, O.B.; Adedeji, O.E.; Chinma, C.E.; Yakubu, C.M.; Ajibo, U.H. Proximate composition, functional, and pasting properties of wheat and groundnut protein concentrate flour blends. *Food Sci. Nutr.* **2018**, *6*, 1173–1178. [CrossRef] [PubMed]
38. Ahmed, H.G.M.D.; LI, M.J.; Khan, S.H.; Kashif, M. Early selection of bread wheat genotypes using morphological and photosynthetic attributes conferring drought tolerance. *J. Integr. Agric.* **2019**, *18*, 2483–2491. [CrossRef]
39. Braun, H.J.; Atlin, G.; Payne, T. Multi-location testing as a tool to identify plant response to global climate change. *Clim. Chang. Crop Prod.* **2010**, *1*, 115–138.
40. Mahloko, L.M.; Silungwe, H.; Mashau, M.E.; Kgatla, T.E. Bioactive compounds, antioxidant activity and physical characteristics of wheat-prickly pear and banana biscuits. *Heliyon* **2019**, *5*, e02479. [CrossRef]
41. Scalzo, J.; Politi, A.; Pellegrini, N.; Mezzetti, B.; Battino, M. Plant genotype affects total antioxidant capacity and phenolic contents in fruit. *Nutrition* **2005**, *21*, 207–213. [CrossRef]
42. Tešević, V.; Nikićević, N.; Milosavljević, S.; Bajic, D.; Vajs, V.; Vučković, I.; Vujisić, L.; Đorđević, I.; Stanković, M.; Velickovic, M. Characterization of volatile compounds of "Drenja", an alcoholic beverage obtained from the fruits of cornelian cherry. *J. Serb. Chem. Soc.* **2009**, *74*, 117–128. [CrossRef]
43. Cakmakci, S.; Tosun, M. Characteristics of mulberry pekmez with cornelian cherry. *Int. J. Food Prop.* **2010**, *13*, 713–722. [CrossRef]
44. Nawirska-Olszańska, A.; Biesiada, A.; Sokół-Łętowska, A.; Kucharska, A.Z. Content of bioactive compounds and antioxidant capacity of pumpkin puree enriched with Japanese quince, cornelian cherry, strawberry and apples. *Acta Sci. Pol. Technol. Aliment.* **2011**, *10*, 51–60. [PubMed]
45. Cerit, İ.; Şenkaya, S.; Tulukoğlu, B.; Kurtuluş, M.; Seçilmişoğlu, Ü.R.; Demirkol, O. Enrichment of functional properties of white chocolates with cornelian cherry, spinach and pollen powders. *Gida/J. Food* **2016**, *41*, 311–316. [CrossRef]
46. Adamenko, K.; Kawa-Rygielska, J.; Kucharska, A.Z.; Piórecki, N. Characteristics of biologically active compounds in Cornelian cherry meads. *Molecules* **2018**, *23*, 2024. [CrossRef]
47. Ivanova, M.; Petkova, N.; Balabanova, T.; Ognyanov, M.; Vlaseva, R. Food design of dairy desserts with encapsulated Cornelian cherry, Chokeberry and Blackberry juices. *Ann. Univ. Dunarea Jos Galati. Fascicle VI—Food Technol.* **2018**, *42*, 137–146.
48. Mantzourani, I.; Nouska, C.; Terpou, A.; Alexopoulos, A.; Bezirtzoglou, E.; Panayiotidis, M.I.; Galanis, A.; Plessas, S. Production of a novel functional fruit beverage consisting of cornelian cherry juice and probiotic bacteria. *Antioxidants* **2018**, *7*, 163. [CrossRef]
49. Adamenko, K.; Kawa-Rygielska, J.; Kucharska, A.Z.; Piórecki, N. Fruit low-alcoholic beverages with high contents of iridoids and phenolics from apple and Cornelian cherry (*Cornus mas* L.) fermented with *Saccharomyces bayanus*. *Pol. J. Food Nutr. Sci.* **2019**, *69*, 307–317. [CrossRef]
50. Kawa-Rygielska, J.; Adamenko, K.; Kucharska, A.Z.; Prorok, P.; Piórecki, N. Physicochemical and antioxidative properties of Cornelian cherry beer. *Food Chem.* **2019**, *281*, 147–153. [CrossRef]
51. Haghani, S.; Hadidi, M.; Pouramin, S.; Adinepour, F.; Hasiri, Z.; Moreno, A.; Munekata, P.E.S.; Lorenzo, J.M. Application of Cornelian cherry (*Cornus mas* L.) peel in probiotic ice cream: Functionality and viability during storage. *Antioxidants* **2021**, *10*, 1777. [CrossRef]
52. Szczepaniak, O.; Jokiel, M.; Stuper-Szablewska, K.; Szymanowska, D.; Dziedziński, M.; Kobus-Cisowska, J. Can cornelian cherry mask bitter taste of probiotic chocolate? Human TAS2R receptors and a sensory study with comprehensive characterisation of new functional product. *PLoS ONE* **2021**, *16*, e0243871. [CrossRef]
53. Tontul, I.; Eroğlu, E.; Topuz, A. Convective and refractance window drying of cornelian cherry pulp: Effect on physicochemical properties. *J. Food Process Eng.* **2018**, *41*, e12917. [CrossRef]
54. Tontul, I.; Topuz, A. Spray-drying of fruit and vegetable juices: Effect of drying conditions on the product yield and physical properties. *Trends Food Sci. Technol.* **2017**, *63*, 91–102. [CrossRef]
55. Syamaladevi, R.M.; Tang, J.; Villa-Rojas, R.; Sablani, S.; Carter, B.; Campbell, G. Influence of water activity on thermal resistance of microorganisms in low-moisture foods: A review. *CRFSFS* **2016**, *15*, 353–370. [CrossRef]
56. Petkova, N.T.; Ognyanov, M.H. Phytochemical characteristics and in vitro antioxidant activity of fresh, dried and processed fruits of Cornelian cherries (*Cornus mas* L.). *Bulg. Chem. Commun.* **2018**, *50*, 302–307.
57. Jaćimović, V.; Božović, D.; Ercisli, S.; Ognjanov, V.; Bosančić, B. Some fruit characteristics of selected cornelian cherries (*Cornus mas* L.) from Montenegro. *Erwerbs-Obstbau* **2015**, *57*, 119–124. [CrossRef]
58. Bijelić, S.; Gološin, B.; Ninić Todorović, J.; Cerović, S.; Bogdanović, B. Promising cornelian cherry (*Cornus mas* L.) genotypes from natural population in Serbia. *Agric. Conspec. Sci.* **2012**, *77*, 5–10.
59. Antoniewska-Krzeska, A.; Ivanišová, E.; Klymenko, S.; Bieniek, A.A.; Šramková, K.F.; Brindza, J. Nutrients content and composition in different morphological parts of Cornelian cherry (*Cornus mas* L.). *Agrobiodivers. Improv. Nutr. Health Life Qual.* **2022**, *6*, 1–10. [CrossRef]
60. Demir, F.; Kalyoncu, I.H. Some nutritional, pomological and physical properties of cornelian cherry (*Cornus mas* L.). *J. Food Eng.* **2003**, *60*, 335–341. [CrossRef]
61. Ercisli.; Orhan, E. Chemical composition of white (*Morus alba*), red (*Morus rubra*) and black (*Morus nigra*) mulberry fruits. *Food Chem.* **2007**, *103*, 1380–1384. [CrossRef]
62. Cosmulescu, S.N.; Trandafir, I.; Cornescu, F. Antioxidant capacity, total phenols, total flavonoids and colour component of cornelian cherry (*Cornus mas* L.) wild genotypes. *Not Bot Horti Agrobot Cluj Napoca* **2019**, *47*, 390–394. [CrossRef]
63. Szczepaniak, O.M.; Kobus-Cisowska, J.; Kusek, W.; Przeor, M. Functional properties of Cornelian cherry (*Cornus mas* L.): A comprehensive review. *Eur. Food Res. Technol.* **2019**, *245*, 2071–2087. [CrossRef]

64. Klymenko, S.; Kucharska, A.Z.; Sokół-Łętowska, A.; Piórecki, N.; Przybylska, D.; Grygorieva, O. Iridoids, Flavonoids, and Antioxidant capacity of *Cornus mas*, *C. officinalis*, and *C. mas* × *C. officinalis* fruits. *Biomolecules* **2021**, *11*, 776. [CrossRef] [PubMed]
65. Dupak, R.; Ivanisova, E.; Grygorieva, O.; Capcarova, M. Antioxidant and biochemical characterisation of Cornelian cherry (*Cornus mas* L.). In *International scientific days 2022: Efficient, Sustainable and Resilient Agriculture and Food Systems—The Interface of Science, Politics and Practice*; Slovak University of Agriculture in Nitra: Nitra, Slovakia, 2022; pp. 659–665. [CrossRef]
66. Szczepaniak, O.M.; Ligaj, M.; Kobus-Cisowska, J.; Maciejewska, P.; Tichoniuk, M.; Szulc, P. Application for novel electrochemical screening of antioxidant potential and phytochemicals in Cornus mas extracts. *CyTA-J. Food* **2019**, *17*, 781–789. [CrossRef]
67. Popović, B.M.; Štajner, D.; Slavko, K.; Sandra, B. Antioxidant capacity of cornelian cherry (*Cornus mas* L.)–Comparison between permanganate reducing antioxidant capacity and other antioxidant methods. *Food Chem.* **2012**, *134*, 734–741. [CrossRef] [PubMed]
68. Gunduz, K.; Saracoglu, O.; Özgen, M.; Serce, S. Antioxidant, physical and chemical characteristics of cornelian cherry fruits (*Cornus mas* L.) at different stages of ripeness. *Acta Sci. Pol. Hortorum Cultus* **2013**, *12*, 59–66.
69. Julkunen-Tiitto, R.; Nenadis, N.; Neugart, S.; Robson, M.; Agati, G.; Vepsäläinen, J.; Jansen, M.A. Assessing the response of plant flavonoids to UV radiation: An overview of appropriate techniques. *Phytochem. Rev.* **2015**, *14*, 273–297. [CrossRef]
70. Kumar, S.; Pandey, A.K. Chemistry and biological activities of flavonoids: An overview. *Sci. World J.* **2013**, *2013*, 162750. [CrossRef]
71. Valková, V.; Ďúranová, H.; Falcimaigne-Cordin, A.; Rossi, C.; Nadaud, F.; Nesterenko, A.; Moncada, M.; Orel, M.; Ivanišová, E.; Chlebová, Z.; et al. Impact of freeze-and spray-drying microencapsulation techniques on β-glucan powder biological activity: A comparative study. *Foods* **2022**, *11*, 2267. [CrossRef]
72. Hager, A.S.; Arendt, E.K. Influence of hydroxypropylmethylcellulose (HPMC), xanthan gum and their combination on loaf specific volume, crumb hardness and crumb grain characteristics of gluten-free breads based on rice, maize, teff and buckwheat. *Food Hydrocoll.* **2013**, *32*, 195–203. [CrossRef]
73. Pancerz, M.; Ptaszek, A.; Sofińska, K.; Barbasz, J.; Szlachcic, P.; Kucharek, M.; Łukasiewicz, M. Colligative and hydrodynamic properties of aqueous solutions of pectin from cornelian cherry and commercial apple pectin. *Food Hydrocoll.* **2019**, *89*, 406–415. [CrossRef]
74. Rosell, C.M.; Rojas, J.A.; De Barber, C.B. Influence of hydrocolloids on dough rheology and bread quality. *Food Hydrocoll.* **2001**, *15*, 75–81. [CrossRef]
75. Collar, C.; Andreu, P.; Martınez, J.C.; Armero, E. Optimization of hydrocolloid addition to improve wheat bread dough functionality: A response surface methodology study. *Food Hydrocoll.* **1999**, *13*, 467–475. [CrossRef]
76. Das, L.; Raychaudhuri, U.; Chakraborty, R. Role of hydrocolloids in improving the physical and textural characteristics of fennel bread. *Int. Food Res. J.* **2013**, *20*, 2253–2259.
77. Kang, N.; Reddy, C.K.; Park, E.Y.; Choi, H.D.; Lim, S.T. Antistaling effects of hydrocolloids and modified starch on bread during cold storage. *LWT* **2018**, *96*, 13–18. [CrossRef]
78. Zhao, F.; Li, Y.; Li, C.; Ban, X.; Cheng, L.; Hong, Y.; Li, Z. Co-supported hydrocolloids improve the structure and texture quality of gluten-free bread. *LWT* **2021**, *152*, 112248. [CrossRef]
79. Wahyono, A.; Tifania, A.Z.; Kurniawati, E.; Kang, W.W.; Chung, S.K. Physical properties and cellular structure of bread enriched with pumpkin flour. In *IOP Conference Series: Earth and Environmental Science*; IOP Publishing: Bristol, UK, 2018; Volume 207, p. 012054. [CrossRef]
80. Jagelaviciute, J.; Cizeikiene, D. The influence of non-traditional sourdough made with quinoa, hemp and chia flour on the characteristics of gluten-free maize/rice bread. *LWT* **2021**, *137*, 110457. [CrossRef]
81. Ivanišová, E.; Drevková, B.; Tokár, M.; Terentjeva, M.; Krajčovič, T.; Kačániová, M. Physicochemical and sensory evaluation of biscuits enriched with chicory fiber. *Food Sci. Technol. Int.* **2020**, *26*, 38–43. [CrossRef] [PubMed]
82. Mishyna, M.; Chen, J.; Benjamin, O. Sensory attributes of edible insects and insect-based foods–Future outlooks for enhancing consumer appeal. *Trends Food Sci. Technol.* **2020**, *95*, 141–148. [CrossRef]
83. Przybylska, D.; Kucharska, A.Z.; Sozański, T. A review on bioactive iridoids in edible fruits–from garden to food and pharmaceutical products. *Food Rev. Int.* **2022**, *38*, 1–31. [CrossRef]
84. Rodríguez, H.; Curiel, J.A.; Landete, J.M.; de las Rivas, B.; de Felipe, F.L.; Gómez-Cordovés, C.; Muñoz, R. Food phenolics and lactic acid bacteria. *Int. J. Food Microbiol.* **2009**, *132*, 79–90. [CrossRef]
85. Deng, S.; West, B.J.; Jensen, C.J. UPLC-TOF-MS characterization and identification of bioactive iridoids in *Cornus mas* fruit. *J. Anal. Chem.* **2013**, *2013*, 710972. [CrossRef]
86. Hobbs, D.A.; Ashouri, A.; George, T.W.; Lovegrove, J.A.; Methven, L. The consumer acceptance of novel vegetable-enriched bread products as a potential vehicle to increase vegetable consumption. *Food Res. Int.* **2014**, *58*, 15–22. [CrossRef]
87. Frutos, M.J.; Guilabert-Antón, L.; Tomás-Bellido, A.; Hernández-Herrero, J.A. Effect of artichoke (*Cynara scolymus* L.) fiber on textural and sensory qualities of wheat bread. *FSTI* **2008**, *14* (Suppl. 5), 49–55. [CrossRef]

Article

Hemp Flour Particle Size Affects the Quality and Nutritional Profile of the Enriched Functional Pasta

Sonia Bonacci [1], Vita Di Stefano [2,*], Fabiola Sciacca [3], Carla Buzzanca [2], Nino Virzì [3], Sergio Argento [4] and Maria Grazia Melilli [4]

[1] Department of Health Sciences, University Magna Græcia of Catanzaro, 88100 Catanzaro, Italy
[2] Department of Biological, Chemical and Pharmaceutical Sciences and Technologies, University of Palermo, 90133 Palermo, Italy
[3] CREA-Council for Agricultural Research and Economics-Research Centre for Cereal and Industrial Crops of Acireale, 00187 Roma, Italy
[4] National Council of Research, Institute of BioEconomy (CNR-IBE), 95126 Catania, Italy
* Correspondence: vita.distefano@unipa.it; Tel.: +39-09123891948

Abstract: The rheological and chemical quality of pasta samples, which were obtained using the durum wheat semolina fortified with the hemp seed solid residue, after oil extraction, sieved at 530 μm (Hemp 1) or 236 μm (Hemp 2) at different percentages of substitution (5%, 7.5%, and 10%, were evaluated. The total polyphenolic content in hemp flour was quantified in the range of 6.38–6.35 mg GAE/g, and free radical scavenging was included in the range from 3.94–3.75 mmol TEAC/100 g in Hemp 1 and Hemp 2, respectively. The phenolic profiles determined by UHPLC-ESI/QTOF-MS showed that cannabisin C, hydroxycinnamic and protocatechuic acids were the most abundant phenolic compounds in both hemp flours. Among the amino acids, isoleucine, glutamine, tyrosine, proline, and lysine were the most abundant in raw materials and pasta samples. Although the hemp seeds were previously subjected to oil extraction, hemp flours retain about 8% of oil, and the fatty acids present in the largest amount were linoleic acid and α-linolenic acid. Characterization of the minerals showed that the concentration of macro and trace elements increased according to fortification percentage. Sensory evaluation and cooking quality indicated that the best performance in terms of process production and consumer acceptance was obtained using Hemp 2 at 7.5%. Hemp supplementation could be a potential option for producing high-quality, nutritionally rich, low-cost pasta with good color and functionality.

Keywords: pasta fortification; hemp flour; durum wheat cultivar; amino acids; fatty acids; mineral fortification

Citation: Bonacci, S.; Di Stefano, V.; Sciacca, F.; Buzzanca, C.; Virzì, N.; Argento, S.; Melilli, M.G. Hemp Flour Particle Size Affects the Quality and Nutritional Profile of the Enriched Functional Pasta. *Foods* **2023**, *12*, 774. https://doi.org/10.3390/foods12040774

Academic Editors: Donatella Bianca Maria Ficco and Grazia Maria Borrelli

Received: 5 January 2023
Revised: 27 January 2023
Accepted: 6 February 2023
Published: 10 February 2023

1. Introduction

In recent years, consumers' eating habits have changed significantly. Food that, in the past, was intended to satisfy hunger and provide the necessary nutrients, today, combined with an active lifestyle, can help with harmonious physical and mental well-being.

Wheat is mainly used for the production of pasta, bread, and sweet and savory baked products. Pasta is one of the basic foods in the Mediterranean diet. Pasta traditionally made with durum wheat semolina can be prepared with "non-wheat flour" or by incorporating by-products from the agro-food industry, in variable percentages, which can increase its nutritional value [1,2]. Food by-products (grape, cereal bran, sunflower, artichoke, etc.) could represent interesting sources of bioactive compounds [3–5]. Fortification is the process by which nutrients with beneficial health effects are added to a food product in order to improve its nutritional quality and to increase its intake levels in the population. Food fortification (or enrichment) often negatively affects the quality of products, in terms of texture, color, cooking quality and sensory properties. Therefore, one of the main challenges in the food industry is to increase the healthiness of food without sacrificing sensory attributes [6].

Hemp (*Cannabis sativa* L.) is a plant used as textile fibers (from vegetative organs) native to the regions north and south of the Himalayas. Its use dates back to the Neolithic, and China is the country where it has been cultivated for the longest time. Its introduction in Europe probably dates back to the second millennium BC. Worldwide, it is cultivated mainly in Asia (China and India), Eastern Europe, and Russia. Today it is mainly used for textile, biocomposite, papermaking, construction, biofuel, and cosmetic purpose.

Cannabis sativa seeds can be used by the agri-food industry to produce flour, pastry, and oil, while the stem through to the canapulo (woody part of the stem) is in the green building sector. Its fiber (the external part of stem) will find new applications in the textile industry. Hemp inflorescences and roots, thanks to the extraction of bioactive molecules, will play an important role in the pharmaceutical and para-pharmaceutical industry [7].

Hemp seeds are mainly used as animal feed, but there is growing interest in their usage for human nutrition as a source of nutrients. They contain 25–35% of oil, 20–25% protein, 20–30% carbohydrates, and 10–15% insoluble fiber, vitamins, and minerals [8]. In particular, hemp seed oil is high in polyunsaturated fatty acids, with an ideal ratio (3:1) of linoleic acid (ω-6) and α-linolenic acid (ω-3) for human nutrition [9,10]. Merlino et al. [4] incorporated hemp seed flour (HSF) as a fortifying ingredient in the production of gnocchi, a typical Italian potato-based fresh pasta, from 5 to 20% HSF in substitution of soft wheat flour. Addition of HSF allowed for the enhancement of the nutritional value of gnocchi as a "source of fiber" in formulations with $\geq$ 10% of HSF. The fortified gnocchi had a high technological quality for cooking loss, cooking resistance, and textural properties, and average sensory quality; however, improving the HSF sensory quality for consumers' satisfaction was necessary in terms of odor and bitter taste. Hemp seeds were used to enrich pasta (15%), and its effects on osteoarticular pain and bone formation markers in patients with osteoarthritis in post-arthroplasty rehabilitation were evaluated. The first results showed that hemp seed can improve pain symptoms in patients with osteoarthritis undergoing arthroplasty surgery and improves bone metabolism both in humans and in vitro [11]. Pasta samples fortified with 5–40% commercially available hemp flour or 2.5–10% of hemp cake obtained from hemp seed oil pressing were studied [12]. The addition of hemp seed raw materials led to an increase in the protein, total dietary fiber (TDF), ash, and fat contents in the pasta samples. Due to its lower granulation and higher nutritional value, hemp flour was found to be a better raw material for the fortification of pasta than hemp cake.

In this study, the rheological and chemical qualities of pasta obtained by using the durum wheat cultivar "Ciclope", fortified with different percentages of hemp flour (cv. Futura 75), were investigated. The influence of semolina replacement at different percentages (5%, 7.5%, and 10%) and using two hemp flours (with different particle sizes) was evaluated, highlighting the effect of the particle size of hemp flours on the sensory properties, cooking quality, mineral composition, and nutritional characteristics of cooked functional pasta.

2. Materials and Methods

2.1. Raw Material

The durum wheat cv Ciclope was chosen among the durum wheat varieties made up by CREA Research Centre for Cereal and Industrial Crops of Acireale (Catania–Sicily, Italy). Durum wheat grain sample was milled using an experimental laboratory mill (Bona Labormill 4RB, Monza, Italy) and fine particles were separated (250 µm), with an extraction rate of about 55–60%.

Hemp flours were obtained from cv Futura seeds provided by Mulino Crisafulli (Caltagirone, Catania, Italy). The hemp seeds were subjected to mechanical extraction of the oil; the residue was milled into powder using a blender model M20 (IKA, Staufen, Germany) and was kept in hermetic bottles at room temperature (20 $\pm$ 2 °C) and sieved at 530 µm (namely Hemp 1) and 236 µm (namely Hemp 2).

2.2. Pasta Making, Cooking Quality and Sensorial Analysis

A Ciclope semolina pasta sample (CTRL) and pasta fortified with variable percentages of hemp flour (Hemp 1 and Hemp 2 at 5, 7.5, and 10% of substitution) were prepared (Table 1).

Table 1. Ciclope and fortified pasta samples.

Material		Fortified Hemp Pasta			
		0	5%	7.5%	10%
Ciclope semolina	(Ciclope)	CTRL			
Hemp 1 flour	(Hemp 1)		Hemp 1_5	Hemp 1_7.5	Hemp 1_10
Hemp 2 flours	(Hemp 2)		Hemp 2_5	Hemp 2_7.5	Hemp 2_10

Hemp 1 is hemp flour sieved at 0.530 mm; Hemp 2 is flour sieved at 0.236 mm. The fortified pastas take into account the different percentages of substitution of semolina with the two different hemp flours.

Pasta samples were prepared using a Pastamatic ARIETE 1591 equipped with a mixer and an extruder (De Longhi Appliances s.r.l., Florence, Italy), mixing 500 g flour (durum wheat semolina + x% w/w of hemp flours) with distilled water for 10.5 min. in order to obtain a dough with 40% moisture. The dough was extruded into a mancherons shape (5 cm long) following the procedures described by Cardullo et al. [13]. The optimal cooking time (OCT), the cooking loss, and the amount of solid substance in the cooking water were evaluated according to the AACC-approved method 66–50 (2000). The swelling index of cooked pasta was determined according to the procedure described in previous papers [2,3,8–10].

In order to evaluate the sensory attributes, a panel of 8 trained tasters (4 men and 4 women, aged between 30 and 64 years) analyzed the cooked pasta in OCT. Panelists' lists were developed on the basis of their sensory skills (ability to accurately determine and communicate sensory attributes such as the appearance, odor, flavor, and texture of a product). They judged bulkiness, firmness, adhesiveness, fibrous, color, odor, and taste. Based on the above-mentioned attributes, panelists were also asked to score the overall quality (OQS) of the product.

A 9-point scale was used: 1—very clear, 9—very dark in terms of color; 1—extremely unpleasant, −9—extremely pleasant in terms of bulkiness, firmness, adhesiveness, and fibrousness; 1—extremely unpleasant, 9—extremely pleasant in terms of odor, taste, and OQS [14–16].

2.3. Chemical Characterization

2.3.1. Polyphenols Extraction

Phenolic compounds (PCs) can be classified as free, conjugated (to sugars and low molecular-weight compounds), and insoluble bound phenolics (BPs); these latter are covalently bound to the structural components of the cell wall [17,18]. BPs are not extractable in aqueous and/or organic solvents; therefore, preventive hydrolysis based on alkaline or acidic treatments is one of the most valuable strategies for targeting these compounds. The alkaline treatment can cleave the ester bonds linking the compounds to the cell wall, thus allowing for the release of PCs (mainly phenolic acids) from the insoluble residues. Free and bound phenolics were extracted using modified methods [19–21].

Eight grams of sample (Ciclope, Hemp 1 and Hemp 2 flours and ground, cooked pasta samples) were homogenized for 45 min in 40 mL 80% methanol solution using an ultrasonic bath. The samples were centrifuged at 5000× g for 15 min, and the supernatant was recovered. The pellet was re-extracted four times (repeating the protocol described above) and the supernatant was collected and evaporated using a rotary evaporator under vacuum at 45 °C. The residue was redissolved in 2 mL of methanol. This solution, containing free phenolic compounds (PCs), was filtered through a 0.22 μm nylon syringe filter into glass vials prior to HPLC-ESI/QTOF-MS analysis. In order to obtain the bound phenolics (BPs) extract, the residues separated after centrifugation were digested in 40 mL of NaOH 4 M for 1 h at room temperature and acidified using hydrochloric acid to pH 2. Subsequently, the acid solution was extracted with ethyl acetate (50 mL) four times and the organic fraction was evaporated in a rotary evaporator at a temperature of 45 °C. The residue was redissolved in 2 mL of methanol and solution filtered through a 0.22 μm nylon syringe filter prior HPLC-ESI/QTOF-MS analysis of bound phenolic (BPs) fraction. Both extractions for free (PCs) and bound phenolics (BPs) in samples were performed in triplicate.

2.3.2. HPLC-ESI/QTOF-MS Analysis of Phenolic Compounds

The phenolic profile of hemp flour and pasta was investigated through an untargeted metabolomics-based approach using a HPLC-ESI/QTOF-MS method previously optimized [21]. The equipment consisted of an Alliance 2695 (Waters) HPLC system equipped with an autosampler, degasser, and column heater coupled with a Quadrupole Time-of-Flight (Waters Q-ToF Premier) mass spectrometer. The compounds were separated using a Phenomenex Luna C18 column (100 cm, 2 mm, 3 μm particle size). The phenolic compounds identified in different flours and pasta samples, were next quantified according to their class and sub-class, using calibration curves in a range of 2.5 μg mL^{-1}–25 μg mL^{-1}, built from pure reference standards (chlorogenic acid, catechin hydrate, rutin, caffeic acid, kaempferol, sinapic acid, and benzoic acid; Supplementary Table S1). When reference compounds were not available, the quantitation was based on structurally related substances. Specifically, rutine in negative mode was the reference compound for the determination of cannabisin B and cannaflavin C. Sinapic acid was the reference compound in negative mode for ferulic acid, chlorogenic acid was the reference compound for N-trans-caffeoyltyramine, and benzoic acid was the reference compound used for protocatechuic acid and vanillic acid semi-quantification.

2.3.3. Total Phenolic Content (TPC)

The content of total phenolics (TPC) was determined using the Folin–Ciocalteau method [22]. A calibration curve was set with gallic acid ranging from 0.001 to 0.25 mg mL^{-1} methanol/water (80:20 v/v) ($y = 10.955x + 0.1405$, $R^2 = 0.992$). The results were expressed as mg gallic acid equivalents per g (mg GAE g^{-1}) of sample. In this method, 5 mL methanol/water (80:20 v/v) was added to 0.5 g of the flour samples (Hemp 1, Hemp 2, Ciclope flours, pasta samples obtained from semolina replacement with the two hemp flours at 5, 7.5, and 10%), then, the obtained mixture was filtered through a 0.45 μm PTFE syringe filter. Next, 125 μL of the solution was mixed with 625 μL of diluted (1:5) Folin–Ciocalteau reagent in water and 120 μL of 7% Na_2CO_3. The samples were left in the dark for 1 h at room temperature. The TPC was measured four times for each sample.

2.3.4. Fatty Acid Composition

After the basic hydrolysis of triglycerides, it was necessary to convert fatty acids into their methyl esters (FAMEs). Quali-quantitative determinations of FAMEs were conducted according to Melilli et al. [23] using a gas chromatography–mass spectrometry (GC/MS) ISQ™ 9000 Quadrupole GC-MS System (Thermo Fisher Scientific, Waltham, MA, USA). The identification of FAMEs was performed by comparing their retention times with those of reference standards (mixture FAME Mix, SUPELCO, which included 37 FAMEs). The results of the FAMEs were expressed as relative percentages (%).

2.3.5. Amino Acids (AAs) Quantification by HPLC-FLD Method

Some procedures are needed for amino acid analysis, such as proteins hydrolysis. The modified procedure employing the acid hydrolysis of protein and the derivatization of the free amino acids using FMOC-Cl (9-fluorenylmethylchloroformate) was required prior to analysis with HPLC-FLD. Five hundred milligrams of the sample were added with 1 mL HCl 6 M in order to support the subsequent hydrolysis of proteins and were incubated in an oven at 110 °C for 24 h. After cooling to room temperature, the sample was diluted with 2 mL of deionized water and filtered with 0.45 μm PTFE syringe filters. The solution was subjected to pre-column derivatization by reaction of the sample with FMOC-Cl: 200 μL of 3 mM FMOC-Cl acetonic solution and 200 μL of borate buffer were added to 50 μL of the solution containing amino acids. The solution was heated at 70 °C for 10 min. Subsequently, 50 μL of a heptylamine solution (3 mL heptylamine, 15 mL ACN, and 175 mL HCl 0.1 M) was added to the solution and mixed for 3 min. Eighty microliters of the latter solution were taken, and 320 μL ACN and 600 μL hexane were added. A volume of 20 μL of this solution was injected into the HPLC-FLD instrument. Derivatized amino acids analyses were carried out using an HPLC Agilent 1100 Series chromatographic system equipped with a G1312A binary gradient pump and a fluorescence FLD detector and controlled by Chemstation software. For the chromatographic separation of derivatized amino, a Discovery HS C18 column was used (4.6 mm × 150 mm. 3.5 μm) (Supelco, Bellefonte, PA, USA), which fitted with guard column. The column operated at 40 °C, the flow rate was maintained at 1mL min^{-1}. Mobile phases were 0.1% formic acid as eluent A and ACN as eluent B. The program of gradient elution was as follows: 0–10 min, 3% B; 3–17 min, linear increase to 10% B; 17–47 min, linear increase to 50% B; 47–57 min, linear increase to 100% B; 57–60 min, hold 100% B; 60–63 min, equilibration and return to the initial conditions. Each derivative eluted from the column was monitored by a fluorometric detector (FLD) set to an excitation wavelength of 254 nm and an emission wavelength of 630 nm. A comparison of the retention times of the standards for peak identification was carried out, and a fortification technique (spiking) was applied. Quantitative determination of the derivatized amino acids was performed using calibration curves. Standard solutions of the derivatized amino acids were prepared at five concentration levels in a range from 0.025 mM to 0.4 mM (Supplementary Table S2). The results were expressed in terms of grams of amino acids in 100 g of sample.

2.3.6. Antiradical Properties of Raw Materials and Functional Pasta

The antiradical activity of samples (flours and fortified pasta) was measured using the DPPH assay.

One gram of each sample was extracted with 4 mL of methanol for 40 min in an ultrasonic bath. The supernatant was filtered using a 0.45 μm PTFE syringe filter. One hundred microliters of the filtrate were mixed with 3 mL DPPH (60 μM in methanol) and placed in the dark for 30 min. Absorbance at 515 nm was measured with a spectrophotometer (Varian Cary® 50 UV-Vis spectrophotometer) using methanol as a blank. Antiradical scavenging activity was expressed as the percentage inhibition of the DPPH radical and was calculated using the following Equation:

$$\text{Scavenging}\% = (A0 - Ai/A0) \times 100$$

where A0 is the absorbance of DPPH without the sample, and Ai is the absorbance of the sample and DPPH. The results were also reported as TEAC (Trolox equivalent antioxidant activity) and expressed in terms of mmol Trolox equivalents (TE)/100 g of sample. Trolox was utilized as the standard, and the calibration curve in a range between 5 and 400 μM was prepared using methanol as solvent (y = 0.0037x + 0.1655 and R2 = 0.987). All of the experiments were carried out in triplicate.

2.3.7. Mineral Profile of Pasta

In order to assess the influence of cooking on the exchange of mineral contents, the elemental composition of the raw and cooked hemp pasta samples and of the different cooking waters was established. Elemental analysis of microelements (As, Be, Cd, Co, Hg, Li, Ni, Sb, Se, Sn, Sr, and V) was performed using an inductively coupled mass spectrometer ICP-MS iCAP RQ, (Thermo Fisher Scientific Inc., Bremen, Germany) operating with argon gas of spectral purity (99.999 sample solutions were pumped by a peristaltic pump from tubes arranged on a CETAC ASX-520 auto-sampler (Thermo Scientific, Omaha, NE, USA). Instrument sensitivity, resolution, and mass calibration were optimized daily with the tuning solution (iCAP Q/RQ Tune aqueous multielement standard solution (Thermo

Scientific, Bremen, Germany) in order to maximize ion signals and minimize interference due to high oxide levels, optimizing torch position, ion lenses, gas output, resolution axis, and background. The optimal parameters are shown in Table 2.

Table 2. Operating conditions and acquisition parameters for ICP-MS.

Parameter	Setting
RF power (W)	500–1700
Reflected power	<10
Plasma gas flow (L min^{-1})	15
Nebulizer gas flow (L min^{-1})	1.00
Auxiliary gas flow (L min^{-1})	0.80
He mode	collision cell mode
He gas flow (mL min^{-1})	5.00
Octopole bias (CCT bias) (V)	−21
Quadrupole bias (pole bias) (V)	−18

The Al, B, Ba, Ca, Cu, Fe, Mg, Mn, Mo, Na, P, and Zn contents were determined using an inductively coupled plasma optical emission spectrometer (ICP-OES Analyzer, iCAP 7400, Thermo Fisher Scientific Inc., Waltham, MA, USA) equipped with a concentric nebulizer and a cyclonic spray chamber. The operating conditions are shown in Table 3.

Table 3. Operating conditions and acquisition parameters for ICP-OES.

Parameter	Setting
Nebulizer	Glass concentric
Nebulizer Gas Flow (L min^{-1})	0.40
Spray chamber	Glass Cyclonic
Purge Gas Flow	Normal
Auxiliary gas flow (L min^{-1})	0.50
Coolant gas flow (L min^{-1})	12
RF Power (W)	1150
Pump Speed (rpm)	50

Sample preparation was carried out using an Anton Paar Multiwave 5000 digestion system equipped with an XF100 rotor. In order to decontaminate PTFE vessels, a cleaning procedure was carried out by adding 4 mL of HNO_3 and 4 mL of H_2O to each vessel under the following conditions: 1100W for 15 min. After cleaning, vessels were rinsed with ultrapure water and dried [24]. Aliquots of 0.5 g of each pooled sample were weighted directly into the PTFE vessel of the microwave system. Digestion was performed by adding 8 mL of HNO_3. The operating conditions used for the microwave digestion were 800 W over 15 min and then hold at this power for 30 min. After digestion, samples were quantitatively transferred to a graduated polypropylene test tube and diluted with ultrapure water to 50 mL and stored at 4 °C until analysis. Each sample's digestion was performed in triplicate. The analytical batch consisted of a set of calibration standard samples and a minimum of three procedural blanks. Each solution was measured in triplicate, and analyses were carried out by a classical external calibration approach. For each element at least six calibration points were considered for calculation. The concentration range was selected based on the expected elemental values and sample dilution. The calibration ranges were: 0.005–100 μg L^{-1} for microelements; 0.002–1 mg L^{-1} for Ba, Cu, Mn, Mo, and Zn; 0.1–100 mg L^{-1} for Ca, Mg, Na, and K; and 0.01–10 mg L^{-1} for Al, B, Fe, and P. Stock solutions of calibration standards were properly diluted with 5% HNO_3.

2.4. Data Analysis

Data were submitted to the Bartlett's test for homogeneity of variance and then analyzed using two-way analysis of variance (ANOVA), based on a factorial combination of particle size (PS) × percentage of substitution (S) for the sensory characteristics of pasta. A separate ANOVA was conducted for each measurement and each main factor (particle size or percentage of substitution); in this case, means were statistically separated based on the Student–Newman–Keuls test. The CTRL was excluded when comparing pasta fortified with the two types of hemp flours. All other data, following Bartlett's test for the homogeneity of variance, were analyzed using one-way ANOVA, and means were compared by *LSD* test when the *F*-test was significant, at least at the 0.05 probability (CoHort Software, CoStat version 6.451).

3. Results

3.1. Chemical Characterization of Raw Material

The chemical characterization of the raw material is shown in Table 4. The total phenolic contents (TPC) of Ciclope durum flour and Hemp 1 and Hemp 2 were examined. For Ciclope flour, total polyphenol content, determined using the Folin–Ciocalteau method, was 2.45 mg GAE/g. In the case of hemp flours, similar total polyphenol values were obtained (6.38 and 6.35 mg GAE/g, respectively, for Hemp 1 and Hemp 2). The DPPH method has been widely used in antiradical activity studies of plant extracts [25]. DPPH radical scavenging activity was also expressed as the % scavenging value. The results of the radical scavenging activity of sample flours showed that all extracts had the ability to scavenge DPPH radical with values of 29.7, 53.2, and 51.5 for Ciclope, Hemp 1, and Hemp 2, respectively. The results confirmed that Hemp 1 and Hemp 2 flours showed similar antiradical activity, which was higher than Ciclope durum wheat flour.

Table 4. Chemical characterization of raw materials.

	Ciclope	Hemp 1	Hemp 2
TPC mgGAE/g	2.45 ± 0.001	6.38 ± 0.002	6.35 ± 0.001
DPPH$_{TEAC}$ mmol TE/100 g	1.35 ± 0.0355	3.94 ± 0.0178	3.75 ± 0.0179
% Scavenging	29.7	53.2	51.5
Fatty acid (Relative percentage %)			
Palmitic acid	17.6 ± 0.30	9.04 ± 0.17	9.10 ± 0.057
Stearic acid	1.54 ± 0.16	3.42 ± 0.025	3.32 ± 0.060
Oleic acid	17.3 ± 0.51	14.6 ± 0.062	14.5 ± 0.081
Linoleic acid (ω-6)	59.9 ± 0.89	53.7 ± 0.35	53.7 ± 0.17
γ linolenic acid (ω-6)	-	2.44 ± 0.075	2.47 ± 0.089
α linolenic acid (ω-3)	3.62 ± 0.53	15.7 ± 0.21	15.5 ± 0.050
$\sum \omega$-6	59.9 ± 0.89	56.1 ± 0.042	56.1 ± 0.259
$\sum \omega$-3	3.62 ±0.53	15.7 ± 0.21	15.5 ± 0.050
Amino acids (g/100 g)			
Arginine	0.34 ± 0.016	0.52 ± 0.020	0.58 ± 0.018
Serine	1.59 ± 0.035	2.00 ± 0.041	1.89 ± 0.038
Glutamine	2.90 ± 0.031	4.96 ± 0.039 b	5.38 ± 0.041 a
Tyrosine	3.37 ± 0.016	3.43 ± 0.014	3.19 ± 0.012
Alanine	0.85 ± 0.009	1.37 ± 0.011	1.84 ± 0.012
Histidine	0.53 ± 0.002	0.92 ± 0.002	0.94 ± 0.003
Proline	1.23 ± 0.025	3.22 ± 0.051	2.99 ± 0.057
Threonine	1.80 ± 0.015	1.97 ± 0.019 b	2.51 ± 0.021 a
Leucine	1.78 ± 0.035	2.75 ± 0.036	2.97 ± 0.035
Methionine	0.14 ± 0.006	0.61 ± 0.011	0.83 ± 0.016
Valine	0.58 ± 0.005	1.17 ± 0.012 b	2.00 ± 0.045 a
Phenylalanine	0.25 ± 0.015	0.40 ± 0.012	0.45 ± 0.013
Isoleucine	3.52 ± 0.055	6.94 ± 0.057	6.21 ± 0.053
Lysine	0.16 ± 0.005	2.07 ± 0.036 b	2.58 ± 0.012 a
$\sum$ AA	19.0 ± 2.39	32.3 ± 3.61 b	34.4 ± 3.76 a
$\sum$ essential AA	8.23 ± 1.03	15.9 b ± 1.78	17.6 ± 1.81 a

Data are expressed as the means ± SD of triplicate experiments. Hemp 1 and Hemp 2 were diluted 10-fold for anti-scavenger determination. Different letters indicated differences at $p < 0.05$ between Hemp 1 and Hemp 2 samples. Values without letters are not significantly different.

Although the hemp flours come from the shredding and sieving of the defatted seeds, a small percentage (about 8%) of oil remained in the solid matrix. The fatty acid profile was evaluated by GC-MS analysis and reported in Table 4. The main fatty acids identified in the lipid fractions of Ciclope flour were palmitic, oleic, and linoleic acids, with lower percentages of α linolenic acid. The lipidic profiles of hemp flours were shown in linoleic and α linolenic acid as major fatty acids, and oleic and palmitic acids as minor. Regarding the quantity of fatty acids, there was no differences in the lipid profile of Hemp 1 and Hemp 2 flours. The fatty acid present in largest amount was linoleic acid in 53% and α linolenic acid with 15.5%. Results were consistent with the analyses of other authors; Siano et al. [26] in agreement with our results, identified a similar fatty acid composition with linoleic acid as prevalent (56.42%), followed by linolenic (14.55%), oleic (12.79%), γ-linolenic (3.03%) and as saturated, palmitic (7.35%) and stearic acids (2.26%). Pojic work [27] highlighted high content of linoleic (54.09%–55.43%), linolenic (17.31–18.42%) and oleic (12.96–13.93%) acids, followed by palmitic (6.48–7.90%), stearic (3.18–3.86%) and γ-linolenic (2.61–2.76%).

There are no studies on the amino acid composition of Ciclope flour and hemp flour. Table 4 moreover summarizes the mean individual and total free amino acid (AAs) contents observed in the studied flours samples. According to the AAs profiles isoleucine, leucine, tyrosine and serine were among the amino acids with the highest content in Ciclope. The total AA content was 19.04 g/100 g in Ciclope. In Hemp 1 and Hemp 2 isoleucine, glutamine, tyrosine, proline and lysine were the most abundant. The particle size affected the amounts of the total AA content with 32.3 g/100 g (Hemp 1) and 34.4 g/100 g (Hemp 2). Particularly interesting was the amount of essential amino acids determined in the hemp flours which resulted in 15.9 g/100 g and 17.5 g/100 g in Hemp 1 and Hemp 2, respectively.

Hydroxycinnamic and protocatechuic acids (Table 5) represented the most abundant phenolic compounds quantified in hemp flour samples. The samples were subjected to alkaline treatment in order to determine the bound phenolic fraction (BPs). Results show a higher amount of bound hydroxycinnamic acid in Hemp 1 than Hemp 2 (1687.4 µg 100 g^{-1} and 1589.0 µg 100 g^{-1}, respectively). The presence of cannaflavin C is highly relevant both in PCs and BPs form. Values ranging from 1384.1 to 3367.0 µg 100 g^{-1} for Hemp 1 and from 1139.2 to 2207.0 µg 100 g^{-1} for Hemp 2.

Table 5. Phenolic compounds (µg 100 g^{-1}): Free and bound quantification (PCs and BPs) of raw materials.

	Ciclope		Hemp 1		Hemp 2	
	PCs	**BPs**	**PCs**	**BPs**	**PCs**	**BPs**
p-Hydroxybenzoic acid	n.d.	n.d.	110.2 ± 1.09	866.0 ± 2.21	73.00 ± 1.39	542.3 ± 2.03
Protocatechuic acid	n.d.	189.5 ± 1.53	45.10 ± 1.48	1210 ± 3.60	79.10 ± 1.87	1361 ± 2.43
Hydroxycinnamic acid	n.d.	329.6 ± 1.33	n.d.	1687 ± 2.02	82.00 ± 1.74	1589 ± 2.69
Vanillic acid	n.d..	75.20 ± 1.37	n.d.	.	n.d.	710.7 ± 1.65
Caffeic acid	n.d.	n.d.	n.d.	849.0 ± 1.23	n.d.	1185 ± 1.97
Ferulic acid	n.d.	n.d.	n.d.	616.2 ± 1.40	30.10 ± 1.11	243.6 ± 1.21
Sinapic acid	n.d.	n.d.	n.d.	550.4 ± 1.23	n.d.	476.5 ± 1.01
Catechin	n.d.	n.d.	n.d.	407.3 ± 0.98	n.d.	n.d.
N-trans-Caffeoyltyramine	n.d.	n.d.	n.d.	n.d.	n.d.	1817 ± 1.42
Chlorogenic acid	n.d.	n.d.	n.d.	353.7 ± 0.89	n.d.	n.d.
Cannabisin B	n.d.	n.d.	n.d.	538.3 ± 0.85	n.d.	n.d.
Cannaflavin C	n.d.	n.d.	1384 ± 1.77	3367 ± 1.15	1139 ± 0.98	2207 ± 1.45

Data are expressed as means ± SD of triplicate experiments. n.d. not detected.

As far as phenol amides were concerned, N-trans-caffeoyltyramine was quantified uniquely in Hemp 2 as bound phenol (1817.1 µg 100 g^{-1}). In Ciclope flour modest amounts of bound phenols have been determined, among these protocatechuic acid, *p*-hydroxycinnamic acid and vanillic acid (189.5, 329.6 and 75.2 µg 100 g^{-1}, respectively). No free-form phenolic compounds were found.

3.2. Chemical Characterization of the Functional Pasta

Total phenolic contents and antiradical activity were determined on cooked Ciclope pasta and the fortified pasta samples as shown in Table 6. The data reveal TPC values in Hemp 1_10% and in Hemp 2_10% pasta samples (4.92 ± 0.31–4.21 ± 0.35 mg GAE/g, respectively), in agreement with the activity of radical scavenging (3.86 ± 0.07–3.14 ± 0.06 mmol TE/100 g, respectively), higher than Ciclope semolina pasta.

Table 6. Chemical characterization of cooked pasta.

	CTRL	Hemp 1_5	Hemp 1_7.5	Hemp 1_10	Hemp 2_5	Hemp 2_7.5	Hemp 2_10	LSD ($p < 0.05$)
TPC mgGAE/g	1.11 ± 0.18	2.50 ± 0.37	4.25 ± 0.18	4.92 ± 0.31	1.95 ± 0.33	2.76 ± 0.40	4.21 ± 0.35	0.31
DPPH$_{TEAC}$ mmol TE/100 g	1.14 ± 0.05	2.30 ± 0.04	3.08 ± 0.08	3.86 ± 0.07	2.08 ± 0.08	2.65 ± 0.09	3.14 ± 0.06	0.08
% Scavenging	27.8	38.3	45.4	52.5	36.4	41.6	45.9	0.08
Fatty acid (Relative percentages%)								
Palmitic acid	15.92 ± 1.09	14.72 ± 0.4	15.11 ± 1.04	15.56 ± 0.3	15.33 ± 0.48	16.40 ± 0.8	16.64 ± 0.89	1.44
Stearic acid	2.72 ± 0.81	1.54 ± 0.26	2.73 ± 0.31	3.13 ± 0.8	3.29 ± 0.34	3.66 ± 0.66	3.88 ± 0.89	0.92
Oleic acid	12.96 ± 0.53	12.30 ± 1.6	13.80 ± 1.4	15.68 ± 0.9	13.42 ± 0.23	15.29 ± 0.46	17.81 ± 0.7	0.97
Linoleic acid	46.40 ± 0.62	41.66 ± 0.87	42.16 ± 0.8	43.05 ± 1.6	46.36 ± 1.23	47.26 ± 0.16	47.94 ± 0.61	0.91
γ linolenic acid	0.86 ± 0.075	3.37 ± 0.30	4.33 ± 0.8	5.26 ± 0.11	4.84 ± 0.04	4.87 ± 0.19	5.71 ± 0.40	1.27
α linolenic acid	4.02 ± 0.63	14.70 ± 0.4	15.51 ± 1.04	15.89 ± 0.3	15.38 ± 0.48	15.61 ± 0.8	15.89 ± 0.86	0.68
$\sum \omega$ 6	47.3	44.9	46.5	48.3	51.2	52.1	53.6	
$\sum \omega$ 3	4.02	14.7	15.5	15.9	15.3	15.6	15.9	
Amino acids (g/100 g)								
Arginine	0.14 ± 0.015	0.17 ± 0.020	0.19 ± 0.006	0.22 ± 0.012	0.19 ± 0.012	0.23 ± 0.020	0.26 ± 0.015	0.03
Serine	0.44 ± 0.015	0.42 ± 0.010	0.52 ± 0.021	0.68 ± 0.012	0.57 ± 0.015	0.67 ± 0.026	0.86 ± 0.015	0.03
Glutamine	0.74 ± 0.020	1.13 ± 0.015	1.19 ± 0.012	1.33 ± 0.020	1.12 ± 0.015	1.24 ± 0.015	1.33 ± 0.021	0.03
Tyrosine	0.44 ± 0.025	0.81 ± 0.015	1.17 ± 0.025	1.53 ± 0.020	0.95 ± 0.021	1.21 ± 0.015	1.49 ± 0.015	0.04
Alanine	0.22 ± 0.020	0.29 ± 0.015	0.32 ± 0.010	0.35 ± 0.015	0.39 ± 0.010	0.43 ± 0.017	0.48 ± 0.015	0.02
Histidine	0.1 ± 0.015	0.20 ± 0.015	0.36 ± 0.015	0.45 ± 0.015	0.26 ± 0.015	0.33 ± 0.020	0.53 ± 0.015	0.03
Proline	0.68 ± 0.010	1.27 ± 0.015	1.43 ± 0.021	1.79 ± 0.021	1.41 ± 0.020	1.65 ± 0.012	1.80 ± 0.021	0.03
Threonine	0.38 ± 0.021	0.49 ± 0.010	0.59 ± 0.015	0.72 ± 0.020	0.90 ± 0.017	1.03 ± 0.021	1.17 ± 0.015	0.03
Leucine	0.16 ± 0.015	0.40 ± 0.015	0.45 ± 0.006	0.57 ± 0.020	0.50 ± 0.021	0.62 ± 0.025	0.69 ± 0.015	0.03
Methionine	0.08 ± 0.001	0.11 ± 0.010	0.15 ± 0.020	0.19 ± 0.010	0.13 ± 0.012	0.16 ± 0.015	0.18 ± 0.015	0.02
Valine	0.09 ± 0.002	0.12 ± 0.010	0.23 ± 0.020	0.30 ± 0.025	0.29 ± 0.010	0.33 ± 0.015	0.37 ± 0.015	0.02
Phenylalanine	0.18 ± 0.012	0.15 ± 0.008	0.20 ± 0.015	0.26 ± 0.012	0.25 ± 0.020	0.29 ± 0.010	0.33 ± 0.015	0.02
Isoleucine	1.00 ± 0.021	1.08 ± 0.020	1.21 ± 0.015	1.41 ± 0.025	1.24 ± 0.015	1.36 ± 0.020	1.52 ± 0.015	0.03
Lysine	0.16 ± 0.020	0.03 ± 0.001	0.52 ± 0.010	0.85 ± 0.020	0.08 ± 0.003	0.22 ± 0.020	0.36 ± 0.021	0.02
$\sum$ AA	4.81 ± 0.070	6.67 ± 0.052	8.53 ± 0.110	10.65 ± 0.045	8.28 ± 0.084	9.77 ± 0.085	11.37 ± 0.035	0.12
$\sum$ essential AA	2.05 ± 0.049	2.38 ± 0.027	3.35 ± 0.055	4.30 ± 0.050	3.39 ± 0.048	4.01 ± 0.081	4.62 ± 0.036	0.09

Data are expressed as means $\pm$ SD of triplicate experiments. Means were separated by LSD test at $p < 0.05$.

Methyl ester fatty acids profile, as shown in Table 6, was also studied on cooked pasta samples. The main differences concern γ linolenic acid and α linolenic acid. In Hemp 1_10% and in Hemp 2_10% pasta samples reported, respectively, 5.26% and 5.71% of γ linolenic acid, and 15.89 of α linolenic acid.

Table 6 also highlights the amino acid content in the cooked pasta sample. Amino acids such as tyrosine, glutamine, proline and isoleucine are among the main in Hemp 1_10% and in Hemp 2_10% pasta. The concentration of essential amino acids in the two different pasta samples was interesting. In particular, there was a concentration of 4.30 g/100 g in Hemp 1_10% and 4.62 g/100 g in Hemp 2_10%. The concentration of total amino acids in the 10% fortified pasta was more than double that of the Ciclope durum wheat pasta.

In hemp-fortified pasta, phenols were detected and quantified (Table 7).

As expected, the alkaline treatment of the samples allowed for the cleavage of the ester bonds that bind the compounds to the cell wall, thus allowing for the release of PCs (mainly phenolic acids) from the insoluble residues.

In the pasta samples fortified, after the alkaline treatment was observed a greater presence of bound phenolics, particularly for the samples obtained with 10% substitution. Cannaflavin C, *p*-hydroxy benzoic acid, protocatechuic acid, hydroxycinnamic acid, and caffeic acid were predominantly found in Hemp 1 pasta in bound form. Conversely, *p*-hydroxycinnamic acid, caffeic acid, *p*-hydroxy benzoic acid, protocatechuic acid, and trans-N-caffeoyl-tyramine resulted in higher quantities in Hemp 2 pasta samples. As expected, the amount of free and bound phenols decrease the percentages of fortification.

Table 7. Phenolic compounds (μg 100 g^{-1}): free and bound quantification in cooked pasta (PCs and BPs).

	Hemp 1_5		Hemp 2_5		Hemp 1_7.5		Hemp 2_7.5		Hemp 1_10		Hemp 2_10	
	PCs	BPs	PCs	BPs	PCs	BPs	PCs	BPs	PCs	BPs	PCs	BPs
p-Hydroxybenzoic acid	n.d.	n.d.	n.d.	164.0 ± 1.02	n.d.	n.d.	n.d.	342.5 ± 1.43	n.d.	496.0 ± 2.01	n.d.	444.2 ± 2.09
Protocatechuic acid	n.d.	320.1 ± 0.89	n.d.	n.d.	n.d.	453.5 ± 1.43	n.d.	n.d.	n.d.	761.0 ± 1.87	n.d.	446.2 ± 1.25
Hydroxycinnamic acid	n.d.	410.3 ± 1.24	n.d.	219.9	n.d.	764.2 ± 1.29	n.d.	802.9 ± 1.98	n.d.	1033 ± 1.78	n.d.	571.3 ± 1.56
Vanillic acid	n.d.	110.0 ± 1.47	n.d.	n.d.	n.d.	335.1 ± 1.06	n.d.	n.d.	n.d.	385.4 ± 1.52	n.d.	n.d.
Caffeic acid	n.d.	n.d.	n.d.	107.9 ± 0.84	n.d.	203.4 ± 1.16	n.d.	367.6 ± 1.26	n.d.	519.3 ± 1.06	n.d.	846.5 ± 1.97
Ferulic acid	n.d.	n.d.	n.d.	n.d.	n.d.	289.4 ± 1.44	n.d.	n.d.	n.d.	372.3 ± 099	n.d.	n.d.
Sinapic acid	n.d.	n.d.	n.d.	67.9 ± 0.89	n.d.	216.4 ± 1.79	n.d.	295.6 ± 1.09	n.d.	410.7 ± 1.32	n.d.	313.4 ± 1.19
Catechin	n.d.	n.d.	n.d.	n.d.	n.d.	n.d.	n.d.	n.d.	n.d.	n.d.	n.d.	n.d.
N-trans-Caffeoyltyramine	n.d.	n.d.	n.d.	n.d.	n.d.	n.d.	n.d.	n.d.	n.d.	n.d.	n.d.	307.3 ± 1.96
Chlorogenic acid	n.d.	n.d.	n.d.	n.d.	n.d.	n.d.	n.d.	n.d.	n.d.	n.d.	n.d.	n.d.
Cannabisin B	n.d.	n.d.	n.d.	n.d.	n.d.	n.d.	n.d.	n.d.	n.d.	n.d.	n.d.	n.d.
Cannaflavin C	132.9 ± 1.45	103.1 ± 1.23	115.9 ± 1.03	110.1 ± 1.78	279.6 ± 1.02	587.4 ± 1.24	176.5 ± 1.80	101.0 ± 1.09	382.0 ± 1.07	855.2 ± 1.69	227.8 ± 1.70	274.8 ± 1.53

Data are expressed as the means ± SD of triplicate experiments. n.d., not detected.

The fortification at 7.5% especially with Hemp 1, showed good values of bound phenols. A similar phenolic profile was also reported by Pannico et al. [28] and Izzo et al. [29].

The quantification of minerals reported in Table 8 was carried out using external calibration curves. The data allow for the assessment of the contribution of hemp enrichment to the macro-element's composition of pasta.

Most of the minerals' concentrations increased according to the fortification percentage. In particular, the addition of hemp to pasta increased the content of iron, potassium, magnesium, and phosphorus. The iron concentration increased from 0.035 mg g^{-1} to 0.051 mg g^{-1}, the potassium concentration increased from 1.953 mg g^{-1} to 2.020 mg g^{-1}, the magnesium concentration increased from 0.873 mg g^{-1} to 1.191 mg g^{-1}, and the phosphorus concentration increased from 2.502 mg g^{-1} to 3.086 mg g^{-1} in Hemp 1_5% pasta and Hemp 1_10% pasta, respectively. As for trace elements, the copper concentration increased from 7.903 ug g^{-1} to 9.669 ug g^{-1} in Hemp 1_5% pasta and Hemp 1_10% pasta, respectively. The same results were recorded in pasta fortified with Hemp 2. The mineral element content in hemp seeds was nutritionally interesting, as reported by Alonso et al., 2022 [30]. Phosphorus, potassium, magnesium, calcium, iron, zinc, manganese, and copper are essential dietary elements for mammals and are involved in many physiological processes [31]. In Figure 1 the percentage of variation is reported for the most important minerals in the hemp-enriched pasta vs the CTRL. Pasta enriched with Hemp_2 yielded the best results for all of the considered minerals except Fe.

Table 8. Mineral content in cooked, fortified pasta.

mg g^{-1}	CTRL	Hemp 1_5	Hemp 1_7.5	Hemp 1_10	Hemp 2_5	Hemp 2_7.5	Hemp 2_10
Al	0.005 ± 0.000	0.034 ± 0.005	0.017 ± 0.000	0.036 ± 0.014	0.012 ± 0.002	0.025 ± 0.004	0.013 ± 0.005
B	<0.002	0.002 ± 0.000	0.002 ± 0.000	0.002 ± 0.000	0.002 ± 0.000	0.002 ± 0.000	0.003 ± 0.000
Ca	0.401 ± 0.0024	0.581 ± 0.021	0.653 ± 0.075	0.621 ± 0.014	0.659 ± 0.067	0.716 ± 0.068	0.764 ± 0.040
Fe	0.022 ± 0.000	0.035 ± 0.001	0.049 ± 0.004	0.051 ± 0.002	0.034 ± 0.001	0.040 ± 0.001	0.047 ± 0.004
K	1.735 ± 0.005	1.953 ± 0.005	1.976 ± 0.024	2.020 ± 0.025	1.907 ± 0.005	2.019 ± 0.005	2.398 ± 0.014
Mg	0.608 ± 0.074	0.873 ± 0.111	1.047 ± 0.130	1.191 ± 0.156	0.976 ± 0.130	1.172 ± 0.156	1.323 ± 0.153
Mn	0.014 ± 0.000	0.020 ± 0.000	0.025 ± 0.001	0.028 ± 0.001	0.023 ± 0.001	0.029 ± 0.000	0.032 ± 0.001
Na	0.092 ± 0.012	0.101 ± 0.015	0.095 ± 0.016	0.101 ± 0.014	0.070 ± 0.011	0.097 ± 0.015	0.087 ± 0.013
P	2.000 ± 0.023	2.502 ± 0.013	2.845 ± 0.000	3.086 ± 0.040	2.707 ± 0.024	3.054 ± 0.063	3.337 ± 0.036
Zn	0.019 ± 0.001	0.024 ± 0.001	0.027 ± 0.000	0.029 ± 0.001	0.026 ± 0.001	0.030 ± 0.001	0.033 ± 0.001
ug g^{-1}							
As	0.006 ± 0.000	0.007 ± 0.000	0.007 ± 0.000	0.008 ± 0.000	0.004 ± 0.000	0.006 ± 0.000	0.006 ± 0.000

Table 8. *Cont.*

mg g^{-1}	CTRL	Hemp 1_5	Hemp 1_7.5	Hemp 1_10	Hemp 2_5	Hemp 2_7.5	Hemp 2_10
Ba	0.777 ± 0.003	1.000 ± 0.003	1.079 ± 0.004	1.276 ± 0.003	0.910 ± 0.003	1.058 ± 0.004	1.009 ± 0.003
Be	<0.01	<0.01	<0.01	<0.01	<0.01	<0.01	<0.01
Cd	0.024 ± 0.000	0.022 ± 0.000	0.022 ± 0.000	0.024 ± 0.000	0.024 ± 0.000	0.025 ± 0.000	0.028 ± 0.000
Co	0.090 ± 0.000	0.094 ± 0.000	0.095 ± 0.000	0.103 ± 0.000	0.099 ± 0.000	0.105 ± 0.000	0.114 ± 0.000
Cr	0.036 ± 0.000	0.039 ± 0.000	0.059 ± 0.000	0.074 ± 0.000	0.035 ± 0.000	0.065 ± 0.000	0.060 ± 0.000
Cu	6.905 ± 0.024	7.903 ± 0.028	8.266 ± 0.029	9.669 ± 0.019	8.320 ± 0.025	9.063 ± 0.032	10.217 ± 0.025
Hg	0.005 ± 0.000	0.004 ± 0.000	0.007 ± 0.000	0.004 ± 0.000	0.004 ± 0.000	0.005 ± 0.000	0.003 ± 0.000
Li	0.054 ± 0.00	0.050 ± 0.00	0.054 ± 0.00	0.085 ± 0.00	0.042 ± 0.00	0.046 ± 0.00	0.062 ± 0.00
Mo	0.490 ± 0.002	0.572 ± 0.002	0.578 ± 0.002	0.626 ± 0.001	0.622 ± 0.002	0.683 ± 0.002	0.760 ± 0.002
Ni	0.279 ± 0.001	0.414 ± 0.001	0.448 ± 0.002	0.566 ± 0.001	0.482 ± 0.001	0.608 ± 0.002	0.763 ± 0.002
Pb	0.552 ± 0.002	0.533 ± 0.002	0.550 ± 0.002	0.532 ± 0.001	0.519 ± 0.002	0.526 ± 0.002	0.575 ± 0.001
Sb	0.004 ± 0.000	0.011 ± 0.000	0.006 ± 0.000	0.003 ± 0.000	0.004 ± 0.000	0.003 ± 0.000	0.006 ± 0.000
Se	0.034 ± 0.000	0.044 ± 0.000	0.046 ± 0.000	0.058 ± 0.000	0.038 ± 0.000	0.039 ± 0.000	0.048 ± 0.000
Sn	0.003 ± 0.000	0.006 ± 0.000	0.014 ± 0.000	0.007 ± 0.000	0.018 ± 0.000	0.017 ± 0.000	0.026 ± 0.000
Sr	4.095 ± 0.014	5.832 ± 0.020	7.334 ± 0.026	8.525 ± 0.017	6.709 ± 0.020	8.808 ± 0.031	8.600 ± 0.021
V	0.016 ± 0.000	0.058 ± 0.000	0.064 ± 0.000	0.086 ± 0.000	0.021 ± 0.000	0.046 ± 0.000	0.045 ± 0.000

Data are expressed as the means ± SD of triplicate experiments.

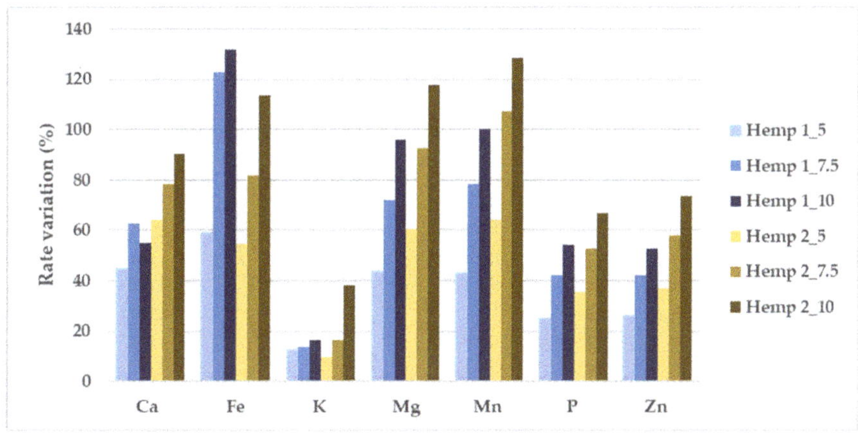

Figure 1. Percentage variation (%) of the hemp-enriched pasta samples vs the CTRL (durum wheat semolina cv. Ciclope).

3.3. Pasta Quality

Semolina particle size is a key quality factor in pasta making. Semolina used for pasta processing typically ranges in particle size from 550 to 150 μm [31]. The semolina used in this study had fine particles (<250 μm), similar to Hemp 2 flours. The addition of hemp flour significantly affected the sensory attributes of cooked pasta (Table 9). Substantial differences were recorded between the two hemp flours for all of the sensory attributes except bulkiness. In general, the use of Hemp 2, with a similar particle size to semolina, to enrich pasta yielded better results than Hemp 1. The absence of proteins such as glutenins and gliadins, responsible for the formation of gluten, has inevitably influenced the characteristics of the product. CTRL recorded the greatest OQS, mainly in terms of firmness and adhesiveness (Table 9); the enrichment of durum wheat pasta with non-gluten flours may affect the parameters, resulting in an increase in adhesiveness. The odor and taste of pasta fortified at different substitution levels resulted similarly to CTRL pasta, suggesting that the particle size of hemp did not affect these traits. The best results in terms of OQS were obtained using Hemp 2 at 7.5% substitution.

Table 9. Sensory characteristics of cooked pasta.

Sample	Bulkiness [I]	Firmness [I]	Adhesiveness [I]	Fibrous [I]	Color [II]	Odor [III]	Taste [III]	OQS [III]
CTRL	3.7 ± 0.012	5.7 ± 0.018	4.3 ± 0.014	5.0 ± 0.016	5.0 ± 0.016	4.7 ± 0.015	6.3 ± 0.021	6.0 ± 0.020
Hemp 1_5	3.0 ± 0.010	4.7 ± 0.015	5.0 ± 0.016	4.7 ± 0.015	4.7 ± 0.015	4.7 ± 0.015	5.3 ± 0.017	5.3 ± 0.017
Hemp 1_7.5	3.7 ± 0.012	4.0 ± 0.016	4.7 ± 0.015	5.7 ± 0.018	5.0 ± 0.016	5.7 ± 0.018	5.7 ± 0.018	5.3 ± 0.017
Hemp 1_10	4.5 ± 0.015	5.0 ± 0.013	3.5 ± 0.011	4.5 ± 0.015	5.0 ± 0.016	4.5 ± 0.015	6.0 ± 0.020	4.0 ± 0.013
Hemp 2_5	3.7 ± 0.012	4.0 ± 0.013	4.3 ± 0.014	5.0 ± 0.016	5.3 ± 0.017	5.3 ± 0.017	5.3 ± 0.017	5.3 ± 0.017
Hemp 2_7.5	5.3 ± 0.017	5.7 ± 0.018	6.0 ± 0.020	5.7 ± 0.018	6.3 ± 0.021	5.7 ± 0.018	6.7 ± 0.022	6.3 ± 0.021
Hemp 2_10	4.3 ± 0.014	5.0 ± 0.016	4.7 ± 0.015	5.7 ± 0.018	5.0 ± 0.016	4.7 ± 0.015	5.3 ± 0.017	5.0 ± 0.016
LSD $(p < 0.05)$	0.034	0.039	0.036	0.040	0.041	0.039	0.045	0.042
Percentage of substitution (S)								
CTRL	3.7 ± 0.012 b	5.7 ± 0.018 a	4.3 ± 0.014 c	5.0 ± 0.016 b	5.0 ± 0.016 b	4.7 ± 0.015 c	6.3 ± 0.021 b	6.0 ± 0.020 a
5%	3.3 ± 0.013 b	4.3 ± 0.014 b	4.6 ± 0.015 b	4.8 ± 0.015 c	5.0 ± 0.016 b	5.0 ± 0.016 b	5.3 ± 0.017 d	5.3 ± 0.017 b
7.5%	4.5 ± 0.015 a	4.8 ± 0.017 a	5.3 ± 0.018 a	5.6 ± 0.018 a	5.6 ± 0.019 a	5.7 ± 0.018 a	6.7 ± 0.020 a	5.8 ± 0.024 a
10%	4.4 ± 0.015 a	5.0 ± 0.015 a	4.0 ± 0.013 c	5.1 ± 0.017 b	5.0 ± 0.016 b	4.6 ± 0.015 c	5.7 ± 0.019 c	4.5 ± 0.015 c
Particle size (PS)								
Hemp 1	3.7 ± 0.012 a	4.5 ± 0.015 a	4.4 ± 0.014 b	4.9 ± 0.016 b	4.9 ± 0.016 b	4.9 ± 0.016 a	5.7 ± 0.018 a	4.9 ± 0.016 b
Hemp 2	4.4 ± 0.014 a	4.9 ± 0.016 b	4.8 ± 0.016 a	5.4 ± 0.018 a	5.5 ± 0.018 a	5.2 ± 0.017 a	5.8 ± 0.019 a	5.5 ± 0.018 a
ANOVA **Main effects**								
S	***	***	***	***	***	**	***	***
PS	***	***	***	ns	***	ns	ns	***
Interaction								
S × PS	***	***	***	***	***	*	***	***

Data are expressed as the means ± SD of triplicate experiments. ***; **, and * indicate significance at $p < 0.001$, $p < 0.01$, and $p < 0.05$, respectively, while "ns" is not significant. The differences among all samples were detected by LSD test at $p < 0.05$. Different small letters in a column for "PS" and "S" factors indicate statistical differences among samples ($p < 0.05$) (Student–Newman–Keuls test). OQS means "over quality score". I: 1—low sensation, -9—high sensation; II: 1—very clear, -9—very dark; III: 1—extremely unpleasant, -9—extremely pleasant; n = 24.

As regards to cooking quality, the replacement of durum wheat semolina with both types of hemp flours in the pasta statistically influenced ($p < 0.05$) the water absorption, most probably due to the high dietary fiber content and resultant strong water absorption capacity [13,15], while the optimal cooking time compared to the CTRL sample increased, particularly when using the Hemp 1 flours (Table 10).

The amount of solid substance lost in the cooking water (cooking loss) did not result as being influenced by the hemp particle size or the percentage of substitution, meaning that the hemp flours were well retained in the pasta. According to Sicignano et al., [31] the hydration of semolina with a wide range of particle sizes affects dough development and pasta quality; the different particle sizes between the semolina and Hemp 1 flours probably led to an over-hydration of the finer fraction and under-hydration of the coarser fraction, affecting the WA and OCT (Table 10).

Table 10. The cooking quality, optimum cooking time (OCT), water absorption capacity (WA), and cooking loss of fortified pasta samples.

	OCT (min)	WA	Cooking Loss
CTRL	12.0	75 ± 1.3	0.99 ± 0.001
Hemp 1_5	13.0 a	100 ± 2.8 a	0.99 ± 0.001
Hemp 2_5	11.5 b	75 ± 1.6 b	1.00 ± 0.001
Hemp 1_7.5	12.5 a	125 ± 3.2 a	0.99 ± 0.001
Hemp 2_7.5	12.0 a	125 ± 3.0 a	0.99 ± 0.001
Hemp 1_10	12.5 a	150 ± 3.9 a	0.99 ± 0.001
Hemp 2_10	12.5 a	100 ± 2.4 b	0.99± 0.001

Different small letters in a column. indicate statistical differences between the hemp flours used for the same concentration at $p < 0.05$; n = 3. Data are expressed as the means ± SD of triplicate experiments.

4. Discussion

The main aim of this study was to develop pasta fortified with variable percentages of hemp flour with different particle sizes. Our results revealed that the incorporation of hemp flour into the pasta formulas led to significant increases in the TPC and DPPH values, AAs, FA composition, and sensory qualities, in addition to obtain satisfactory properties and good cooking qualities, related to the percentage of hemp substitution used in the production recipe. The fibrous sensation recorded was probably due to the different particle sizes between the semolina and Hemp 1 flour and to the different percentages of water required for the dough development detected by farinograph analysis. Blends of semolina and Hemp 2 required less water (on average 58%) vs Hemp 1 (on average 61%) (data not shown).

The Folin–Ciocalteau method was used for raw material; similar polyphenol content values were obtained for Hemp 1 and Hemp 2 (6.38 and 6.35 mg GAE/g, respectively), while the TPC value was lower for Ciclope flour (2.45 mg GAE/g). TPC values increased with the addition of hemp flour in pasta. The highest increase in TPC was observed in pasta samples containing 10% hemp flour (4.92 ± 0.31 mg/GAE and 4.21 ± 0.35 mg/GAE for Hemp 1 and Hemp 2, respectively) while the lowest was recorded for the CTRL pasta sample (1.11 ± 0.18 mg/GAE).

However, despite the loss in amino acids and phenolics during cooking (about 40%), the enriched pasta still showed good antioxidant activity. The improvements to DPPH values were found to be higher in pasta formulas with the addition of hemp flour than in CTRL samples prepared with 100% Ciclope wheat flour. In fact, the supplementation of 10% hemp flour also enhanced the antioxidant activity (3.86 ± 0.07 mmol TE/100 g and 3.14 ± 0.06 mmol TE/100 g for Hemp 1 and Hemp 2 pasta samples, respectively) compared to the CTRL samples (1.14 ± 0.05 mmol TE/100 g).

This study also focused on the AAs composition of fortified pasta. The contents of some amino acids considered essential in the human diet can be low in wheat proteins, especially lysine and threonine. The preparation of a functional pasta enriched with variable percentages of hemp flour could affect the content of these two amino acids.

From observed data, the lysine content was found to be 0.16, 0.85–0.36 g/100 g in the CTRL, Hemp 1_10, and Hemp 2_10 flours respectively, while threonine was found to be 0.38, 0.72–1.17 g/100 g in the CTRL, Hemp 1_10, and Hemp 2_10 flours respectively). The total essential amino acids in the fortified pasta samples were 4.30 and 4.62 g/100 g (in Hemp 1_10 and in Hemp 2_10, respectively), while in the durum wheat pasta they were 2.05 g/100 g.

An increase in mono- and polyunsaturated fatty acids was also observed in fortified pasta. The total ω 3 contents varied between 15.9 and 15.3% in Hemp 2 pasta samples and between 15.9–14.7% in Hemp 1 pasta samples. Linoleic was the fatty acid present in the largest amount in Hemp 2 pasta samples (47.94%). The control sample had a lower amount of total ω 3 (4.02%).

Phenols represent the most relevant compounds found in hemp, including some phenylamides, phenolic acids, lignanamides, and flavonoids, such as flavonols, flavones, and flavanols. Current literature suggests that the long-term consumption of diets rich in phenolic compounds protects against certain cancers, cardiovascular diseases, type 2 diabetes, osteoporosis, lung damage, and neurodegenerative diseases [32,33].

An untargeted metabolomics-based approach was used to comprehensively screen and profile phenols in different hemp flours and pasta samples through UHPLC-ESI/QTOF-MS analysis. A total of 12 phenols were identified (Table 8) by comparison with the retention time, MS spectra, and accurate mass measurement obtained from the literature data [17–21] and by phenolic reference standards (Supplementary Table S3).

As is known, phenolic compounds are contained in plant materials in the free (PCs) but also in the insoluble bound form (BPs); these latter are covalently linked to the structural components of the cell wall. Particularly interesting are the implications of BPs in foods, in terms of bioaccessibility, transformation during digestion, and modulation of the gut microbiota [34]. For this reason, the study of the bound polyphenolic fraction (BPs) in pasta samples was of interest.

In raw samples, the contents of free phenolic compounds (PCs) were rather low in Ciclope flour; hydroxybenzoic and protocatechuic acids represented the most abundant phenolic compounds quantified in hemp flour samples. The results of the phenolic component after alkaline treatment of the matrices showed a higher amount of these in flour. Regarding fortified pastas, the phenol content is good, especially in the bound form. As could easily be predicted, the content of these increases with the percentage of replacement. Cannabisin B was only found in Hemp 1 as a bound phenol and at lower levels with respect to cannaflavin C. Cannaflavin A, cannaflavin B, and cannabisins A, B, and C are non-psychoactive molecules exclusively present in hemp plants that suppress PGE2 production in synovial membrane cells, exhibiting anti-inflammatory power 30-times stronger than acetylsalicylic acid [35]. Among hemp's exclusive lignanamides, these exhibit remarkable beneficial effects on human health [36].

Caffeoyltyramine and its phenolic amides, including cis-N-caffeoyltyramine and trans-N-caffeoyltyramin, are known to have anti-fungal, antioxidant, anti-inflammatory, and antihyperlipidemic activities [28]. As observed from Table 8, the phenols cannaflavin C, *p*-hydroxybenzoic acid, protocatechuic acid, hydroxycinnamic acid, and caffeic acid were predominantly found in Hemp 1 pasta in bound form.

In the pasta samples obtained by replacing the semolina with Hemp 2 flour, *p*-hydroxycinnamic acid, caffeic acid, *p*-hydroxybenzoic acid, protocatechuic acid, and trans-N-caffeoyltyramine always resulted to a greater extent in the bound form. As expected, the amount of free and bound phenols is strongly related to the fortification rate.

The incorporation of hemp flours to produce pasta has been studied by other research teams, but the contribution of mineral content was not investigated. Our results represent the first finding of the addition of hemp flours producing fortified pasta rich in minerals. However, in this study the presence of phytates in hemp flours was not detected, and even if the pasta samples could be an excellent source of mineral elements, their nutritional quality could be reduced.

Each increase in the addition of hemp flour resulted in an increase in pasta quality (water absorption, due to gluten dilution) and in the satisfactory organoleptic properties (until 7.5%); the fibrous sensation recorded at higher percentages of substitution was probably due to the different particle sizes between semolina and Hemp 1 flour and to the different percentages of water required for the dough development detected by farinograph analysis. Blends of semolina and Hemp 2 required less water (on average, doses of 58%) compared to Hemp 1 (on average, doses of 61%). Additionally, the contribution of hemp enrichment improved the micro-elemental composition of pasta (iron, potassium, magnesium, and phosphorus) compared to the control sample. The number of health claims relating to mineral elements which could be used for hemp flours could be high, but further studies are needed on the bioaccessibility and bioavailability in order to clarify the role of hemp flour as a dietary source of mineral elements due to the presence of phytates in the raw material.

5. Conclusions

The present study concluded that the incorporation of hemp flours with different particle sizes represents the best compromise between pasta properties and nutrient content; the addition of hemp flour to durum semolina cv Ciclope flour effectively increased the anti-radical potential. The best results in terms of overall quality score (OQS) were obtained using flour with a minor particle size (Hemp 2) for the preparation of the pasta. The maximum substitution level, which showed the best performance during the production process, was 7.5% for both types of hemp flour used. However, some differences can be highlighted: specifically, the pasta obtained by the replacement of 7.5% of the Ciclope semolina with Hemp 2 flour showed a better profile in terms of mineral salts and amino acids and a greater quantity of polyunsaturated fatty acids, while the pasta obtained from the same percentage of substitution but with Hemp 1 flour showed a better phenolic profile and TCP and better anti-radical activity. Hemp supplementation could be a potential option for the production of high-quality, nutritionally rich, low-cost pasta with good organoleptic properties.

In summary, the enrichment of wheat pasta with hemp flours is a very interesting future trend that enables more attractive pasta products to be obtained for consumers in terms of increased nutritional and pro-health value. On the basis of these results, hemp-enriched pasta may have great potential in the industry for the development of functional products. Further studies on the bioavailability of nutrients, the glycemic index, and the effects on the intestinal microbiota of Hemp 1- and Hemp 2-fortified pasta will be conducted in the near future.

Supplementary Materials: The following supporting information can be downloaded at: https://www.mdpi.com/article/10.3390/foods12040774/s1. Table S1: Retention time (min.), coefficient of determination (R2) and linear regression model of external standards used for phenolic compounds calibration; Table S2: Retention time (min.), coefficient of determination (R2) and linear regression model of external standards used for amino acids calibration; Table S3: Retention time (min.), formula, MS spectra, and accurate mass measurement of phenolic compound identified.

Author Contributions: Conceptualization, M.G.M. and V.D.S.; methodology, V.D.S. and M.G.M.; formal analysis, S.B., V.D.S., C.B. and S.A.; investigation, S.B. and F.S.; resources, N.V.; data curation, V.D.S., C.B., and M.G.M.; writing—original draft preparation, M.G.M., F.S., C.B. and S.B.; writing—review and editing, V.D.S., C.B. and M.G.M.; supervision, V.D.S. and M.G.M.; project administration, M.G.M.; funding acquisition, M.G.M. All authors have read and agreed to the published version of the manuscript.

Funding: CNR-DISBA project NutrAge (project number 7022).

Data Availability Statement: The data presented in this study are available on request from the corresponding author.

Acknowledgments: We would like to thank Giuseppe and Francesco Sammartino, project leaders of "The hemp supply chain-from hemp seed to table" SEMINCANTA", for supplying hemp flours.

Conflicts of Interest: The authors declare no conflict of interest.

References

1. Melilli, M.G.; Di Stefano, V.; Sciacca, F.; Pagliaro, A.; Bognanni, R.; Scandurra, S.; Virzì, N.; Gentile, C.; Palumbo, M. Improvement of Fatty Acid Profile in Durum Wheat Breads Supplemented with *Portulaca Oleracea* L. Quality Traits of Purslane-Fortified Bread. *Foods* **2020**, *9*, 764. [CrossRef] [PubMed]
2. Melilli, M.G.; Pagliaro, A.; Scandurra, S.; Gentile, C.; Di Stefano, V. Omega-3 Rich Foods: Durum Wheat Spaghetti Fortified with Portulaca Oleracea. *Food Biosci.* **2020**, *37*, 100730. [CrossRef]
3. Branca, F.; Arena, D.; Argento, S.; Frustaci, F.; Ciccarello, L.; Melilli, M.G. Recovery of Healthy Compounds in Waste Bracts of Globe Artichoke Heads. *Acta Hortic.* **2022**, 361–366. [CrossRef]
4. Merlino, M.; Tripodi, G.; Cincotta, F.; Prestia, O.; Miller, A.; Gattuso, A.; Verzera, A.; Condurso, C. Technological, Nutritional, and Sensory Characteristics of Gnocchi Enriched with Hemp Seed Flour. *Foods* **2022**, *11*, 2783. [CrossRef]
5. Del Gatto, A.; Mengarelli, C.; Foppa Pedretti, E.; Duca, D.; Pieri, S.; Mangoni, L.; Signor, M.; Raccuia, S.A.; Melilli, M.G. Adaptability of Sunflower (*Helianthus Annuus* L.) High Oleic Hybrids to Different Italian Areas for Biodiesel Production. *Ind. Crops Prod.* **2015**, *75*, 108–117. [CrossRef]
6. Garbetta, A.; D'Antuono, I.; Melilli, M.G.; Sillitti, C.; Linsalata, V.; Scandurra, S.; Cardinali, A. Inulin Enriched Durum Wheat Spaghetti: Effect of Polymerization Degree on Technological and Nutritional Characteristics. *J. Funct. Foods* **2020**, *71*, 104004. [CrossRef]
7. Sorrentino, G. Introduction to Emerging Industrial Applications of Cannabis (*Cannabis Sativa* L.). *Rend. Lincei* **2021**, *32*, 233–243. [CrossRef]
8. Callaway, J.C. Hempseed as a Nutritional Resource: An Overview. *Euphytica* **2004**, *140*, 65–72. [CrossRef]
9. Baldini, M.; Ferfuia, C.; Piani, B.; Sepulcri, A.; Dorigo, G.; Zuliani, F.; Danuso, F.; Cattivello, C. The Performance and Potentiality of Monoecious Hemp (*Cannabis Sativa* L.) Cultivars as a Multipurpose Crop. *Agronomy* **2018**, *8*, 162. [CrossRef]
10. Kriese, U.; Schumann, E.; Weber, W.E.; Beyer, M.; Brühl, L.; Matthäus, B. Oil Content, Tocopherol Composition and Fatty Acid Patterns of the Seeds of 51 Cannabis Sativa L. Genotypes. *Euphytica* **2004**, *137*, 339–351. [CrossRef]
11. Maurotti, S.; Mare, R.; Pujia, R.; Ferro, Y.; Mazza, E.; Romeo, S.; Pujia, A.; Montalcini, T. Hemp Seeds in Post-Arthroplasty Rehabilitation: A Pilot Clinical Study and an in Vitro Investigation. *Nutrients* **2021**, *13*, 4330. [CrossRef] [PubMed]
12. Teterycz, D.; Sobota, A.; Przygodzka, D.; Lysakowska, P. Hemp Seed (*Cannabis Sativa* L.) Enriched Pasta: Physicochemical Properties and Quality Evaluation. *PLoS One* **2021**, *16*, e0248790. [CrossRef] [PubMed]

13. Cardullo, N.; Muccilli, V.; Di Stefano, V.; Bonacci, S.; Sollima, L.; Melilli, M.G. Spaghetti Enriched with Inulin: Effect of Polymerization Degree on Quality Traits and α-Amylase Inhibition. *Molecules* **2022**, *27*, 2482. [CrossRef] [PubMed]

14. Di Stefano, V.; Pagliaro, A.; Del Nobile, M.A.; Conte, A.; Melilli, M.G. Lentil Fortified Spaghetti: Technological Properties and Nutritional Characterization. *Foods* **2021**, *10*, 4. [CrossRef] [PubMed]

15. Padalino, L.; Costa, C.; Conte, A.; Melilli, M.G.; Sillitti, C.; Bognanni, R.; Raccuia, S.A.; Del Nobile, M.A. The Quality of Functional Whole-Meal Durum Wheat Spaghetti as Affected by Inulin Polymerization Degree. *Carbohydr. Polym.* **2017**, *173*, 84–90. [CrossRef] [PubMed]

16. Petitot, M.; Boyer, L.; Minier, C.; Micard, V. Fortification of Pasta with Split Pea and Faba Bean Flours: Pasta Processing and Quality Evaluation. *Food Res. Int.* **2010**, *43*, 634–641. [CrossRef]

17. Shahidi, F.; Yeo, J. Insoluble-Bound Phenolics in Food. *Molecules* **2016**, *21*, 1216. [CrossRef]

18. Acosta-Estrada, B.A.; Gutiérrez-Uribe, J.A.; Serna-Saldívar, S.O. Bound Phenolics in Foods, a Review. *Food Chem.* **2014**, *152*, 46–55. [CrossRef]

19. Gong, E.S.; Li, B.; Li, B.; Podio, N.S.; Chen, H.; Li, T.; Sun, X.; Gao, N.; Wu, W.; Yang, T.; et al. Identification of Key Phenolic Compounds Responsible for Antioxidant Activities of Free and Bound Fractions of Blackberry Varieties' Extracts by Boosted Regression Trees. *J. Sci. Food Agric.* **2022**, *102*, 984–994. [CrossRef]

20. Gong, E.S.; Gao, N.; Li, T.; Chen, H.; Wang, Y.; Si, X.; Tian, J.; Shu, C.; Luo, S.; Zhang, J.; et al. Effect of In Vitro Digestion on Phytochemical Profiles and Cellular Antioxidant Activity of Whole Grains. *J. Agric. Food Chem.* **2019**, *67*, 7016–7024. [CrossRef]

21. Di Stefano, V.; Buzzanca, C.; Melilli, M.G.; Indelicato, S.; Mauro, M.; Vazzana, M.; Arizza, V.; Lucarini, M.; Durazzo, A.; Bongiorno, D. Polyphenol Characterization and Antioxidant Activity of Grape Seeds and Skins from Sicily: A Preliminary Study. *Sustainability* **2022**, *14*, 6702. [CrossRef]

22. Slinkard, K.; Singleton, V.L. Total Phenol Analysis: Automation and Comparison with Manual Methods. *Am. J. Enol. Vitic.* **1977**, *28*, 49–55. [CrossRef]

23. Melilli, M.G.; Pagliaro, A.; Bognanni, R.; Scandurra, S.; Di Stefano, V. Antioxidant Activity and Fatty Acids Quantification in Sicilian Purslane Germplasm. *Nat. Prod. Res.* **2019**, *34*, 26–33. [CrossRef]

24. Chiesa, L.M.; Ceriani, F.; Procopio, A.; Bonacci, S.; Malandra, R.; Panseri, S.; Arioli, F. Exposure to Metals and Arsenic from Yellow and Red Tuna Consumption. *Food Addit. Contam. Part A* **2019**, *36*, 1228–1235. [CrossRef] [PubMed]

25. Rice-Evans, C.; Miller, N.; Paganga, G. Antioxidant Properties of Phenolic Compounds. *Trends Plant Sci.* **1997**, *2*, 152–159. [CrossRef]

26. Siano, F.; Moccia, S.; Picariello, G.; Russo, G.; Sorrentino, G.; Di Stasio, M.; La Cara, F.; Volpe, M. Comparative Study of Chemical, Biochemical Characteristic and ATR-FTIR Analysis of Seeds, Oil and Flour of the Edible Fedora Cultivar Hemp (*Cannabis Sativa* L.). *Molecules* **2018**, *24*, 83. [CrossRef]

27. Pojić, M.; Mišan, A.; Sakač, M.; Dapčević Hadnađev, T.; Šarić, B.; Milovanović, I.; Hadnađev, M. Characterization of Byproducts Originating from Hemp Oil Processing. *J. Agric. Food Chem.* **2014**, *62*, 12436–12442. [CrossRef]

28. Pannico, A.; Kyriacou, M.C.; El-Nakhel, C.; Graziani, G.; Carillo, P.; Corrado, G.; Ritieni, A.; Rouphael, Y.; De Pascale, S. Hemp Microgreens as an Innovative Functional Food: Variation in the Organic Acids, Amino Acids, Polyphenols, and Cannabinoids Composition of Six Hemp Cultivars. *Food Res. Int.* **2022**, *161*, 111863. [CrossRef]

29. Izzo, L.; Castaldo, L.; Narváez, A.; Graziani, G.; Gaspari, A.; Rodríguez-Carrasco, Y.; Ritieni, A. Analysis of Phenolic Compounds in Commercial *Cannabis Sativa* L. Inflorescences Using UHPLC-Q-Orbitrap HRMS. *Molecules* **2020**, *25*, 631. [CrossRef]

30. Alonso-esteban, I.; Torija-isasa, M.E.; S, M.D.C. Mineral Elements and Related Antinutrients, in Whole and Hulled Hemp (*Cannabis Sativa* L.) Seeds. *J. Food Compos. Anal.* **2022**, *109*. [CrossRef]

31. Sicignano, A.; Masi, P.; Cavella, S. From Raw Material to Dish: Pasta Quality Step by Step. *J. Sci. Food Agric.* **2015**, *95*, 2579–2587. [CrossRef] [PubMed]

32. Andre, C.; Larondelle, Y.; Evers, D. Dietary Antioxidants and Oxidative Stress from a Human and Plant Perspective: A Review. *Curr. Nutr. Food Sci.* **2010**, *6*, 2–12. [CrossRef]

33. Prociuk, M.A.; Edel, A.L.; Richard, M.N.; Gavel, N.T.; Ander, B.P.; Dupasquier, C.M.C.; Pierce, G.N. Cholesterol-Induced Stimulation of Platelet Aggregation Is Prevented by a Hempseed-Enriched Diet, in the Special Issue Bridging the Gap: Where Progress in Cardiovascular and Neurophysiologic Research Me. *Can. J. Physiol. Pharmacol.* **2008**, *86*, 153–159. [CrossRef] [PubMed]

34. Rocchetti, G.; Gregorio, R.P.; Lorenzo, J.M.; Barba, F.J.; Oliveira, P.G.; Prieto, M.A.; Simal-Gandara, J.; Mosele, J.I.; Motilva, M.; Tomas, M.; et al. Functional Implications of Bound Phenolic Compounds and Phenolics–Food Interaction: A Review. *Compr. Rev. Food Sci. Food Saf.* **2022**, *21*, 811–842. [CrossRef] [PubMed]

35. Barrett, M.L.; Scutt, A.M.; Evans, F.J. Cannaflavin A and B, Prenylated Flavones From *Cannabis Sativa* L. *Experientia* **1986**, *42*, 452–453. [CrossRef] [PubMed]

36. Chen, T.; He, J.; Zhang, J.; Li, X.; Zhang, H.; Hao, J.; Li, L. The Isolation and Identification of Two Compounds with Predominant Radical Scavenging Activity in Hempseed (Seed of *Cannabis Sativa* L.). *Food Chem.* **2012**, *134*, 1030–1037. [CrossRef] [PubMed]

Article

Functional Bread Produced in a Circular Economy Perspective: The Use of Brewers' Spent Grain

Antonietta Baiano *, Barbara la Gatta, Mariacinzia Rutigliano and Anna Fiore

Dipartimento di Scienze Agrarie, Alimenti, Risorse Naturali e Ingegneria (DAFNE), University of Foggia, 71122 Foggia, Italy
* Correspondence: antonietta.baiano@unifg.it; Tel.: +39-(0)881589249

Abstract: Brewers' spent grain (BSG) is the main by-product of the brewing industry, corresponding to ~85% of its solid residues. The attention of food technologists towards BSG is due to its content in nutraceutical compounds and its suitability to be dried, ground, and used for bakery products. This work was aimed to investigate the use of BSG as a functional ingredient in bread-making. BSGs were characterised for formulation (three mixtures of malted barley and unmalted durum (Da), soft (Ri), or emmer (Em) wheats) and origin (two cereal cultivation places). The breads enriched with two different percentages of each BSG flour and gluten were analysed to evaluate the effects of replacements on their overall quality and functional characteristics. Principal Component Analysis homogeneously grouped BSGs by type and origin and breads into three sets: the control bread, with high values of crumb development, a specific volume, a minimum and maximum height, and cohesiveness; Em breads, with high values of IDF, TPC, crispiness, porosity, fibrousness, and wheat smell; and the group of Ri and Da breads, which have high values of overall smell intensity, toasty smell, pore size, crust thickness, overall quality, a darker crumb colour, and intermediate TPC. Based on these results, Em breads had the highest concentrations of nutraceuticals but the lowest overall quality. Ri and Da breads were the best choice (intermediate phenolic and fibre contents and overall quality comparable to that of control bread). Practical applications: the transformation of breweries into biorefineries capable of turning BSG into high-value, low-perishable ingredients; the extensive use of BSGs to increase the production of food commodities; and the study of food formulations marketable with health claims.

Keywords: brewers' spent grain; bread-making; circular economy; common wheat; dietary fibre; durum wheat; emmer; phenolics; proteins; sustainable food production

Citation: Baiano, A.; la Gatta, B.; Rutigliano, M.; Fiore, A. Functional Bread Produced in a Circular Economy Perspective: The Use of Brewers' Spent Grain. *Foods* **2023**, *12*, 834. https://doi.org/10.3390/foods12040834

Academic Editors: Donatella Bianca Maria Ficco and Grazia Maria Borrelli

Received: 30 January 2023
Revised: 11 February 2023
Accepted: 13 February 2023
Published: 15 February 2023

1. Introduction

Topics related to a circular and sustainable economy are gaining more and more attention as means to combat climate change, reduce the use of fossil fuels and other natural resources, recover all products from resources without generating waste, and thus creating new opportunities for long-term economic growth [1]. In this perspective, the sustainable management of by-products and waste is an integral aspect of the circular economy.

Brewers' spent grain (BSG) is a lignocellulosic material representing the main by-product generated from the brewing industry, since it corresponds to about 85% of the total solid residues [2]. It has been estimated that between 14 and 20 kg of wet BSG are obtained per 100 L of finished beer [3]. The average annual global production is estimated to be around 39 million tons, with ~3.4 million tons (which could become 8.5 be the end of 2030) produced in the European Union [4]. In Italy, a production of 188 thousand tons/year of BSG is estimated. Thirty percent of them are reused, mainly in the livestock sector. Furthermore, the BSG market value as cattle feed is very low, being around ~€35/ton [5]. The remaining 70% must be disposed of by breweries, with costs between 75 and 100 €/t (in the EU) and environmental impacts deriving from the release of 513 kg of CO_2 equivalent of greenhouse gases per ton of landfilled BSG [6]. The research of Bolwig et al. [7] highlighted that an increasing number of farmers are refusing to use BSG as animal feed. This situation, together with the absence of on-site storage facilities, slows down the BSG disposal, sometimes forcing breweries to halt production.

The composition of BSG paves the way for the possibility of recovering and recycling them as secondary starting materials to produce a variety of value-added goods, included functional foods. Although BSG composition depends on several factors including the type of grains (mainly malted barley and wheat, but also unmalted grains and other starchy raw materials) and the malting and mashing methods [8], it includes on average: hemicellulose (~20%), cellulose (15.2–28.7%), lignin (3.35–21%), proteins (18.5–24.7%), lipids (8.4%), starch (5.3%), and ash (3.7%) [9]. From a nutritional standpoint, the interest towards BSG is mainly due to its content in phenolic compounds and soluble and insoluble fibre components such as β-glucans, and the less known arabinoxylans. Since most of the phenolic compounds of cereals are contained in the husk, BSG is rich in valuable antioxidant compounds such as phenolic acids (both benzoic and cinnamic acid derivatives), flavonoids, tannins, proanthocyanidins, and amino-phenolics and, for this reason, can be considered a suitable target for exploitation as a health promoting food supplement [10,11]. β-glucans and arabinoxylans are known to exert physiological functions in the gastro-intestinal tract. In particular, β-glucans are implicated in lowering plasma cholesterol and reducing glycaemic index and the risk of colon cancer, while arabinoxylans exert positive effects such as the production of short-chain fatty acids, the reduction of serum cholesterol, the enhancement of calcium and magnesium adsorption and, because of their esterification with hydroxycinnamic acids, they may also have some antioxidant properties [12]. Regardless of their health effects, the incorporation of arabinoxylans or β-glucans into wheat flour can also negatively influence bread-making. Loaf volume is depressed as a consequence of the effects of gluten dilution; the disruption of the gluten network related to the presence of fibres; the ability of β-glucans to bind water thus reducing its availability for the gluten network development; and, consequently, the gas holding capacity [13,14]. Although the use of BSG in bread-making has been extensively investigated, most research has been focused on the effects of the level of addition of spent grains on the characteristics of doughs and breads [15–18]. Topics related to the effects of nature (composition and brewing style) and geographical origin of BSG on concentrations of functional compounds in both spent grains and final products have been thoroughly investigated in very few studies and generally not together [19,20].

Our work was aimed at investigating the feasibility of using BSGs from the brewing of Belgian style white beers as functional ingredients in bread-making from a perspective of the sustainable use of natural resources, the reduction of waste disposal, and the re-use of by-products. BSGs were firstly characterised in order to evaluate their content in nutraceutical compounds as a function of both formulation (three different mixtures of malted barley and unmalted cereals) and geographical origin (two different places of cereal cultivation) of the starting brewing materials. To our knowledge, this is the first time that such an approach has been addressed. Successively, the bread samples obtained by replacing the Manitoba wheat flour with two different percentages of each BSG flour were submitted to physical, chemical, image, and sensory analyses with the aim of evaluating the effects of such replacements on their overall quality and functional characteristics.

2. Materials and Methods

2.1. Production of Cereals Used in Brewing

Barley malt cv. Fortuna was supplied by Agroalimentare Sud (Melfi, Potenza, Italy). The unmalted cereals, i.e., durum wheat (*Triticum durum*) cv. Dauno III, soft wheat (*Triticum aestivum*) cv. Risciola, and emmer (*Triticum dicoccum*) were produced from seeds supplied by CREA-CI Research Centre for Cereal and Industrial Crops (Foggia, Italy) cultivated under the same agricultural practices in two areas of Puglia region, namely Daunia (Soc. Cooperativa Agricola Valleverde, Bovino, Foggia, Italy) and Salento (Birra Salento

Societa' Semplice Agricola Di Leo Consolata and Co, Leverano, Lecce, Italy). The two areas strongly differ for geographical position and climate as follows:

Bovino: Height above sea level: 620 m; Latitude 41.250944 N; Longitude 15.342204 E; Climate zone: E; Degree days: 2243;

Leverano: Height above sea level: 37 m; Latitude 40.289737 N; Longitude 18.001176 E; Climate zone: C; Degree days: 1197.

2.2. Brewing Procedures and Obtainment of BSG Flour

BSGs were collected after mashing during brewing of craft Belgian style white beers performed in a 30 L Braumeister system (Speidel Tank-und Behälterbau GmbH, Ofterdingen, Germany) as follows: cereal mixtures made of 65% barley malt and 35% unmalted cereals were crushed with a 2-roller mill (Albrigi Luigi, Stallavena, Verona, Italy) under mill gaps of 0.5 ± 0.1 mm; water-to-cereal ratio was set to 20:5 (v/w); mash-in temperature was 52 °C, followed by a 13 min stand at 55 °C, a 35 min stand at 64 °C, a 20 min stand at 72 °C, and a final mash-off at 78 °C for 15 min; 7 L of sparge water at 80 °C was passed through the grain bed. Table 1 lists the resulting six types of BSGs differing for type and geographical origin of the unmalted cereals in the mixtures.

Table 1. BSG composition (%) and geographical origin.

BSGs	Origin: Daunia			Barley Malt cv. Fortuna	Origin: Salento		
	Soft Wheat cv. Risciola	Durum Wheat cv. Dauno III	Emmer		Soft Wheat cv. Risciola	Durum Wheat cv. Dauno III	Emmer
Ri-D	35			65			
Ri-S				65	35		
Da-D		35		65			
Da-S				65		35	
Em-D			35	65			
Em-S				65			35

Since freezing techniques are not appropriate to preserve BSG quality as they affect its arabinose content [21], after separation of wort, BSG samples were dried at 20 °C for 24 h through a Forced-air-drying system. The dried BSGs were ground (particle size distribution: <15% higher than 500 μm; 35–45% between 500 and 250 μm; 30–40% between 250 and 125 μm; <15% between 125 and 63 μm; <1% lower than 63 μm) with a blade mill, taking care to keep the temperature below 35 °C. The resulting BSG flours were sealed in polyethylene bags and stored at −20 °C until further use and analysis.

The following ingredients were used in bread formulations: BSGs, Manitoba soft wheat flour type 0 (COOP, Casalecchio di Reno, Italy), water, extra-virgin olive oil (Pazienza, Foggia, Italy), gluten (Elgranero, Madrid, Spain), sodium chloride (Compagni Europea Sali, Margherita di Savoia, Italy), and dehydrated *Saccharomyces cerevisiae* yeast (Cameo, Desenzano del Garda, Italy).

2.3. Formulations and Bread-Making

Thirteen types of bread were produced according to the formulations reported in Table 2: a control, made of 100% Manitoba soft wheat flour type 0 (Figure 1a); and twelve samples, obtained by replacing Manitoba flour with two different amounts (20 and 25%) of each of the six BSG flours (Figure 1b). When 20% of Manitoba flour was replaced, gluten was also added in an amount corresponding to 3% of the total weight of flours (BSG + Manitoba), while at 25% of flour substitution, a corresponding 4% of gluten was added, always referred to in the total amount of flours. Relying on the results of preliminary tests, we decided to use the same water amount for all formulations, since it was sufficient to ensure the right level of hydration and the correct development of the dough without changing the production times. For each type of breads, three technological replicates were performed.

Table 2. Ingredients (g) used in bread formulations.

Bread Samples	BSG Flour [1]	Gluten [2]	Manitoba Flour (g)	Water	EVOO	NaCl	Dehydrated *Saccharomyces cerevisiae* Yeast
Control	-	-	500	350	40	12	7
Ri-D-20G3	100 (20%)	15 (3%)	385	350	40	12	7
Ri-D-25G4	125 (25%)	20 (4%)	355	350	40	12	7
Ri-S-20G3	100 (20%)	15 (3%)	385	350	40	12	7
Ri-S-25G4	125 (25%)	20 (4%)	355	350	40	12	7
Da-D-20G3	100 (20%)	15 (3%)	385	350	40	12	7
Da-D-25G4	125 (25%)	20 (4%)	355	350	40	12	7
Da-S-20G3	100 (20%)	15 (3%)	385	350	40	12	7
Da-S-25G4	125 (25%)	20 (4%)	355	350	40	12	7
Em-D-20G3	100 (20%)	15 (3%)	385	350	40	12	7
Em-D-25G4	125 (25%)	20 (4%)	355	350	40	12	7
Em-S-20G3	100 (20%)	15 (3%)	385	350	40	12	7
Em-S-25G4	125 (25%)	20 (4%)	355	350	40	12	7

[1] The percentage of replacement of Manitoba flour with BSG is reported in brackets. [2] The percentage of gluten reported in brackets is referred to the total amount of flours (Manitoba + BSG).

Figure 1. Slices of (**a**) control bread and (**b**) the 12 functional breads produced.

Loaves of regular shape were produced using a mould of 20 cm length and 11 cm width. In order to standardise the production, the entire procedure was performed in a bread-making machine (Zero-Glu Pro, Imetec, Azzano S. Paolo, Italy) using the program number 14 (total duration 3 h and 12 min) that includes the following steps: kneading, 22 min; 1st leavening, 40 min; 1st stirring, 5 s; 2nd leavening, 73 min; 2nd stirring, 5 s; 3rd leavening, 50 min; cooking, 47 min.

2.4. Analysis of BSGs

Moisture and ash contents were determined according to the AACC Methods 44-15.02 and 08-01.01, respectively, and expressed as % [22].

BSG proteins were extracted and analysed through electrophoretic and chromatographic techniques. Total proteins from BSG samples (~1 g) were extracted for 3 h with 10 mL of a buffer containing Tris-HCl 0.0625 M (pH 6.8), SDS 2%, and glycerol 10% (v/v) and then the extracts were centrifuged at $13{,}000\times g$ for 10 min at 25 °C. Supernatants were carefully removed and stored at -20 °C until the analysis. Protein content was determined using Quick Start™ Bradford Protein Assay (Bio-Rad, Hercules, CA, USA), according to the supplier instructions. Electrophoretic separation (SDS-PAGE) was performed through a Mini-PROTEAN Tetra system electrophoresis cell (Bio-Rad, Hercules, CA, USA), according to la Gatta et al. [23] as follows: loading of 10 μg of sample onto the gel; running buffer consisting of Tris 0.025 M, Glycine 20 mM, 1% SDS; application of a potential of 200 V for one hour; use of a Prestained SDS-PAGE standard (Bio-Rad, Richmond, CA, USA). The gels were stained with a 0.25% (w/v) solution of Coomassie Brilliant Blue G-250 (CBB), fixed with a 7% (v/v) solution of acetic acid and 40% (v/v) of methanol, and destained with water. The polymeric components of BSG proteins were analysed using a two-step extraction procedure. An aliquot of each sample (20 μL) was analysed according to la Gatta et al. [23] using a Biosep SEC-S4000 column (300 × 7.8 mm, Phenomenex, Torrence, CA, USA). The fractions of total unextractable polymeric proteins (tUPP%) and of large unextractable polymeric proteins (lUPP%) were calculated according to Kuktaite et al. [24]. In addition, the incidence of unextractable (both monomeric and polymeric forms) proteins on the total protein content was also calculated (UP%).

The extraction of total phenolics was performed according to Gandolpho et al. [25] with some modifications. Briefly, 1 g of each BSG was suspended in a hydroalcoholic solution (ethanol 58%, solid-to-liquid ratio 1 g to 30 mL) and treated in an ultrasonic bath according to the following conditions: 37 kHz, 30 °C, 30 min. After extraction, the solutions were centrifuged ($2000\times g$, 25 min, 20 °C) and the supernatants were removed and filtered through nylon filter (0.45 μm). The filtered extracts were analysed for their phenolic content and antioxidant activity. TPC was determined using the Folin–Ciocalteu reagent [26] and expressed as mg of gallic acid/g of dry matter. The phenolic profile was analysed by a HPLC system equipped with a diode array detector (Agilent 1100 Liquid Chromatograph, Santa Clara, CA, USA) according to Aliakbarian et al. [27] using a 100 × 4.6 mm × 3 μm RP-C18 Gemini column (Phenomenex, Aschaffenburg, Germany). The wavelengths used were 250, 280, and 320 nm. Identification of phenolics was made taking into account their retention times and comparing their spectra with those of standard materials while their quantification (mg/g dm) was obtained by comparing their peak areas with those of standard curves. The antioxidant activity was measured using 2,2-diphenyl-1-picrylhydrazyl (DPPH) free radical scavenging [28] and results were expressed as mg of Trolox (6-hydroxy-2,5,7,8-tetramethylchroman2-carboxylic acid) per g of dry matter.

Soluble and insoluble dietary fibres were determined according to AACC Methods 32-05.01 and 32-21.01 [22] using the K-TDFR-200A Megazyme kits (Neogen Europe Ltd., Ayr, Scotland) and expressed as %.

2.5. Analysis of Functional Bread

Previously ground bread slices were submitted to the determination of moisture, ash, TPC, antioxidant activity, phenolic profile, and dietary fibre, and to the protein characterization as already described for BSG in Section 2.3.

In order to evaluate the chromatic and structural characteristics of the bread samples, the images of the slices were acquired with an Epson scanner (mod. XP-3100, Cinisello Balsamo, Italy) at a resolution of 1200 dpi and saved in the tiff format. A ruler was placed next to the samples in order to convert pixels to centimetres through the measurement tools of the image analysis software. The image processing was performed with ImageJ ver. 1.52a, a free public domain software developed at the National Institutes of Health (USA).

The crust and crumb colours were expressed according to the coordinates of the colour space defined by the International Commission on Illumination: L*, lightness/brightness from black to white on a scale of zero to 100 and a* and b*, which represent chromaticity with no specific numeric limits. Negative and positive a* values correspond with green and red, respectively, while negative and positive b* values correspond with blue and yellow, respectively. The following structural characteristics were determined: slice height; crumb specific volume (cm^3/g); number of pore/cm^2; average pore size (mm^2); porosity %, i.e., the surface of the slice occupied by pores; pore circularity, calculated as 4π*area/$perimeter^2$ (it ranges between 1.0 and 0.0, with a value of 1.0 indicating a perfect circle and value near to 0.0 indicating increasingly elongated shapes).

A panel of 10 trained judges (5 females and 5 males) between 20 and 65 years of age, experienced in the sensory evaluation of baked foods, carried out a Quantitative Descriptive Analysis (QDA) in a sensory laboratory equipped with booths at 23 ± 1 °C, under a white light. All samples were assessed 2 h after baking. A slice of each bread sample was randomly labelled with a three-digit numeric code and provided to assessors in a double-blind presentation to avoid any expectation error. The attributes were selected among those found in literature and generated by the same panel so as to give an overall description of the products without overlapping. As a result of this selection, the profile sheet used by panellists for the QDA of bread samples included 5 visual (for crust: colour and thickness; for crumb: colour, pore size, and development), 6 olfactory (on crust and crumb together: overall, freshly baked bread, wheat, and malty smell; toasty smell on crust; yeast smell on crumb), 4 gustatory (sweetness, saltiness, and acidity/sourness of crumb; crust bitterness), and 6 tactile (for crust: hardness and crispiness; for crumb: resistance to chewing, cohesiveness, graininess/fibrousness, stickiness) attributes to evaluate on numeric category scales. Unipolar category scales were used for all the attributes with the exception of tactile attributes, which were evaluated on bipolar category scales. The attribute definitions were retrieved from Callejo [29] and supplied to the panellists that were also asked to evaluate the overall quality of each bread, i.e., the comprehensive likeness/dislikeness of the sample expressed considering all the sensory attributes previously evaluated [30]. The panellists rated the intensity of each parameter from 0 (minimum scale) to 9 (maximum scale). Judges were instructed to rinse their mouths between samples with natural water. To prevent sensory fatigue, the samples were divided into two subsets, one completely evaluated in the morning tasting session and another one completely evaluated in the session of the afternoon of the same day.

2.6. Statistical Analysis

Each analysis was replicated at least three times, with the exception of the chromatographic analyses of proteins, which were performed in duplicates, and the image analysis, with five acquisitions for each sample. The averages and the standard deviations were calculated. Analysis of Variance (ANOVA) and LSD test were applied (p-value < 0.01) to study the single and interactive effects of type of BSG and geographical origin (of the unmalted cereal in the mixtures used in brewing) on the characteristics of BSGs. The same statistical analyses were applied to evaluate the single and interactive effects of type of BSG, geographical origin (of the unmalted cereal in the mixtures used in brewing), and amounts of BSG-gluten used in the formulation of the characteristics of the functional breads. Principal Component Analysis (PCA) was applied to evaluate the possibility of homogeneously grouping both BSGs and breads according to the experimental data. Pearson correlation coefficients at p-value < 0.01 was applied to individuate significant correlations among pairs of characteristics of bread samples. The statistical analyses were carried out using Excel software V. 14.0.0 for Mac and Statistica for Windows V. 7.0. (Statsoft, Tulsa, OK, USA).

3. Results and Discussion

3.1. BSGs

3.1.1. Physicochemical Characteristics of BSGs

Moisture was determined to check that drying was successful and to verify that the amount of water indicated in the bread formulations (Table 2) ensured the right dough hydration. According to the results of Table 3, neither type of BSGs nor their origin significantly affected the residual moisture. The mean moistures ranged from 2.9 to 3.5%, thus ensuring the inhibition of microbial growth [31]. In agreement with the findings of previous studies, BSG is a good source of ash with a great variability among samples depending on both the type and geographical origin of BSG [32,33]. The highest ash content was observed in Em spent grains coming from Daunia area while the lowest concentrations were detected in **Ri** samples (Table 3). The differences among BSG types depended on the use of not de-hulled emmer while the influence exerted by the geographical origin was due to the substantial differences in soil and climate conditions between Daunian sub-Apennines (D) and Salento peninsula (S), being the same agricultural practices adopted [34,35].

Table 3. Single and interactive effects of type of BSG and geographical origin of the cereal mixtures used in brewing on physical and chemical characteristics of BSGs.

BSGs	Moisture (%)	Ash (%)	% tUPP	% IUPP	UP%	IDF%	SDF%
			Interactive effects (Type of BSG Geographical Origin of Cereal Mixtures)*				
Ri-D	3.3 ± 0.7 [a]	3.0 ± 0.0 [a]	34.54 ± 1.30 [bc]	52.54 ± 3.45 [c]	28.23 ± 0.77 [cd]	25.25 ± 0.05 [c]	2.06 [f]
Ri-S	3.4 ± 0.0 [a]	3.1 ± 0.0 [a]	36.97 ± 2.24 [c]	42.13 ± 5.16 [b]	25.33 ± 1.30 [ab]	27.25 [c]	0.64 [b]
Da-D	3.3 ± 0.3 [a]	4.8 ± 0.3 [b]	28.82 ± 0.21 [a]	25.96 ± 0.46 [a]	26.05 ± 0.55 [bc]	23.97 [b]	1.04 [d]
Da-S	3.5 ± 0.4 [a]	2.9 ± 0.0 [a]	30.45 ± 0.39 [ab]	20.91 ± 0.46 [a]	23.30 ± 0.01 [a]	21.58 [a]	0.42 [a]
Em-D	3.0 ± 0.5 [a]	6.7 ± 0.4 [d]	37.07 ± 1.03 [c]	49.96 ± 0.50 [bc]	28.79 ± 0.35 [d]	28.39 [d]	1.43 [e]
Em-S	2.9 ± 0.4 [a]	5.6 ± 0.5 c	90.82 ± 0.22 [d]	87.65 ± 1.40 [d]	32.23 ± 0.33 [e]	34.44 [e]	0.90 [c]
Significance	n.s.	*	*	*	*	*	*
			Single Effect (Type of BSG)				
Ri	3.3 [a]	3.1 [a]	35.76 [b]	23.43 [b]	26.78 [b]	27.25 [b]	1.35 [c]
Da	3.4 [a]	3.9 [b]	29.64 [a]	47.33 [a]	24.67 [a]	22.78 [a]	0.73 [a]
Em	2,9 [a]	6.2 [c]	63.94 [c]	68.80 [c]	30.51 [c]	31.41 [c]	1.17 [b]
Significance	n.s.	*	*	*	*	*	*
			Single Effect (Geographical Origin of Cereal Mixtures)				
D	3.2 [a]	4.8 [b]	33.48 [a]	42.82 [a]	27.69 [a]	26.54 [a]	1.51 [b]
S	3.3 [a]	3.9 [a]	52.75 [b]	50.23 [b]	26.95 [a]	27.76 [b]	0.65 [a]
Significance	n.s.	*	*	*	n.s.	*	*

In column, different letters indicate significant differences at $p < 0.01$; * $p < 0.01$; n.s.: not significant.

BSG showed the common chromatographic profiles (Figure 2) of SDS-extractable and SDS-unextractable cereal proteins with the elution of four main peaks between 7 and 14 min: peak one, related to the elution of Large Polymeric Proteins (LPP), peak two related to the elution of Small Polymeric Proteins (SPP), peak three related to the elution of Large Monomeric proteins (LMP), and peak four related to the elution of Small Monomeric Proteins (SMP) [23,24]. The importance of studying the size distribution of polymeric proteins is due to its influence on the technological properties of flour [36]. Since UPP% of wheat cultivars is affected by both genetic and environment, this index has been suggested as an effective evaluation parameter in breeding programs of locally adapted wheats intended for bread-making, even considering the effects of environmental factors such as temperature, and nitrogen application and timing [37–39]. In agreement with the finding of a previous work [36], the proportion of the total unextractable polymeric proteins (tUPP%) in BSG samples was significantly affected by genotype and environment, whether considered individually or not. In addition, our results highlighted that the interactions of the two factors was also statistically significant, and that some considerations can be extended to the fate of the large unextractable polymeric fraction (lUPP%). The highest tUPP% and lUPP% were quantified in Em spent grains from Salento while the lowest values were found in Da spent grains from Daunia. Regarding the effects of BSG type,

Table 4. Effects of type of BSG and geographical origin of the cereal mixtures used in brewing on phenolic contents and antioxidant activity of BSGs.

BSGs	TPC (mg/g d.m.)	Antioxidant Activity (mg Trolox/g d.m.)	Phenolics (mg/g d.m.)											
			Gallic Acid	Vanillic Acid	Caffeic Acid	p-Coumaric Acid	4-Hydroxybenzoic Acid	Ferulic Acid	Epicatechin	Epicatechin Gallate	Resveratrol	Quercetin	Kaempferol
					Interactive effects (Type of BSG* Geographical origin of Cereal Mixtures)									
Ri-D	3.129 ± 0.107 [a]	2.88 ± 0.14 [c]	0.016 [c]	0.010	0.178 [bc]	0.018 [b]	0.001 [b]	0.047 [d]	0.164 [bc]	0.0004 [e]	0.031 [c]	0.029 [b]	1.018 ± 0.022 [c]
Ri-S	3.114 ± 0.326 [a]	2.89 ± 0.11 [c]	0.017 [d]	0.010	0.016 [abc]	0.018 [b]	n.d. [a]	0.046 [a]	0.160 [a]	0.002 [b]	0.030 [b]	0.029 [a]	0.914 ± 0.069 [b]
Da-D	3.664 ± 0.085 [a]	1.24 ± 1.00 [ab]	0.016 [c]	0.010	0.014 ± 0.002 [a]	0.017 [a]	0.001 [b]	0.046 [bc]	0.165 [c]	0.003 [c]	0.307 [a]	0.029 [a]	1.382 ± 0.020 [d]
Da-S	3.274 ± 0.187 [a]	1.39 ± 0.20 [ab]	0.015 [a]	0.010	0.018 [cd]	0.017 [a]	0.003 [d]	0.046 [c]	0.163 [b]	0.001 [a]	0.031 [b]	0.029 [b]	0.900 ± 0.179 [b]
Em-D	4.868 ± 0.821 [b]	0.64 ± 0.10 [a]	0.017 [d]	0.010	0.020 [d]	0.019 [c]	0.005 [e]	0.048 [e]	0.168 [d]	0.002 [c]	0.032 [e]	0.031 [d]	1.807 ± 0.016 [e]
Em-S	4.789 ± 0.191 [b]	1.96 ± 0.11 [bc]	0.015 [b]	0.006 ± 0.004	0.015 [ab]	0.018 [c]	0.002 [c]	0.048 [e]	0.1645 [c]	0.003 [d]	0.032 [d]	0.031 [c]	0.795 ± 0.039 [a]
Significance	*	*	*	n.s.	*	*	*	*	*	*	*	*	*
					Single Effect (Type of BSG)									
Ri	3.122 [a]	2.88 [b]	0.017 [c]	0.010	0017 [ab]	0.018 [b]	0.001 [a]	0.047 [b]	0.162 [a]	0.003 [b]	0.031 [b]	0.030 [a]	0.966 [a]
Da	3.469 [b]	1.32 [a]	0.015 [a]	0.010	0.016 [a]	0.017 [a]	0.002 [b]	0.046 [a]	0.164 [b]	0.002 [a]	0.030 [a]	0.030 [a]	1.116 [b]
Em	4.829 [c]	1.30 [a]	0.016 [b]	0.010	0.018 [b]	0.019 [c]	0.003 [c]	0.048 [c]	0.166 [c]	0.003 [b]	0.032 [c]	0.031 [b]	1.301 [c]
Significance	*	*	*	n.s.	*	*	*	*	*	*	*	*	*
					Single Effect (Geographical Origin of Cereal Mixtures)									
D	3.8870	1.59	0.017 [b]	0.010	0.017	0.018	0.002 [b]	0.047 [b]	0.165 [b]	0.003 [b]	0.032 [b]	0.031 [b]	1.386 [b]
S	3.726	2.08	0.016 [a]	0.010	0.017	0.018	0.001 [a]	0.046 [a]	0.162 [a]	0.002 [a]	0.031 [a]	0.030 [a]	0.870 [a]
Significance	n.s.	n.s.	*	n.s.	n.s.	n.s.	*	*	*	*	*	*	*

In column, different letters indicate significant differences at $p < 0.01$; * $p < 0.01$; n.s.: not significant; n.d.: not detected.

3.2. Functional Breads

3.2.1. Physicochemical Characteristics of the Functional Breads

Colour, moisture, ash, and the protein distribution of breads enriched with BSGs are reported in Table 5 and compared with the characteristics of the control bread.

Regarding colour indices, the crust of the control bread showed the highest L* (together with Em-D breads) and b* (with Da-S-20G3) values and the lowest a* indices (with Da-D-25G4, Da-S-20G3, and Em-D-20G3). Em-S-25G4 had the lowest L* and b* values, while the highest a* was measured on Da-S-25G4. The crumb of the control bread also showed the highest L* and b* (with Ri-D-20G3) values and the lowest a* indices.

The type of BSG exerted significant effects on all colour indices. Concerning crust, the highest brightness and yellow index were measured on Da samples and the lowest red index was detected in Ri breads. The crumb of Ri breads was the darkest, the reddest (with Da samples), and showed the lowest yellow index (together with Em). The different behaviour observed for the various types of BSG was correlated to the colour of the starting spent grains.

The level of BSG-gluten addition significantly affected the colorimetric indices of bread with the exception of the crust a* values, but the effects were opposite between crust and crumb. The increase in BSG-gluten content generally decreased the L* and b* values of the crust, i.e., determined the formation of a darker colour. This behaviour is explained by the higher protein content in the formulation which caused intensive Maillard reactions [53]. Contrary to what happened to the crust, the increase in BSG-gluten content corresponded to a slightly lighter crumb (due to the clear colour of the fibre added) but also to a more intense golden colour (increased a* and b* values). In most previous works, a darker crumb was observed as a consequence of BSG addition [15] but Gómez et al. [54] pointed out that the effects depend on the colour of fibres contained in the added ingredient.

Remarkable water content was quantified in the control bread. However, other bread samples showed higher crust moisture (Em-D-20G3) and similar amounts of water in crumb (Em-D-25G4). The type of BSG did not affect crust moisture while the crumb of breads produced with Da spent grains contained generally higher water in their composition. The crumb moisture was significantly and positively affected by the BSG-gluten level, and this behaviour may be attributed to the higher fibre and protein contribution [17].

Regarding ash, the lowest and the highest amounts were quantified in control (3.1%) and Em-S-25G4 (3.6%) breads, respectively. The addition of all the BSG types significantly increased bread ash content but, consistently with the BSG composition reported in Table 3, the greatest ash amount was contributed by the Em type. Since BSG is mainly composed of the husk of grain and minerals are present in a greater amount in their outer layer, the amount of BSG-gluten added also had a significant and positive effect on the bread ash content [17,55].

Concerning protein size distribution in bread samples, the chromatographic profiles highlighted the presence of only two main peaks, eluted between 10 and 12.5 min (i.e., the range of large and small monomeric proteins) and the total lack of peaks in the elution range of polymeric protein aggregates. This result could be due to a variety of factors including technological process; BSG diluting effect on the protein network; and possible accumulation of low-molecular weight metabolites (mainly glutathione) deriving from the lysed yeast cell. Glutathione (GSH) was found to be responsible for the chemical modification of the gluten protein structure, leading to its depolymerization [56] and the formation of lower molecular weight gluten proteins [57,58]. Consequently to these changes, unextractable proteins included only monomeric forms (Table 5). From the comparison of data in Tables 3 and 5, a remarkable increase in UP% from BSGs to the corresponding breads can be inferred. It could be due to baking, whose high temperatures are known to induce interactions among different class of proteins, thus leading to the formation of larger aggregates or a supramolecular structure. The average unextractable monomeric proteins ranged from a minimum of 35.91% in Em-S-20G3 to a maximum of 42.07% in Ri-S-20G3. Percentages higher than 40% were also detected in Ri-D-25G4, Ri-S-25G4, and the control bread. Significant single effects of BSG type were observed, with UP% higher in Ri breads and lower in Em breads.

Table 5. Effects of type of BSG and amounts of BSG-gluten in the formulation on physical and chemical characteristics of the functional breads.

Bread Samples	Crust			Crumb			Crust Moisture %	Crumb Moisture %	Ash (%)	UP%
	L*	a*	b*	L*	a*	b*				
Control	57.0 ± 2.5 [def]	16.6 ± 1.2 [a]	47.0 ± 1.1 [g]	69.2 ± 5.6 [g]	2.6 ± 0.5 [a]	29.4 ± 1.6 [ab]	10.5 ± 1.1 [cde]	24.5 ± 2.4 [c]	3.1 [a]	41.00 ± 0.68 [ef]
	Interactive effects (Type of BSG* Geographical origin of Cereal Mixtures* Amount BSG-gluten)									
Ri-D-20 G3	43.8 ± 3.3 [a]	19.3 ± 0.8 [d]	37.9 ± 1.5 [bc]	46.7 ± 2.8 [a]	6.7 ± 0.4 [b]	29.2 ± 1.2 [a]	7.8 ± 0.7 [ab]	16.9 ± 1.8 [a]	3.3 [b–e]	38.98 ± 0.84 [cde]
Ri-D-25 G4	53.8 ± 2.3 [bc]	17.3 ± 0.7 [b]	39.6 ± 0.8 [bc]	58.0 ± 5.1 [ef]	10.4 ± 0.5 [g]	35.1 ± 1.8 [g]	11.2 ± 1.1 [de]	23.6 ± 0.8 [bc]	3.4 [efg]	40.26 ± 0.52 [def]
Ri-S-20 G3	54.5 ± 4.1 [cd]	16.3 ± 1.4 [ab]	40.0 ± 2.4 [cd]	53.7 ± 2.7 [bd]	8.1 ± 2.6 [de]	33.7 ± 1.5 [d–g]	10.1 ± 1.0 [b–e]	21.8 ± 2.7 [bc]	3.2 [b]	42.07 ± 0.15 [f]
Ri-S-25 G4	60.0 ± 3.9 [f]	16.1 ± 1.9 [a]	42.8 ± 2.9 [ef]	53.0 ± 2.4 [b]	9.0 [def]	33.8 ± 0.9 [d–g]	9.7 ± 0.9 [a–e]	20.7 ± 1.3 [abc]	3.3 [bcd]	41.02 ± 0.45 [ef]
Da-D-20 G3	55.1 ± 1.7 [cd]	20.7 ± 0.5 [e]	41.5 ± 0.5 [de]	56.4 ± 5.1 [b–f]	8.7 ± 1.6 [de]	33.3 ± 2.7 [def]	8.3 ± 1.8 [abc]	18.9 ± 1.5 [ab]	3.3 [c–f]	39.19 ± 0.56 [de]
Da-D-25 G4	56.7 ± 3.2 [cde]	16.9 ± 1.1 [a]	40.6 ± 0.8 [cd]	58.3 ± 2.1 [f]	10.0 ± 0.5 [fg]	37.0 ± 1.1 [g]	9.7 ± 1.7 [a–e]	23.6 ± 0.2 [bc]	3.4 [def]	36.47 ± 1.29 [ab]
Da-S-20 G3	62.4 ± 1.8 [f]	16.8 ± 0.6 [a]	45.9 ± 1.0 [g]	53.1 ± 2.7 [bc]	7.9 ± 0.3 [ce]	32.0 ± 1.0 [cd]	10.4 ± 1.5 [b–e]	24.8 ± 1.5 [c]	3.2 [bc]	39.18 ± 0.06 [de]
Da-S-25 G4	51.3 ± 2.0 [b]	21.1 ± 1.4 [e]	40.5 ± 1.1 [cd]	58.4 ± 2.7 [f]	9.0 ± 1.2 [ef]	34.8 ± 1.9 [fg]	9.2 ± 0.4 [a–d]	23.2 ± 2.7 [bc]	3.4 ± 0.1 [def]	39.67 ± 1.09 [de]
Em-D-20 G3	59.3 ± 2.4 [ef]	16.9 ± 0.6 [a]	44.0 ± 1.8 [f]	57.1 ± 1.8 [def]	6.8 ± 0.8 [bc]	33.8 ± 1.9 [efg]	12.1 ± 1.6 [e]	20.9 ± 0.5 [abc]	3.4 [fg]	38.18 ± 0.01 [bcd]
Em-D-25 G4	54.2 ± 1.6 [bcd]	19.5 ± 07 [d]	39.1 ± 1.5 [b]	56.9 ± 2.9 [b–f]	7.8 ± 0.4 [bcd]	32.5 ± 1.0 [cde]	10.7 ± 0.2 [c–e]	24.6 ± 4.0 [c]	3.5 [h]	36.82 ± 1.20 [abc]
Em-S-20 G3	59.1 ± 1.9 [ef]	18.4 ± 0.5 [c]	43.1 ± 1.8 [ef]	57.0 ± 2.2 [c–f]	7.9 ± 0.3 [ce]	33.8 ± 0.4 [efg]	9.2 ± 0.7 [a–d]	17.0 ± 0.3 [a]	3.5 ± 0.1 [gh]	35.91 ± 0.38 [a]
Em-S-25 G4	44.5 ± 1.9 [a]	19.3 ± 0.5 [d]	34.5 ± 1.4 [a]	54.1 ± 2.3 [b–e]	7.9 ± 0.3 [ce]	31.0 ± 0.9 [bc]	7.3 ± 1.3 [a]	18.7 ± 2.0 [ab]	3.6 [i]	38.26 ± 0.55 [bcd]
Significance	*	*	*	*	*	*	*	*	*	*
	Single Effect (Type of BSG)									
Ri	53.0 [a]	17.2 [a]	40.1 [a]	52.8 [a]	8.6 [b]	32.6 [a]	9.7	20.8 [ab]	3.3 [a]	40.58 [c]
Da	56.4 [b]	18.5 [b]	42.1 [b]	56.6 [b]	8.9 [b]	34.3 [b]	9.4	22.6 [b]	3.3 [a]	38.63 [b]
Em	54.3 [a]	18.8 [b]	40.2 [a]	56.3 [b]	7.6 [a]	32.8 [a]	9.8	20.3 [a]	3.5 [b]	37.29 [a]
Significance	*	*	*	*	*	*	n.s.	*	*	*
	Single Effect (Geographical Origin of Cereal Mixtures)									
D	53.8	18.4	40.4	55.6	8.4	33.5	10.3	21.4	3.4	38.32
S	55.3	18.0	41.1	54.8	8.3	33.2	9.3	21.0	3.4	39.35
Significance	n.s.	n.s.	n.s.	n.s.	n.s.	n.s.	n.s.	n.s.	n.s.	n.s.
	Single Effect (Amount BSG-gluten)									
20 G3	55.7 [b]	18.1	41.1 [b]	54.0 [a]	7.7 [a]	32.6 [a]	9.6	20.1 [a]	3.3 [a]	38.92
25 G4	53.4 [a]	18.2	39.5 [a]	56.4 [b]	9.0 [b]	34.0 [b]	9.6	22.4 [b]	3.4 [b]	38.75
Significance	*	n.s.	*	*	*	*	n.s.	*	*	n.s.

In column, different letters indicate significant differences at $p < 0.01$; * $p < 0.01$; n.s.: not significant.

The geographical origin of the starting cereal mixtures did not significantly influence any of the parameters of Table 5.

3.2.2. Phenolic Concentration, Antioxidant Activity, and Fibre Content of the Functional Breads

The amounts of antioxidants and fibres are among the most important factors to consider in assessing the nutraceutical quality of breads.

The total phenolic contents of functional breads were always higher than those of control bread (1.555 ± 0.158 mg/g d.m.) and comprised between 1.793 ± 0.183 mg/g d.m. (Ri-D-20G3) and 2.833 ± 0.772 mg/g d.m. (Da-S-25G4) (Table 6). These were the results of the interactions among the type of BSG, the geographical origin of cereal mixtures, and the amount of BSG-gluten, although the single effects of these variables were also statistically significant (with the exception of the geographical origin). In particular, the highest TPCs were detected in Em breads, consistently with the high phenolic concentrations of the starting spent grains. Moreover, the bread antioxidant content increased with the increase in BSG-gluten amount in the bread formulation. Finally, it is appropriate to consider that the phenolic content of the final breads was also affected by bread-making and, although the same process was applied for all samples, the magnitude of these changes could not be the same. According to a recent research of Tian et al. [59], bread-making generally improved the potential health benefits of whole wheat products. The authors pointed out that fermentation and baking generally increased soluble phenolic content and its antioxidant activity due to the contribution of Maillard reaction products, and that those steps only slightly increased the insoluble phenolic fraction and its antioxidant activity. The TPCs of our experimental functional breads were considerable higher than those (0.47 ± 0.06 mg/g d.m.) retrieved in the recent literature [60].

The antioxidant activity values of the functional breads showed trends similar to those of TPCs. They were always higher than those of the control bread (0.36 ± 0.02 mg Trolox/g d.m.) and comprised between 0.67 ± 0.16 mg/g d.m. (Ri-D-20G3) and 3.45 ± 0.47 mg/g d.m. (Em-D-25G4) (Table 6). The single effects of the type of BSG and the amount of BSG-gluten were statistically significant. In particular, the highest and the lowest antioxidant activity values were detected in breads produced with Em and Ri spent grains, respectively. Furthermore, the bread antioxidant activity increased with the increase in BSG-gluten amount in the bread formulation. TPC and antioxidant activity showed a strong correlation (R = 0.885, *p*-value < 0.01).

The bread phenolic profiles were simpler than those of the starting BSGs. Five phenolic compounds were identified in all the functional breads, including four phenolic acids (gallic, vanillic, caffeic, and sinapic) and a flavanol (epicagallotechin gallate). Nevertheless, their concentrations depended on BSG type, the geographical origin of the stating cereal mixtures, and BSG-gluten amount. *p*-Coumaric acid was detected only in the control bread that, instead, did not contain phenolics such as epigallocatechin gallate, and vanillic, caffeic, and sinapic acids. The interactive effects of the three factors were also statistically significant. The type of BSG showed significant single effects, with the highest concentrations of almost all compounds detected in Em breads. The higher level of BSG-gluten supplementation allowed to obtain breads with a higher content of almost all compounds. The single effects of the geographical origin of the starting mixtures were less significant.

Table 6. Effects of type of BSG, geographical origin, and amounts of BSG-gluten in the formulation on antioxidant and fibre contents of the functional breads.

Bread Samples	TPC (mg/g d.m.)	Antioxidant Activity (mg Trolox/g d.m.)	Gallic Acid	Vanillic Acid	Caffeic Acid	*p*-Coumaric Acid	Sinapic Acid	Epigallocatechin Gallate	IDF%	SDF%
Control	1.555 ± 0.158 [a]	0.36 ± 0.02 [a]	0.028 [a]	n.d. [a]	n.d. [a]	0.027 [b]	n.d. [a]	n.d. [a]	2.05 [a]	0.63 [k]
Interactive effects (Type of BSG* Geographical origin of Cereal Mixtures* Amount BSG-gluten)										
Ri-D-20 G3	1.793 ± 0.183 [b]	0.67 ± 0.16 [b]	0.040 [de]	n.d. [a]	0.018 [f]	n.d.	n.d.[a]	n.d. [a]	4.17 [b]	0.33 [f]
Ri-D-25 G4	2.066 ± 0.159 [bcd]	0.97 ± 0.12 [bc]	0.029 [ab]	n.d. [a]	0.018 [f]	n.d.	n.d.[a]	n.d. [a]	5.98 [g]	0.05 [b]
Ri-S-20 G3	1.881 ± 0.117 [bc]	1.48 ± 0.18 [cd]	0.029 [ab]	n.d. [a]	0.018 [f]	n.d.	n.d.[a]	n.d. [a]	5.08 [d]	0.20 [d]
Ri-S-25 G4	2.300 ± 0.135 [b-e]	2.12 ± 0.46 [ef]	0.049 [f]	n.d. [a]	0.018 [f]	n.d.	n.d.[a]	n.d. [a]	6.76 [k]	0.004 [a]
Da-D-20 G3	1.888 ± 0.086 [bc]	2.00 ± 0.10 [de]	0.031 ± 0.003 [ab]	0.015 [b]	0.017 [d]	n.d.	0.017 [b]	n.d. [a]	5.25 [e]	0.30 [e]
Da-D-25 G4	2.488 ± 0.583 [b-e]	2.53 ± 0.14 [e-g]	0.035 [bc]	0.017 [b-d]	0.017 [d]	n.d.	0.033 [c]	n.d. [a]	6.15 [h]	0.43 [h]
Da-S-20 G3	1.951 ± 0.106 [b-d]	1.51 ± 0.09 [cd]	0.028 [a]	0.016 [bc]	0.015 [b]	n.d.	0.018 [b]	n.d. [a]	4.49 [c]	0.52 [i]
Da-S-25 G4	2.833 ± 0.772 [e]	2.93 ± 0.39 [gh]	0.040 [de]	0.018 [de]	0.017 [d]	n.d.	0.035 ± 0.002 [cd]	0.010 ± 0.001 [b]	5.36 [f]	1.56 [n]
Em-D-20 G3	2.169 ± 0.246 [b-e]	2.61 ± 0.37 [fg]	0.041 [e]	0.018 [de]	0.016 [c]	n.d.	0.037 ± 0.001 [d]	0.022 ± 0.002 [c]	6.30 [i]	0.78 [m]
Em-D-25 G4	2.584 ± 0.289 [de]	3.45 ± 0.47 [h]	0.048 ± 0.005 [f]	0.020 [e]	0.018 [f]	n.d.	0.042 [e]	0.030 ± 0.004 [d]	6.84 [l]	0.08 [c]
Em-S-20 G3	2.550 ± 0.089 [c-e]	2.87 ± 0.43 [g]	0.046 [ef]	0.025 [g]	0.017 [d]	n.d.	0.019 ± 0.001 [b]	n.d.[a]	6.76 [k]	0.40 [g]
Em-S-25 G4	2.453 ± 0.139 [b-e]	2.63 ± 0.06 [fg]	0.044 ± 0.004 [ef]	0.018 ± 0.001 [ef]	0.017 [d]	n.d.	0.019 [b]	0.002 [a]	7.15 [m]	0.70 [l]
Significance	*	*	*	*	*		*	*	*	*
Single Effect (Type of BSG)										
Ri	2.009 [a]	1.310 [a]	0.037 [b]	n.d. [a]	0.018 [c]	n.d.	n.d. [a]	n.d. [a]	5.50 [b]	0.15 [a]
Da	2.290 [ab]	2.243 [b]	0.033 [a]	0.017 [b]	0.016 [a]	n.d.	0.026 [b]	0.003 [b]	5.31 [a]	0.70 [c]
Em	2.439 [b]	2.891 [c]	0.045 [c]	0.020 [c]	0.017 [b]	n.d.	0.029 [c]	0.013 [c]	6.76 [c]	0.49 [b]
Significance	*	*	*	*	*		*	*	*	*
Single Effect (Geographical Origin of Cereal Mixtures)										
D	2.165	2.039	0.037	0.012	0.017	n.d.	0.021 [b]	0.009 [b]	5.78 [a]	0.33 [a]
S	2.327	2.257	0.039	0.013	0.017	n.d.	0.015 [a]	0.002 [a]	5.93 [b]	0.56 [b]
Significance	n.s.	n.s.	n.s.	n.s.	n.s.		*	*	*	*
Single Effect (Amount BSG-gluten)										
20 G3	2.039 [a]	1.859 [a]	0.036 [a]	0.012	0.016 [a]	n.d.	0.015 [a]	0.004 [a]	5.34 [a]	0.42 [a]
25 G4	2.453 [b]	2.437 [b]	0.041 [b]	0.012	0.017 [b]	n.d.	0.021 [b]	0.007 [b]	6.37 [b]	0.47 [b]
Significance	*	*	*	n.s.	*		*	*	*	*

In column, different letters indicate significant differences at $p < 0.01$; * $p < 0.01$; n.s.: not significant; n.d.: not detected.

The interest towards the fibre content of BSG-enriched bread is due to the possibility, established by Reg. (EU) N°. 432/2012 [61], to use the following two health claims: "Barley grain fibre contributes to an increase in faecal bulk" for foods which are high in that fibre, i.e., for those foods that contain at least 6 g of fibre per 100 g or at least 3 g of fibre per 100 kcal; and "Beta-glucans contribute to the maintenance of normal blood cholesterol levels" for food which contains at least 1 g of beta-glucans from barley/barley bran per quantified portion. [62]. The increase of faecal bulk is related to the ingestion of insoluble dietary fibre. Other long-known benefits of dietary fibre intake include the modulation of glycaemic index and potential prebiotic capacity, which are known to be linked to arabinoxylan and arabinoxylan-oligosaccharides, the latter primarily deriving from wheat [63,64]. USDA recommends daily intakes of fibres equal to 25 g for women and 38 g for men up to 50 years old and to 21 and 30 g for elder women and men, respectively [65]. Since the actual intake is generally lower, especially in Western countries, the regular consumption of BSG-enriched breads could help consumers to meet such recommendations. As can be inferred from Table 6, the amounts of IDF and SDF in BSG- enriched breads were significantly lower than in the corresponding spent grains, due to the dilution effect of Manitoba flour. IDF% and SDF% felt both interactive and single effects of the factors. More in depth, IDF% ranged from 2.05% in the control bread to 7.15% of Em-S-25G4 while SDF% was in the 0.004% (Ri-S-25G4)–1.56% (Da-S-25G4) range and the percentages of soluble fibre on the total fibre were appreciably higher than in the BSG samples, ranging from 0.06% (RiS25-G4) to 22.51% (DaS25-G4), thus contributing beneficial effects that go beyond the simple increase in faecal bulk. A first reason for this behaviour is the high percentages of soluble fibre in the Manitoba flour. In fact, the control bread had a percentage of soluble fibre on the total fibres equal to 23.49%. Another reason is that, during bread-making, a decomposition of dietary fibre (first hemicellulose and afterwards cellulose) occurred, reducing the fibre molecular weight [66]. The reason for this degrading action could be both the yeast, since some *Saccharomyces cerevisiae* strains are able to produce cellulase and xylanase [66], and the first step of the baking process when both temperatures and moisture are elevated, thus simulating the conditions of an autoclave treatment [67]. In the control bread, the incidence of soluble fibre on the total amount of fibre was equal to 23.49%. However, while data indicate that the single effects exerted by BSG type and the geographical origin on bread IDF% were similar to those observed for spent grains, they also describe the opposite effects of BSG type and geographical origin on the soluble fibre fraction between breads and spent grains. Significant increases in both IDF% and SDF% were evaluated by increasing the amount of BSG-gluten added. The total dietary fibres ranged between 4.50% of RiD20-G3 and 7.85% of EmS25-G4, resulting as significantly higher than those quantified in the control bread (2.68%) and slightly higher than the ranges found in recent literature (3.32–6.37%) [44]. According to the data concerning the dietary fibre contents of the enriched breads, the inclusion of Em spent grains in percentages equal or higher than those used in these experiments could allow the use of one or both of the health claims mentioned above [61].

3.2.3. Structural Characteristics of the Functional Breads

Consumers are more likely to purchase well-leavened and regularly shaped breads. For this reason, height and specific volume are often considered as the key quality parameters.

In our work, the minimum and maximum slice heights were used as indices of bread shape regularity. The highest values of the minimum (9.36 ± 0.24 cm) and maximum (10.14 ± 0.17 cm) height were measured on the control bread, while the lowest values were found for Da-S-20G3 (7.72 ± 0.84 cm) and Ri-D-25G4 (8.39 ± 0.20 cm), respectively (Table 7). As can be inferred from these data, slice height values were significantly affected by interactions among BSG type, origin, and the amount of BSG-gluten, but the BSG type also exerted significant single effects, with the highest values observed in Em breads. The crumb specific volume is another key parameter since superior bread quality is often characterised by high specific volumes [44]. Crumb specific volume (Table 7) was only slightly reduced by the BSG addition, ranging from 2.11 ± 0.9 cm^3/g of Da-D-20G3 to 2.64 cm^3/g of Em-D-25G4 and control bread. These results were significantly better than those observed by Amoriello et al. [15] in breads produced with medium or strong wheat flour supplemented with 5 or 10% of BSG. In that work, the authors attributed the limited dough development of BSG-enriched breads to a reduction of extensibility and the gas-retention ability of gluten, in turn caused by dilution with non-gluten proteins and disruption due to the interference of fibres. The better specific volume of supplemented breads obtained in our work was determined by the concurrent addition of gluten. As always in our work, only the interactive effects of the three factors were significant, but it is interesting to point out that, although not significant, the highest specific volumes were quantified in Em breads, especially at increasing amounts of BSG-gluten added, i.e., in the samples with the highest fibre contents and deriving from the spent grains that had the highest amounts of total and large unextractable polymeric proteins.

Pores are created within the dough structure as a consequence of CO_2 production during leavening. Their characteristics are described in Table 7. The control bread had a high pore density (0.73 ± 0.02 pores/cm^2) but not the highest, since that index ranged from ~0.3 pores/cm^2 (Ri-S breads, Em-D-25G4, and Em-S-25G4) to ~0.82 pores/cm^2 (Em-D-20G3). This result was not in agreement with the findings of Neylon et al. [44], who observed a significant decrease in the number of cells in BSG-fortified breads. BSG type, origin, and the amount of BSG-gluten exerted statistically significant single and interactive effects, with the higher density measured on breads produced with Da-D spent grains at the lowest level of addition. The average pore size was inversely correlated ($R = 0.938$, p-value < 0.01) with pore density, ranging from 0.10 mm^2 (control bread and Da-S-25G4) to 0.30 mm^2 (Em-D-25G4 and Ri-S-20G3). As for pore density, the results were not in agreement with the findings of Neylon et al. [44], who observed a decrease in the cell diameter in BSG-fortified breads. Only the BSG type showed a significant effect on pore size, with the lowest values observed in Da breads. Porosity % ranged from ~35% (control bread, Da-D-20G3, and Ri-D-25G4) to ~47% (Em-S-25G4) and, as for the average pore size, only the BSG type showed a significant effect, with the highest values observed in Em breads. According to these results, the crumb of Em breads had a less compact structure than those of all other breads. Porosity % was well correlated with the total pore surface (pore density × verage pore size), showing an R value of 0.694 (p-value < 0.01). Circularity ranged from 0.774 ± 0.029 (Ri-S-20G3) to 0.829 ± 0.029 (Em-S-25G4), thus indicating a predominantly circular shape. Nevertheless, data showed a remarkable variability already within the samples of each type of bread and only the interactive effects of the three factors were statistically significant, making it difficult to understand the weight of each independent variable.

3.2.4. Sensorial Characteristics of the Functional Breads

Control bread and breads enriched with the BSGs were evaluated by a trained panel through a Quantitative Descriptive Analysis and the results are reported in Table 8. Two of the characteristics that judges were requested to evaluate, namely crust bitterness and crumb stickiness, were not detected for any of the experimental breads. The interactive effects of BSG type, the origin of cereal mixtures, and the amount of BSG-gluten added were significant for all the other sensorial parameters.

Table 7. Effects of type of BSG, geographical origin, and amounts of BSG-gluten in the formulation on structural characteristics of the functional breads.

Bread Samples	Minimum Slice Height (cm)	Maximum Slice Height (cm)	Crumb Specific Volume (cm³/g)	N. pores/cm²	Average Pore Size (mm²)	Porosity %	Pore Circularity
Control	9.36 ± 0.24 [d]	10.14 ± 0.17 [d]	2.64 [b]	0.73 ± 0.02 [cd]	0.10 [a]	34.9 ± 1.3 [a]	0.800 ± 0.007 [a-d]
Interactive effects (Type of BSG* Geographical origin of Cereal Mixtures* Amount BSG-gluten)							
Ri-D-20 G3	8.01 ± 0.6 [ab]	8.97 ± 0.22 [abc]	2.37 [ab]	0.59 ± 0.07 [bc]	0.18 ± 0.04 [abc]	39.5 [a-d]	0.813 ± 0.007 [b-e]
Ri-D-25 G4	7.94 ± 0.24 [ab]	8.39 ± 0.20 [a]	2.29 [ab]	0.70 ± 0.07 [cd]	0.14 ± 0.05 [ab]	35.8 ± 1.8 [a]	0.835 ± 0.013 [e]
Ri-S-20 G3	8.52 ± 0.65 [a-d]	9.11 ± 0.59 [abc]	2.47 [ab]	0.34 ± 0.03 [a]	0.30 [d]	36.0 ± 1.9 [ab]	0.774 ± 0.029 [a]
Ri-S-25 G4	8.38 ± 0.57 [abc]	9.11 ± 0.55 [abc]	2.50 [ab]	0.39 ± 0.07 [a]	0.24 ± 0.05 [bcd]	36.6 ± 1.8 [ab]	0.788 ± 0.012 [ab]
Da-D-20 G3	8.43 ± 0.50 [a-d]	9.26 ± 0.29 [bc]	2.11 ± 0.9 [a]	0.69 ± 0.10 [cd]	0.14 ± 0.05 [ab]	35.2 ± 2.5 [a]	0.816 ± 0.005 [b-e]
Da-D-25 G4	8.14 ± 0.26 [abc]	9.21 ± 0.20 [abc]	2.45 [ab]	0.66 ± 0.09 [cd]	0.16 ± 0.05 [ab]	38.4 ± 2.0 [a-d]	0.825 ± 0.010 [de]
Da-S-20 G3	7.72 ± 0.84 [a]	8.97 ± 0.71 [abc]	2.32 [ab]	0.69 ± 0.07 [cd]	0.16 ± 0.05 [ab]	38.2 ± 2.2 [abc]	0.835 ± 0.026 [e]
Da-S-25 G4	7.99 ± 0.45 [ab]	8.48 ± 0.22 [ab]	2.35 [ab]	0.75 ± 0.04 [cd]	0.10 [a]	36.4 ± 3.8 [ab]	0.817 ± 0.013 [cde]
Em-D-20 G3	8.58 ± 0.43 [a-d]	8.89 ± 0.82 [abc]	2.42 [ab]	0.82 ± 0.17 [d]	0.16 ± 0.05 [ab]	43.1 ± 2.9 [de]	0.819 ± 0.006 [cde]
Em-D-25 G4	8.84 ± 1.21 [bcd]	9.67 ± 0.92 [cd]	2.64 [b]	0.33 ± 0.08 [a]	0.32 ± 0.08 [d]	42.6 ± 0.1 [cde]	0.807 ± 0.015 [b-e]
Em-S-20 G3	8.46 ± 0.31 [a-d]	9.72 ± 0.41 [cd]	2.52 [ab]	0.47 ± 0.22 [ab]	0.24 ± 0.11 [bcd]	40.8 ± 3.6 [bcd]	0.791 ± 0.017 [abc]
Em-S-25 G4	9.08 ± 0.18 [cd]	9.58 ± 0.21 [cd]	2.27 [ab]	0.40 ± 0.11 [a]	0.28 ± 0.08 [cd]	46.7 ± 6.2 [e]	0.829 ± 0.029 [e]
Significance	*	*	*	*	*	*	*
Single Effect (Type of BSG)							
Ri	8.21 [a]	8.89 [a]	2.41	0.50 [a]	0.21 [b]	36.9 [a]	0.80
Da	8.01 [a]	8.98 [a]	2.30	0.70 [b]	0.14 [a]	37.0 [a]	0.82
Em	8.74 [b]	9.46 [b]	2.46	0.50 [a]	0.25 [b]	43.3 [b]	0.81
Significance	*	*	n.s.	*	*	*	n.s.
Single Effect (Geographical Origin of Cereal Mixtures)							
D	8.31	9.06	2.38	0.63 [b]	0.18	39.1	0.82
S	8.36	9.16	2.40	0.51 [a]	0.22	39.1	0.81
Significance	n.s.	n.s.	n.s.	*	n.s.	n.s.	n.s.
Single Effect (Amount BSG-gluten)							
20 G3	8.29	9.15	2.37	0.60 [b]	0.20	38.8	0.81
25 G4	8.39	9.07	2.42	0.54 [a]	0.20	39.4	0.82
Significance	n.s.	n.s.	n.s.	*	n.s.	n.s.	n.s.

In column, different letters indicate significant differences at $p < 0.01$; * $p < 0.01$; n.s.: not significant.

Table 8. Effects of type of BSG, geographical origin, and amounts of BSG-gluten in the formulation on sensorial parameters of the functional breads.

| Bread Samples | Visual Characteristic | | | | | | Smell | | | | | Taste | | | Tactile Characteristics/Texture | | | | | Overall Quality |
| | Crust | | Crumb | | | | Crust and Crumb | | | Crust | Crumb | Crumb | | | Crust | | | Crumb | | |
	Colour	Thickness	Colour	Pore Size	Development	Overall Intensity	Freshly Baked Bread	Wheat	Malty	Toasty	Yeast	Sweetness	Saltiness	Acidity	Hardness	Crispiness	Resistance to Chewing	Cohesiveness	Fibrousness	
Control	0.5 ± 0.1 a	4.5 ± 0.6 ab	0.5 ± 0.1 a	2.5 ± 0.6 ab	8.5 ± 0.6 c	4.5 ± 0.6 a	4.5 ± 0.6 de	0.5 ± 0.1 a	1.5 ± 0.3 bc	0.5 ± 0.1 a	0.5 ± 0.1 a	6.5 ± 0.6 ab	7.5 ± 0.6 f	2.5 ± 0.2 c	3.5 ± 0.6 a	4.5 ± 0.6 bc	0 a	7.5 ± 0.6 b	0 a	7.5 ± 0.6 c
							Interactive effects (Type of BSG* Geographical origin of Cereal Mixtures* Amount BSG-gluten)													
Ri-D-20 G3	6.5 ± 0.6 b	4.5 ± 0.6 ab	6.5 ± 0.6 cd	3.5 ± 0.6 b	6.5 ± 0.6 ab	7.5 ± 0.6 c	2.5 ± 0.6 bc	2.5 ± 0.6 bc	3.5 ± 0.6 d	2.5 ± 0.1 b	0.5 ± 0.1 a	6.5 ± 0.6 ab	4.5 ± 0.6 de	1.5 ± 0.3 bc	4.5 ± 0.6 ab	1.5 ± 0.2 a	1.5 ± 0.1 bc	6.5 ± 0.6 b	3.5 ± 0.6 de	7.5 ± 0.6 c
Ri-D-25 G4	6.5 ± 0.6 b	3.5 ± 0.6 a	7.5 ± 0.6 d	3.5 ± 0.6 b	6.5 ± 0.6 ab	6.5 ± 0.6 bc	2.5 ± 0.6 bc	3.5 ± 0.6 cd	3.5 ± 0.6 d	1.5 ± 0.1 ab	2.5 ± 0.3 bc	5.5 ± 0.6 a	4.5 ± 0.6 de	0.5 ± 0.2 ab	3.5 ± 0.6 a	1.5 ± 0.1 a	1.5 ± 0.2 bc	6.5 ± 0.6 b	2.5 ± 0.2 cd	7.5 ± 0.6 c
Ri-S-20 G3	7.5 ± 0.6 bc	4.5 ± 0.6 ab	7.5 ± 0.6 d	4.5 ± 0.6 c	6.5 ± 0.6 ab	7.5 ± 0.6 c	2.5 ± 0.6 bc	1.5 ± 0.6 ab	1.5 ± 0.6 bc	1.5 ± 0.1 ab	1.5 ± 0.1 ab	6.5 ± 0.6 ab	3.5 ± 0.5 cd	0.5 ± 0.1 ab	3.5 ± 0.6 a	4.5 ± 0.6 bc	1.5 ± 0.4 bc	7.5 ± 0.6 b	2.5 ± 0.2 cd	7.5 ± 0.6 c
Ri-S-25 G4	6.5 ± 0.6 b	3.5 ± 0.6 a	7.5 ± 0.6 d	3.5 ± 0.6 b	7.5 ± 0.6 bc	6.5 ± 0.6 bc	5.5 ± 0.6 e	4.5 ± 0.6 de	3.5 ± 0.6 d	1.5 ± 0.2 ab	5.5 ± 0.6 d	5.5 ± 0.6 a	3.5 ± 0.6 cd	1.5 ± 0.2 bc	3.5 ± 0.5 a	1.5 ± 0.2 a	2.5 ± 0.2 cd	7.5 ± 0.6 b	0.5 ± 0.1 ab	7.5 ± 0.6 c
Da-D-20 G3	6.5 ± 0.6 b	5.5 ± 0.6 b	6.5 ± 0.6 cd	4.5 ± 0.6 c	5.5 ± 0.6 a	6.5 ± 0.6 bc	5.5 ± 0.6 e	0.5 ± 0.1 a	0.5 ± 0.1 ab	1.5 ± 0.2 ab	1.5 ± 0.1 ab	5.5 ± 0.6 a	2.5 ± 0.3 bc	1.5 ± 0.3 bc	5.5 ± 0.6 bc	0.5 ± 0.2 a	3.5 ± 0.6 d	4.5 ± 0.1 a	0.5 ± 0.1 ab	5.5 ± 0.6 ab
Da-D-25 G4	6.5 ± 0.6 b	4.5 ± 0.6 ab	6.5 ± 0.6 cd	2.5 ± 0.6 ab	6.5 ± 0.6 ab	6.5 ± 0.6 bc	1.5 ± 0.3 ab	2.5 ± 0.6 bc	0.5 ± 0.1 ab	1.5 ± 0.5 ab	2.5 ± 0.6 bc	5.5 ± 0.6 a	3.5 ± 0.6 cd	0.5 ± 0.2 ab	5.5 ± 0.6 bc	3.5 ± 0.6 b	1.5 ± 0.6 bc	6.5 ± 0.6 b	0.5 ± 0.1 ab	7.5 ± 0.6 c
Da-S-20 G3	6.5 ± 0.6 b	5.5 ± 0.6 b	6.5 ± 0.6 cd	3.5 ± 0.6 b	5.5 ± 0.6 a	6.5 ± 0.6 bc	3.5 ± 0.6 cd	0.5 ± 0.1 a	1.5 ± 0.4 bc	2.5 ± 0.5 b	0.5 ± 0.1 a	5.5 ± 0.6 a	4.5 ± 0.5 de	0.5 ± 0.1 ab	4.5 ± 0.6 ab	5.5 ± 0.6 cd	2.5 ± 0.3 cd	6.5 ± 0.6 b	0.5 ± 0.1 ab	7.5 ± 0.6 c
Da-S-25 G4	7.5 ± 0.6 bc	5.5 ± 0.6 b	6.5 ± 0.6 cd	2.5 ± 0.6 ab	5.5 ± 0.6 a	7.5 ± 0.6 c	2.5 ± 0.4 bc	2.5 ± 0.3 bc	2.5 ± 0.2 cd	1.5 ± 0.3 ab	2.5 ± 0.5 bc	7.5 ± 0.6 b	5.5 ± 0.6 e	1.5 ± 0.2 bc	5.5 ± 0.5 bc	1.5 ± 0.6 a	2.5 ± 0.6 cd	6.5 ± 0.6 b	1.5 ± 0.6 bc	6.5 ± 0.5 bc
Em-D-20 G3	7.5 ± 0.6 bc	3.5 ± 0.6 a	4.5 ± 0.6 b	2.5 ± 0.6 ab	7.5 ± 0.6 bc	4.5 ± 0.6 a	1.5 ± 0.1 ab	3.5 ± 0.6 cd	0.5 ± 0.1 ab	0.5 ± 0.1 a	1.5 ± 0.3 ab	6.5 ± 0.6 ab	3.5 ± 0.6 cd	0.5 ± 0.1 ab	3.5 ± 0.5 a	6.5 ± 0.6 d	0.5 ± 0.1 ab	6.5 ± 0.1 b	5.5 ± 0.6 f	5.5 ± 0.5 ab
Em-D-25 G4	8.5 ± 0.6 c	3.5 ± 0.6 a	7.5 ± 0.6 d	2.5 ± 0.6 ab	7.5 ± 0.6 bc	6.5 ± 0.6 bc	5.5 ± 0.6 e	5.5 ± 0.6 e	0.5 ± 0.6 ab	1.5 ± 0.2 ab	1.5 ± 0.2 ab	5.5 ± 0.4 a	1.5 ± 0.3 ab	1.5 ± 0.3 bc	6.5 ± 0.5 c	4.5 ± 0.4 bc	5.5 ± 0.6 e	6.5 ± 0.6 b	3.5 ± 0.6 de	4.5 ± 0.4 a
Em-S-20 G3	6.5 ± 0.6 b	4.5 ± 0.6 ab	5.5 ± 0.6 bc	1.5 ± 0.4 a	6.5 ± 0.6 ab	5.5 ± 0.6 ab	3.5 ± 0.6 cd	5.5 ± 0.6 e	0.5 ± 0.2 ab	0.5 ± 0.1 a	3.5 ± 0.2 c	6.5 ± 0.4 ab	1.5 ± 0.3 ab	1.5 ± 0.3 bc	4.5 ± 0.5 ab	6.5 ± 0.6 d	1.5 ± 0.2 bc	7.5 ± 0.6 b	3.5 ± 0.3 de	6.5 ± 0.5 bc
Em-S-25 G4	7.5 ± 0.6 bc	4.5 ± 0.6 ab	5.5 ± 0.6 bc	1.5 ± 0.3 a	6.5 ± 0.6 ab	4.5 ± 0.6 a	0.5 ± 0.1 a	4.5 ± 0.6 de	0 a	0.5 ± 0.1 a	1.5 ± 0.3 ab	5.5 ± 0.5 a	0.5 ± 0.1 a	0 a	3.5 ± 0.4 a	5.5 ± 0.5 cd	1.5 ± 0.3 bc	6.5 ± 0.6 b	4.5 ± 0.5 ef	4.5 ± 0.6 a
Significance	*	*	*	*	*	*	*	*	*	*	*	*	*	*	*	*	*	*	*	*
							Single Effect (Type of BSG)													
Ri	6.7 a	4.0 a	7.2 c	3.7 c	6.7 b	7.0 b	3.2	3.0 b	3.0 c	1.7 b	2.5 b	6.0	4.0 b	1.0	3.7 a	2.5 a	1.7 a	7.0 b	2.2 b	7.5 c
Da	6.7 a	5.2 b	6.5 b	3.2 b	5.7 a	6.7 b	3.2	1.5 a	1.2 b	1.7 b	1.7 a	6.0	4.0 b	1.0	5.2 c	2.7 a	2.5 b	6.0 a	0.7 a	6.7 b
Em	7.5 b	4.0 a	5.7 a	2.0 a	7.0 b	5.2 a	2.7	4.7 c	0.4 a	0.7 a	2.0 ab	6.0	1.7 a	0.9	4.5 b	5.7 b	2.2 bc	6.7 b	4.2 c	5.2 a
Significance	*	*	*	*	*	*	n.s.	*	*	*	*	n.s.	*	n.s.	*	*	*	*	*	*
							Single Effect (Geographical Origin of Cereal Mixtures)													
D	7	4.2	6.5	3.2	6.7	6.3	3.1	3.0	1.5	1.5	1.7 a	5.8	3.3	1.0	4.8 b	3.0 a	2.3	6.2 a	2.7	6.3 a
S	7	4.7	6.5	2.8	6.3	6.3	3.0	3.2	1.6	1.3	2.5 b	6.2	3.2	0.9	4.2 a	4.2 b	2.0	7 b	2.2	6.7 b
Significance	n.s.	n.s.	n.s.	n.s.	n.s.	n.s.	n.s.	n.s.	n.s.	n.s.	*	n.s.	n.s.	n.s.	*	*	n.s.	*	n.s.	*
							Single effect (Amount BSG-gluten)													
20 G3	6.8	4.7	6.2	3.3	6.3	6.3	3.2	2.3 a	1.3	1.5	1.5 a	6.2	3.3	1.0	4.3 a	4.2 b	1.8 a	6.5	2.7	6.7 b
25 G4	7.2	4.2	6.8	2.7	6.7	6.3	3.0	3.8 b	1.7	1.3	2.7 b	5.8	3.2	0.9	4.7 b	3.0 a	2.5 b	6.7	2.2	6.3 a
Significance	n.s.	n.s.	n.s.	n.s.	n.s.	n.s.	n.s.	*	n.s.	n.s.	*	n.s.	n.s.	n.s.	*	*	*	n.s.	n.s.	*

In column, different letters indicate significant differences at $p < 0.01$; * $p < 0.01$; n.s.: not significant.

Bread colour is the first variable evaluated by consumers and strongly affects their willingness to purchase and the product acceptability. In agreement with the findings of Ginindza et al. [68], the addition of BSG always resulted in a significant colour change compared with the control bread and the reason is that BSG contributed a higher amount of aminoacids, thus favouring the non-enzymatic browning reactions [69]. The Em breads were evaluated as the darkest. The single effects of geographical origin and the amount of BSG-gluten added were not significant either for crust or for crumb colour. Regarding the other visual characteristics, the type of BSG had significant single effects on crust thickness (higher in Da breads while control bread had intermediate scores), crumb pore size (higher in Ri breads while control bread had intermediate scores), and crumb development (higher in Em samples among the BSG-enriched breads but lower than that of the control bread).

The fortified breads generally showed the highest intensity of wheat, malt, and yeast flavours. According to Ktenioudaki et al. [16], these flavours were due to volatile compounds already present in BSG (2-heptane, butanal, 2-methylbutanale, benzene, and 2,3-butanedione) and arising from fermentation (ethanol, butanol, and other acholic compounds) and Maillard reactions (furfural, pyrazine). The type of BSG exerted significant effects on overall intensity, malty, and toasty flavours (lower in Em breads), and on wheat and yeast smells (higher in Da breads). The amount of BSG-gluten added exerted significant single effects only on wheat and yeast smells, which increased with the level of replacement.

Concerning crumb taste, the single effects of the three factors were not statistically significant. The saltiness and acidity/sourness of the functional breads were evaluated as lower or equal to those of the control bread. The latter obtained intermediate scores for sweetness, perhaps as a consequence of the maltose and glucose contained in BSG. Moreover, the sweet taste was also enhanced by the volatile compounds responsible for sweet and malty flavours.

Regarding bread texture, the addition of BSG-gluten always increased resistance to chewing and the fibrousness of the functional breads with respect to the control. The control bread obtained intermediate scores for crispness. The highest crust hardness and resistance to chewing and the lowest cohesiveness and fibrousness were attributed to breads fortified with Da spent grains. The amount of BSG-gluten added significantly affected the bread texture, except for crumb cohesiveness. The hardness of the crust and the resistance to chewing of crumb increased with BSG-gluten content. This behaviour could be explained by both the higher fibre and protein contents of functional breads, which caused a greater water absorption [17,70], and also the presence of pentopans, a fibre BSG component, responsible for the gluten protein cross linking [71]. Crust crispness decreased with the increase in BSG-gluten content due to the increase in water absorbed, while crumb cohesiveness and fibrousness did not show significant changes. These results were only partially in agreement with those of Yitayew et al. [17], who found that the hardness and breakage of crumb increased with BSG level.

Finally, the best and the worse overall ratings were attributed to Ri (together with control) and Em breads, respectively, and the panellists accorded their preferences to breads produced with the lowest amount of BSG-gluten added only for DaS and Em breads. In the other case, the % of BSG added did not affect the overall sensory quality. This is an interesting finding since the percentages of BSG in our products (20 or 25%) was much higher than those ($\leq$10%) indicated as optimal by Yitayew et al. [17] and Ginindza et al. [68] who found the maximum and minimum acceptability score for the control sample and the bread supplemented with the highest amount of BSG, respectively. The interest of our finding is also due to the fact that, in other research, the BSG-fortified foods exerted a higher appeal [42] than the not-enriched counterparts as a consequence of consumer interest for health issues, characteristics that overshadows the product sensory properties [72].

3.2.5. Principal Component Analysis and Pearson Coefficients

Principal Component Analysis was carried out to verify the possibility of grouping BSGs by type and/or by place of origin on the basis of their physical characteristics and the content of antioxidants, fibres, and non-extractable proteins. The ability of PCA to group BSG samples belonging to the same base malts group with a high percentage of explained variance (75%) was already highlighted by di Matteo et al. [73]. Figure 3 shows a biplot of Factors 1 and 2 that accounted for 67.63% of the variance in the whole data set. Regarding the projection of the BSG samples on the factorial plan (Figure 3a), Em samples differed from the others for their negative loading of Factor 1, accounting for 43.19% of the explained variance. Em samples were furtherly divided into two geographically homogeneous groups, one including spent grains from Daunia and another including spent grains from Salento, characterised by the negative and positive loading of Factor 2, respectively. The negative loadings of this factor were associated with high amounts of ash and most of the phenolic compounds while the positive loadings of these factors were associated with high amounts of insoluble fibres, unextractable proteins, *p*-coumaric acid, and epigallocatechingallate (Figure 3b). Ri-D and Ri-S spent grains were homogeneously grouped but close to each other in the upper right quadrant, the first with loadings near to 0, the latter with more positive loadings of Factor 1 (Figure 3a), characterised by high antioxidant activity values and low phenolic concentrations (Figure 2b). Da-D and Da-S spent grains were homogeneously grouped but close to each other in the lower right quadrant (Figure 3a). This quadrant was associated with low-to-intermediate values of all variables (Figure 3b).

Principal Component Analysis was also performed to highlight the relationship between bread samples and their chemical, physical, structural, and sensory characteristics. Figure 4a,b shows the projection on the factorial plan of breads and analytical information, respectively. Factors 1 and 2 only explained 44.21% of the variance in the whole data set and made it possible to group the thirteen types of samples into just three sets that stood out for their position in the factorial plan: one including the control bread, placed in the lower left quadrant and associated with high values of crumb development, specific volume, minimum and maximum height, cohesiveness, and amount of *p*-coumaric acid; another group, comprising all the Em breads distributed within the lower right quadrant and characterised by high concentrations of IDF, TPC, gallic acid, vanillic acid, sinapic acid, epigallocatechingallate, and high values of crispiness, porosity, fibrousness, and wheat smell; and the last group, including Ri and Da breads, which were partially overlapped and concentrated straddling the two upper quadrants, associated with high values of overall smell intensity, a toasty smell, pore size, crust thickness, overall quality, a high concentration of caffeic acid, a darker crumb colour, and intermediate TFC. As can be inferred from Figure 4a, PCA of the overall data set was not able to discriminate bread samples on the base of the two different percentages of the BSG-gluten used in bread-making. In the paper of Ktenioudaki et al. [16], PCA analysis of volatile compounds was able to clear separate snack samples in homogeneous clusters for the percentage of BSG added but it is appropriate to point out that they worked with a single type of BSG instead of three types as we did.

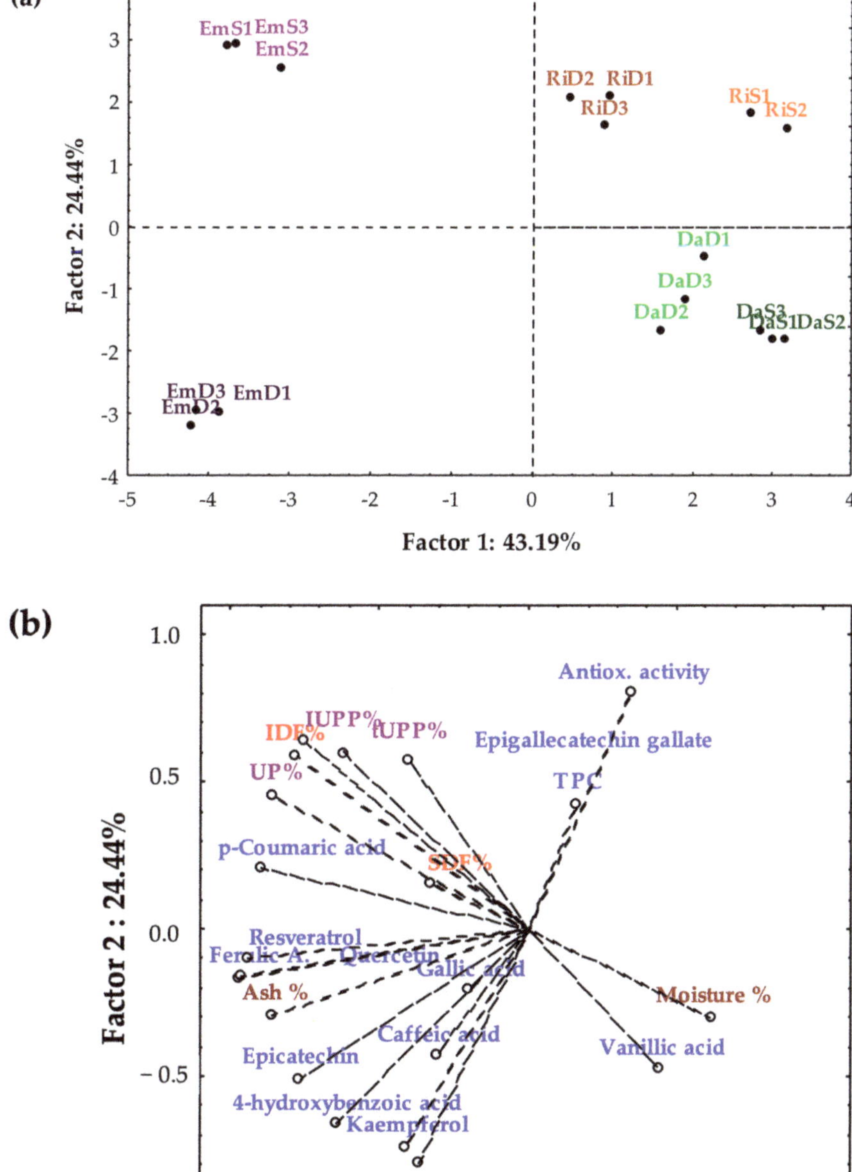

Figure 3. Projections of (**a**) BSG samples and (**b**) the related experimental data on the factorial plane. Phenolics are in blue font; Insoluble and Soluble Dietary Fibres are in red; tUPP%, lUPP%, and UP% are in purple; Moisture % and Ash % are in brown.

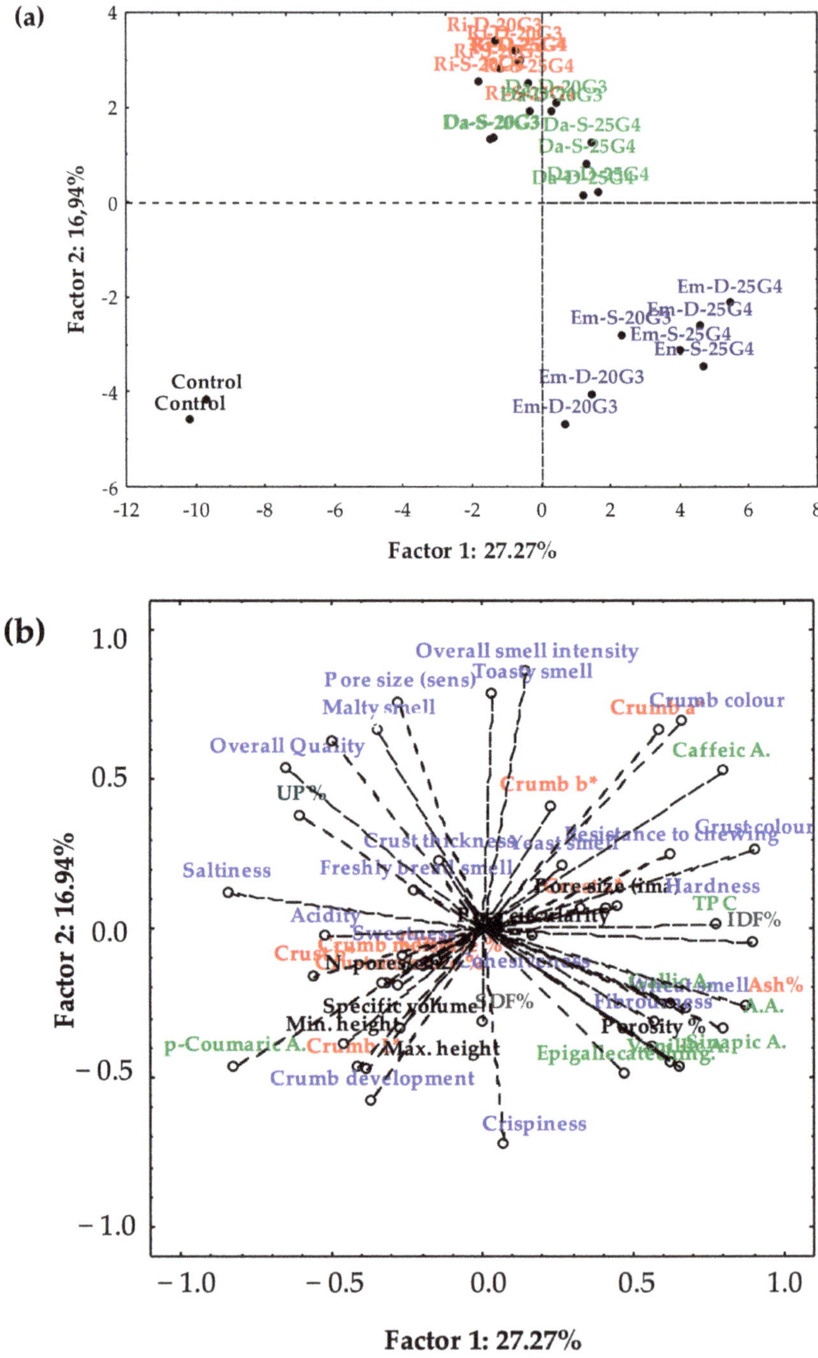

Figure 4. Projections of (**a**) control and functional breads and (**b**) the related experimental data on the factorial plane. Phenolics are in green font; Insoluble and Soluble Dietary Fibres are in purple; UP% is in dark green; Moisture %, Ash %, and colorimetric indices are in red; structural parameters are in black.

Pearson correlation coefficients were calculated to individuate correlations among the quality characteristics of breads and measured variables. For the sake of synthesis, only the main statistically significant correlations ($p < 0.01$) are discussed.

The percentage of unextractable proteins was positively correlated with the crumb pore size (+0.61) and negatively correlated with porosity (−0.56), sensory crust hardness (−0.53), and antioxidant activity (−0.60). A possible explanation of this behaviour is that the formation of larger protein aggregates made the crumb structure more compact and entrapped phenolic compounds through protein-polyphenol complexation while increasing the water holding capacity of the crust structure.

The insoluble dietary fibre content was positively correlated with ash content (+0.83), porosity (+0.47), and crumb fibrousness (+0.51) while the soluble dietary fibre content was positively correlated with crust thickness (+0.51) and crumb sweetness (+0.70) as a consequence of the depolymerization of hemicellulose and cellulose and the production of simple sugars. To confirm the significant effects of the addition of BSGs in terms of colour changes, IDF% resulted in a significant correlation with crust b* (−0.54), crumb a* (+0.58), and the colour of the crust (+0.79) and crumb (+0.62). However, contrary to what Ginindza et al. [68] highlighted, the darkening of the colour did not negatively affect the overall quality of the bread. IDF% as an index of the amount of BSG added also affected bread taste by reducing saltiness (−0.82) and olfactory characteristics, and enhancing the intensity of the wheat (+0.80) and yeast (+0.58) smell. Contrary to what Ginindza et al. [68] highlighted, the flavour changes caused by the addition of BSG were not correlated with the overall sensory quality of bread.

TPC was positively correlated with ash content (+0.66), IDF% (+0.74), antioxidant activity (+0.70), and with the individual concentration of most of the phenolic compounds, thus demonstrating that wholegrain derivatives can be considered good sources of phenolic antioxidants.

Further significant correlations were found between the physical, structural, and sensory characteristics of breads, as in the following cases: crumb specific volume and crumb development (+0.43); porosity and crumb pore size (−0.48); crust b* and crust colour (−0.51); and crumb L*, a*, and b* and crumb colour (−0.38, +0.68, +0.43).

Finally, the overall quality of breads was positively correlated with a malty smell (0.68) and crumb saltiness (0.70), and negatively correlated with ash and IDF% (−0.72 and −0.52), and with crumb fibrousness and porosity (−0.56 and −0.58). Our results were only partially in agreement with the findings of Ktenioudaki et al. [16], who found that taste and texture were the attributes that mostly affected the overall acceptability, and with the results of Combest and Warren [74], who found significant correlations only between taste and overall liking. In fact, in our study, some smell, taste, and texture attributes are correlated with the overall sensory quality of breads.

4. Conclusions

The partial replacement of wheat flour with BSG resulted in significant increases in phenolic content, and insoluble and soluble dietary fibres of the enriched breads with respect to the control thus confirming the nutraceutical and functional nature of BSGs and BSG-enriched breads, respectively. Concerning BSG samples, single and interactive effects of the type and the geographical origin of the starting cereal mixtures were highlighted, to point out that their composition is mainly determined by genetics but can be significantly influenced by environmental conditions.

The highest amounts of phenolic compounds were detected in Em spent grains, followed by Da, and Ri. Em spent grains also showed the highest concentrations of both insoluble and soluble dietary fibres, followed by Ri, and Da. The supplementation with the highest percentage of BSGs exerted a positive influence on the contents of phenolics and dietary fibres without worsening the structural and sensory attributes. These results were probably due to the simultaneous addition of gluten in amounts able to counterbalance the negative effects exerted by the fibres added on the gluten network. Moreover, the increase in soluble fibres from BSGs to the corresponding breads may be partially responsible for the good structural and sensorial performance of the enriched breads compared with the control bread.

The breads enriched with Em spent grains showed the highest percentage of phenolics and insoluble fibres and are therefore the best breads in terms of the content of nutraceuticals. However, since these desirable characteristics were accompanied by a worsening of the overall sensory quality evaluated by a trained panel, the production of bread supplemented with Ri or Da spent grains represented the optimal choice, since their intermediate phenolic and fibre contents and their overall quality scores were comparable to that of the control bread.

As a practical application, the large amount of BSG produced annually in the world makes the transformation of breweries into biorefineries capable of turning this by-product into high-value, low-perishable ingredients for food and feed industries economically convenient. Since BSG composition depends on their type and origin, they should be offered for sale with labels showing the precise information on their composition, nutritional value, and the suggested percentages of use in the formulation of the finished products. On the other hand, food companies should explore the huge variety of uses of BSG, in particular to a) increase the production of bread and other food commodities to meet the needs of the growing world population, and b) study food formulations that can be labelled with health claims.

Author Contributions: Conceptualization, A.B.; methodology, A.B., A.F., B.l.G. and M.R.; software, A.B.; validation, A.B.; formal analysis, A.B.; investigation, A.B., A.F., B.l.G. and M.R.; resources, A.B.; data curation, A.B., A.F.; writing—original draft preparation, A.B.; writing—review and editing, A.B.; visualization, A.B.; supervision, A.B.; project administration, A.B.; funding acquisition, A.B. All authors have read and agreed to the published version of the manuscript.

Funding: PSR PUGLIA 2014/2020—Misura 16—Cooperazione—Sottomisura 16.2—"Sostegno a progetti pilota e allo sviluppo di nuovi prodotti, pratiche, processi e tecnologie"—Project entitled "Dal Campo al Boccale" (BE^2R). Grant Number: DDS 94250036020.

Data Availability Statement: No new data were created or analyzed in this study. Data sharing is not applicable to this article.

Abbreviations

BSG	Brewers's spent grain;
Ri	BSG deriving from mashing of a mixture of 65% barley malt and 35% unmalted soft wheat cv. Risciola;
Da	BSG deriving from mashing of a mixture of 65% barley malt and 35% unmalted durum wheat cv. Dauno III;
Em	BSG deriving from mashing of a mixture of 65% barley malt and 35% unmalted emmer;
D	Geographical origin from Daunia;
S	Geopraphical origin from Salento;
20G3 and 25G4	The first number is referred to the percentages of substitution of the Manitoba flour with BSG flour while the second one is referred to the percentages of gluten added calculated on the total amount of flours;
EVOO	Extra-virgin olive oil;
TPC	Total Phenolic Content;
tUPP%	Percentage of total unextractable polymeric proteins;
lUPP%	Percentage of large unextractable polymeric protein;
UP%	Percentage of unextractable (both monomeric and polymeric forms) proteins;
TDF	Total Dietary Fibres;
SDF	Soluble Dietary Fibres;
IDF	Insoluble Dietary Fibres.

References

1. European Commission. *A Sustainable Bioeconomy for Europe: Strengthening the Connection between Economy, Society and the Environment: Updated Bioeconomy Strategy*; Publications Office of the European Union: Brussels, Belgium, 2018. Available online: https://knowledge4policy.ec.europa.eu/publication/sustainable-bioeconomy-europe-strengthening-connection-between-economy-society_en (accessed on 9 January 2023).
2. Montusiewicz, A.; Pasieczna-Patkowska, S.; Lebiocka, M.; Szaja, A.; Szymanaska-Chargot, M. Hydrodynamic cavitation of brewery spent grain diluted by wastewater. *Chem. Eng. J.* **2017**, *313*, 946–956. [CrossRef]
3. Gupta, M.; Abu-Ghannam, N.; Gallaghar, E. Barley for Brewing: Characteristic changes during malting, brewing and applications of its by-products. *Compr. Rev. Food Sci. Food Saf.* **2010**, *9*, 318–328. [CrossRef] [PubMed]
4. Lynch, K.M.; Steffen, E.J.; Arendt, E.K. Brewers' spent grain: A review with an emphasis on food and health. *J. Inst. Brew.* **2016**, *122*, 553–568. [CrossRef]
5. Buffington, J. The economic potential of brewer's spent grain (BSG) as a biomass feedstock. *Adv. Chem. Eng. Sci.* **2014**, *4*, 308–318. [CrossRef]
6. Kavalopoulos, M.; Stoumpou, V.; Christofi, A.; Mai, S.; Barampouti, E.M.; Moustakas, K.; Malamis, D.; Loizidou, M. Sustainable valorisation pathways mitigating environmental pollution from brewers' spent grains. *Environ. Pollut.* **2021**, *270*, 116069. [CrossRef]
7. Bolwig, S.; Mark, M.S.; Happel, M.K.; Brekke, A. Beyond animal feed? The valorisation of brewers' spent grain. In *From Waste to Value: Valorisation Pathways for Organic Waste Streams in Circular Bioeconomies*; Klitkou, A., Fevolden, A.M., Capasso, M., Eds.; Routledge: London, UK, 2019; pp. 107–126.
8. Villacreces, S.; Blanco, C.A.; Caballero, I. Developments and characteristics of craft beer production processes. *Food Biosci.* **2022**, *45*, 101495. [CrossRef]
9. Zeko-Pivač, A.; Tišma, M.; Žnidaršič-Plazl, P.; Kulisic, B.; Sakellaris, G.; Hao, J.; Planinić, M. The potential of brewer's spent grain in the circular bioeconomy: State of the art and future perspectives. *Front. Bioeng. Biotechnol.* **2022**, *10*, 870744. [CrossRef] [PubMed]
10. Hernanz, D.; Nuñez, V.; Sancho, A.I.; Faulds, C.B.; Williamson, G.; Bartolomé, B.; Gómez-Cordovés, C. Hydroxycinnamic acids and ferulic acid dehydrodimers in barley and processed barley. *J. Agric. Food Chem.* **2001**, *49*, 4884–4888. [CrossRef]
11. McCarthy, A.; O'Callaghan, Y.; Piggott, C.; Fitzgerald, R.; O'Brien, N. Brewers' spent grain; bioactivity of phenolic component, its role in animal nutrition and potential for incorporation in functional foods: A review. *Proc. Nutr. Soc.* **2012**, *72*, 117–125. [CrossRef]
12. Izydorczyk, M.S.; Dexter, J.E. Barley β-glucans and arabinoxylans: Molecular structure, physicochemical properties, and uses in food products–A Review. *Food Res. Int.* **2008**, *41*, 850–868. [CrossRef]
13. Gill, S.; Vasanthan, T.; Ooarikul, B.; Rossnagel, B. Wheat bread quality as influenced by the substitution of waxy and regular barley flours in their native and extruded forms. *J. Cereal Sci.* **2002**, *36*, 219–237. [CrossRef]
14. Gill, S.; Vasanthan, T.; Ooarikul, B.; Rossnagel, B. Wheat bread quality as influenced by the substitution of waxy and regular barley flours in their native and cooked forms. *J. Cereal Sci.* **2002**, *36*, 239–252. [CrossRef]
15. Amoriello, T.; Mellara, F.; Galli, V.; Amoriello, M.; Ciccoritti, R. Technological properties and consumer acceptability of bakery products enriched with brewers' spent grains. *Foods* **2020**, *9*, 1492. [CrossRef] [PubMed]
16. Ktenioudaki, A.; Chaurin, V.; Reis, S.F.; Gallagher, E. Brewer's spent grain as a functional ingredient for breadsticks. *Int. J. Food Sci. Technol.* **2012**, *47*, 1765–1771. [CrossRef]
17. Yitayew, T.; Moges, D.; Satheesh, N. Effect of brewery spent grain level and fermentation time on the quality of bread. *Int. J. Food Sci.* **2022**, *2022*, 8704684. [CrossRef] [PubMed]
18. Vriesekoop, F.; Haynes, A.; van der Heijden, N.; Liang, H.; Paximada, P.; Zuidberg, A. Incorporation of Fermented Brewers Spent Grain in the Production of Sourdough Bread. *Fermentation* **2021**, *7*, 96. [CrossRef]
19. Czubaszek, A.; Wojciechowicz-Budzisz, A.; Spychaj, R.; Kawa-Rygielska, J. Effect of added brewer's spent grain on the baking value of flour and the quality of wheat bread. *Molecules* **2022**, *27*, 1624. [CrossRef]
20. Farcas, A.C.; Socaci, S.A.; Chis, M.S.; Pop, O.L.; Fogarasi, M.; Păucean, A.; Igual, M.; Michiu, D. Reintegration of Brewers Spent Grains in the Food Chain: Nutritional, Functional and Sensorial Aspects. *Plants* **2021**, *10*, 2504. [CrossRef]
21. Faccenda, A.; Zambom, M.A.; Avila, A.S.; Castagnara, D.D.; Dri, R.; Fischer, M.L.; Tinini, R.C.R.; Dessbesell, J.G.; Almeida, A.-R.E.; Almeida, K.V. Influence of the storage period on the nutritional and microbiological value of sun-dried brewer's grains. *Rev. Colomb. Cienc. Pecu.* **2021**, *34*, 254–266. [CrossRef]
22. AACC; American Association of Cereal Chemists. Official Methods 46–30.01 Crude Protein—Combustion Method; 44-15.02 Moisture—Air-Oven Methods; 08-01.01 Ash—Basic Method; 32-05.01 Total Dietary Fiber Method; 32-21.01 Insoluble and Soluble Dietary Fiber in Oat Products. In *Enzymatic-Gravimetric Method*, 11th ed.; AACC International, Inc.: St. Paul, MN, USA, 2012.
23. la Gatta, B.; Rutigliano, M.; Rusco, G.; Petrella, G.; Di Luccia, A. Evidence for different supramolecular arrangements in pasta from durum wheat (Triticum durum) and einkorn (Triticum monococcum) flours. *J. Cereal Sci.* **2017**, *73*, 76–83. [CrossRef]
24. Kuktaite, R.; Johansson, E.; Juodeikiene, G. Composition and concentration of proteins in Lithuanian wheat cultivars: Relationships with bread-making quality. *Cereal Res. Commun.* **2000**, *28*, 195–202. [CrossRef]
25. Gandolpho, B.C.G.; Almeida, A.R.; Gandolpho, G.M.; Freitas, D.Z.; Gasparini, O.C.; Machado, M.H.; Barreto, P.L.M. Optimization of brewing waste's (trub) phenolic compounds extraction by ultrasound assisted using response surface methodology. *Quim. Nova* **2021**, *44*, 478–483. [CrossRef]
26. Almeida da Rosa, A.; Ferreira Geraldo, M.R.; Ribeiro, L.F.; Vieira Silva, M.; de Oliveira Brisola Maciel, M.V.; Haminiuk, C.W.I. Bioactive compounds from brewer's spent grain: Phenolic compounds, fatty acids and in vitro antioxidant capacity. *Acta Sci.* **2017**, *39*, 269–277. [CrossRef]

27. Aliakbarian, B.; Casazza, A.A.; Perego, P. Valorization of olive oil solid waste using high pressure–high temperature reactor. *Food Chem.* **2011**, *128*, 704–710. [CrossRef]
28. Brand-Williams, W.; Cuvelier, M.E.; Berset, C. Use of a free radical method to evaluate antioxidant activity. *Food Sci.Technol.* **1995**, *28*, 25–30. [CrossRef]
29. Callejo, M.J. Present situation on the descriptive sesnory analysis of bread. *J. Sens. Stud.* **2011**, *26*, 255–268. [CrossRef]
30. Puri, R.; Khamrui, K.; Khetra, Y.; Malhotra, R.; Devraja, H.C. Quantitative descriptive analysis and principal component analysis for sensory characterization of Indian milk product cham-cham. *J. Food Sci. Technol.* **2016**, *53*, 1238–1246. [CrossRef] [PubMed]
31. Thai, S.; Avena-Bustillos, R.J.; Alves, P.; Pan, J.; Osorio-Ruiz, A.; Miller, J.; Tam, C.; Rolston, M.R.; Teran-Cabanillas, E.; Yokoyama, W.H.; et al. Influence of drying methods on health indicators of brewers spent grain for potential upcycling into food products. *Appl. Food Res.* **2022**, *2*, 100052. [CrossRef]
32. Santos, M.; Jiménez, J.J.; Bartolomé, B.; Gómez-Cordovés, C.; Del Nozal, M.J. Variability of brewer's spent grain within a brewery. *Food Chem.* **2003**, *80*, 17–21. [CrossRef]
33. Senthilkumar, S.; Viswanathan, T.V.; Mercy, A.D.; Gangadevi, P.; Ally, K.; Shyama, K. Chemical composition of brewery waste. *Indian J. Vet. Anim. Sci.* **2010**, *6*, 49–51.
34. Cotecchia, V.; Simeone, V.; Gabriele, S. Caratteri Climatici. Available online: https://www.google.it/url?sa=i&rct=j&q=&esrc=s&source=web&cd=&cad=rja&uact=8&ved=0CAMQw7AJahcKEwjwi-ajosz8AhUAAAAAHQAAAAAQAg&url=https%3A%2F%2Fwww.isprambiente.gov.it%2Ffiles2017%2Fpubblicazioni%2Fperiodici-tecnici%2Fmemorie-descrittive-della-carta-geologica-ditalia%2Fvolume-92%2Fmemdes_92_1_7_caratteri_climatici.pdf&psig=AOvVaw1UKtSAbEyCgo80B8HdWks4&ust=1673957330501947 (accessed on 16 January 2023).
35. L'Abate, G.; Costantini, E.; Barbetti, R.; Fantappiè, M.; Lorenzetti, R.; Magini, S. *Carta dei Suoli D'italia 1:1,000,000 (Soil Map of Italy, Scale 1:1,000,000)*; Società Geografica: Firenze, Italy, 2015; Available online: https://www.researchgate.net/publication/283906875_Carta_dei_Suoli_d%27Italia_11000000_Soil_map_of_Italy_scale_11000000 (accessed on 16 January 2023).
36. He, G.Y.; Rooke, L.; Steel, S.; Bekes, F.; Gras, P.; Tatham, A.S.; Fido, R.; Barcelo, P.; Shewry, P.R.; Lazzeri, P.A. Transformation of pasta wheat (Triticum turgidum L. var. durum) with high-molecular-weight glutenin subunit genes and modification of dough functionality. *Mol. Breed.* **1999**, *5*, 377–386. [CrossRef]
37. Malik, A.H.; Kuktaite, R.; Johansson, E. Combined effect of genetic and environmental factors on the accumulation of proteins in the wheat grain and their relationship to bread-making quality. *J. Cereal Sci.* **2003**, *57*, 170–174. [CrossRef]
38. Zhang, P.; He, Z.; Zhang, Y.; Xia, X.; Chen, D.; Zhang, Y. Association between % SDS-unextractable polymeric protein (%UPP) and end-use quality in Chinese bread wheat cultivars. *Cereal Chem.* **2008**, *85*, 696–700. [CrossRef]
39. Hussain, A.; Larsson, H.; Kuktaite, R.; Prieto-Linde, M.L.; Johansson, E. Amount and size distribution of monomeric and polymeric proteins in the grain of organically produced wheat. *Cereal Chem.* **2003**, *90*, 80–86. [CrossRef]
40. Johansson, E.; Prieto-Linde, M.L.; Svensson, G.; Jönsson, J.Ö. Influence of cultivar, cultivation year and fertilizer rate on amount of protein groups and amout and size distribution of mono- and polymeric proteins in wheat. *J. Agric. Sci.* **2003**, *140*, 275–284. [CrossRef]
41. Lynch, K.M.; Strain, C.R.; Johnson, C.; Patangia, D.; Stanton, C.; Koc, F.; Gil-Martinez, J.; O'Riordan, P.; Sahin, A.W.; Ross, R.P.; et al. Arendt Extraction and characterisation of arabinoxylan from brewers spent grain and investigation of microbiome modulation potential. *Eur. J. Nutr.* **2021**, *60*, 4393–4411. [CrossRef]
42. Shih, Y.; Wang, W.; Hasenbeck, A.; Stone, D.; Zhao, Y. Investigation of physicochemical, nutritional, and sensory qualities of muffins incorporated with dried brewer's spent grain flours as a source of dietary fiber and protein. *J. Food Sci.* **2020**, *85*, 3943–3953. [CrossRef] [PubMed]
43. Waters, D.M.; Jacob, F.; Titze, J.; Arendt, E.K.; Zannini, E. Fibre, protein and mineral fortification of wheat bread through milled and fermented brewer's spent grain enrichment. *Eur. Food Res. Technol.* **2012**, *235*, 767–778. [CrossRef]
44. Neylon, E.; Arendt, E.K.; Zannini, E.; Sahin, A.W. Fermentation as a tool to revitalise brewers' spent grain and elevate techno-functional properties and nutritional value in high fibre bread. *Foods* **2021**, *10*, 1639. [CrossRef]
45. Shewry, P.R.; Hey, S. Do "ancient" wheat secies differ from modern bread wheat in their contents of bioactive compounds? *J. Cereal Sci.* **2015**, *65*, 236–243. [CrossRef]
46. Bento-Silva, A.; Koistinen, V.M.; Mena, P.; · Bronze, M.R.; Hanhineva, K.; Sahlstrøm, S.; Kitrytė, V.; Moco, S.; Aura, A.M. Factors affecting intake, metabolism and health benefits of phenolic acids: Do we understand individual variability? *Eur. J. Nutr.* **2020**, *59*, 1275–1293. [CrossRef]
47. Rao, S.; Schwarz, L.J.; Santhakumar, A.B.; Chinkwo, K.A.; Blanchard, C.L. Cereal phenolic contents as affected by variety and environment. *Cereal Chem.* **2018**, *95*, 589–602. [CrossRef]
48. Langos, D.; Granvogl, M. Studies on the simultaneous formation of aroma-active and toxicologically relevant vinyl aromatics from free phenolic acids during wheat beer brewing. *J. Agric. Food Chem.* **2016**, *64*, 2325–2332. [CrossRef] [PubMed]
49. Schwarz, K.J.; Boitz, L.I.; Methner, F.-J. Release of phenolic acids and amino acids during mashing dependent on temperature, ph, time, and raw materials. *J. Am. Soc. Brew. Chem.* **2012**, *70*, 290–295. [CrossRef]
50. Moreira, M.M.; Morais, S.; Carvalho, D.O.; Barros, A.A.; Delerue-Matos, C.; Guido, L.F. Brewer's spent grain from different types of malt: Evaluation of the antioxidant activity and identification of the major phenolic compounds. *Food Res. Int.* **2013**, *54*, 382–388. [CrossRef]
51. Birsan, R.I.; Wilde, P.; Waldron, K.W.; Rai, D.K. Recovery of polyphenols from brewer's spent grains. *Antioxidants* **2019**, *8*, 380. [CrossRef] [PubMed]

52. Baiano, A.; Fiore, A.; la Gatta, B.; Tufariello, M.; Gerardi, C.; Savino, M.; Grieco, F. Single and interactive effects of unmalted cereals, hops, and yeasts on quality of white-inspired craft beers. *Beverages* **2023**, *9*, 9. [CrossRef]
53. Rufián-Henares, J.A.; Delgado-Andrade, C.; Morales, F.J. Assessing the Maillard reaction development during the toasting pro-cess of common flours employed by the cereal products industry. *Food Chem.* **2009**, *114*, 93–99. [CrossRef]
54. Gómez, M.; Ronda, F.; Blanco, C.A.; Caballero, P.A.; Apesteguia, A. Effect of dietary fibre on dough rheology and bread quality. *Eur. Food Res. Technol.* **2003**, *216*, 51–56. [CrossRef]
55. Musa, R.; Elawad, O.; Yang, T.A.; Ahmed, A.H.R.; Ishag, K.E.A.; Mudawi, H.A.; Abdelrahim, S.M.K. Chemical composition and functional properties of wheat bread containing wheat and legumes bran. *Int. J. Food Sci. Nutr.* **2016**, *1*, 10–15.
56. Guo, L.; Xu, D.; Fanf, F.; Jin, Z.; Xu, X. Effect of glutathione on wheat dough properties and bread quality. *J. Cereal Sci.* **2020**, *96*, 103116. [CrossRef]
57. Delcour, J.A.; Hoseney, R.C. Yeast-leavened products. In *Principles of Cereal Science and Technology*; AACC International, Inc.: St. Paul, MN, USA, 2010; pp. 177–206.
58. Heitmann, M.; Zannini, E.; Arendt, E. Impact of Saccharomyces cerevisiae metabolites produced during fermentation on bread quality parameters: A review. *Crit. Rev. Food Sci. Nutr.* **2018**, *58*, 1152–1164. [CrossRef] [PubMed]
59. Tian, W.; Chen, G.; Tilley, M.; Li, Y. Changes in phenolic profiles and antioxidant activities during the whole wheat bread-making process. *Food Chem.* **2021**, *345*, 128851. [CrossRef] [PubMed]
60. Baáez, J.; Fernaández-Fernaández, A.; Briozzo, F.; Díaz, S.; Dorgans, A.; Tajam, V. Effect of enzymatic hydrolysis of brewer's spent grain on bioactivity, techno-functional properties and nutritional value when added to a bread formulation. *Biol. Life Sci. Forum* **2021**, *6*, 100.
61. European Union. Regulation (EU) N°. 432/2012. Commission Regulation (EU) No 43272012 of 16 May 2012 establishing a list of permitted health claims made on foods, other than those referring to the reduction of disease risk and to children's development and health. *Off. J. Eur. Union* **2012**, *L136*, 1–40.
62. European Union. Regulation (EC) N.° 1924/2006. Regulation (EC) No 1924/2006 of the European Parliament and of the Council of 20 December 2006 on nutrition and health claims made on foods. *Off. J. Eur. Union* **2006**, *L404*, 9–25.
63. Depeint, F.; Lecerf, J.-M.; Pouillart, P.; Beckers, Y.; Khorsi-Cauet, H. Bifidogenic effect of a wheat arabinoxylan (MC-AX) is observed across two animal models, a simulated human intestine model (SHIME®) and a clinical study. *Med. Res. Arch.* **2017**, *5*. Available online: https://esmed.org/MRA/mra/article/view/938\T1\textgreater{} (accessed on 25 January 2023).
64. Malunga, L.N.; Izydorczyk, M.; Beta, T. Antiglycemic effect of water extractable arabinoxylan from wheat aleurone and bran. *J. Nutr. Metab.* **2017**, *2017*, 5784759. [CrossRef] [PubMed]
65. Bliss, R. Online Nutrition Resources at Your Fingertips. Available online: https://www.usda.gov/media/blog/2015/03/31/online-nutrition-resources-your-fingertips (accessed on 25 January 2023).
66. Onyetugo, C.; Egong, E.J.; Nwagu, T.N.; Okpala, G.; Onwosi, C.O.; Chukwu, G.C.; Okolo, B.N.; Agu, R.C.; Moneke, A.N. Process optimization for simultaneous production of cellulase, xylanase and ligninase by Saccharomyces cerevisiae SCPW 17 under solid state fermentation using Box-Behnken experimental design. *Heliyon* **2020**, *6*, e04566.
67. Naibaho, J.; Korzeniowska, M.; Wojdyło, A.; Figiel, A.; Yang, B.; Laaksonen, O.; Foste, M.; Vilu, R.; Viiard, E. Fiber modification of brewers' spent grain by autoclave treatment to improve its properties as a functional food ingredient. *LWT-Food Sci. Technol.* **2021**, *149*, 111877. [CrossRef]
68. Ginindza, A.; Solomon, W.K.; Shelembe, J.S.; Nkambule, T.P. Valorisation of brewer's spent grain flour (BSGF) through wheat-maize-BSGF composite flour bread: Optimization using D-optimal mixture design. *Heliyon* **2022**, *8*, e09514. [CrossRef]
69. Guo, M.; Du, J.; Zhang, Z.; Jin, Y. Optimization of brewer's spent grain-enriched biscuits processing formula. *J. Food Proc. Eng.* **2014**, *37*, 122–130. [CrossRef]
70. Petrović, J.S.; Pajin, B.S.; Kocić Tanackov, S.D.; Pejin, J.D.; Fišteš, A.Z.; Bojanić, N.D.; Lončarević, I.S. Quality properties of cookies supplemented with fresh brewer's spent grain. *Food Feed Res.* **2017**, *44*, 57–63. [CrossRef]
71. Haruna, M.; Udobi, C.; Ndife, J. Effect of added brewers dry grain on the physico-chemical, microbial and sensory quality of wheat bread. *Am. J. Food Nutr.* **2011**, *1*, 39–43. [CrossRef]
72. Crofton, E.C.; Scannell, A.G. Snack foods from brewing waste: Consumer-led approach to developing sustainable snack options. *Br. Food J.* **2020**, *122*, 3899–3916. [CrossRef]
73. Di Matteo, P.; Bortolami, M.; Curulli, A.; Feroci, M.; Gullifa, G.; Materazzi, S.; Risoluti, R.; Petrucci, R. Phytochemical Characterization of Malt Spent Grain by Tandem Mass Spectrometry also Coupled with Liquid Chromatography: Bioactive Compounds from Brewery By-Products. *Front. Biosci.* **2023**, *28*, 3. [CrossRef]
74. Combest, S.; Warren, C. The effect of upcycled brewers' spent grain on consumer acceptance and predictors of overall liking in muffins. *J. Food Qual.* **2022**, *2022*, 6641904. [CrossRef]

 **foods**

Article

A Functional End-Use of Avocado (cv. Hass) Waste through Traditional Semolina Sourdough Bread Production

Enrico Viola [1], Carla Buzzanca [2], Ilenia Tinebra [1,*], Luca Settanni [1], Vittorio Farina [1,3], Raimondo Gaglio [1] and Vita Di Stefano [2]

[1] Department of Agricultural, Food and Forest Sciences (SAAF), Università degli Studi di Palermo, Viale delle Scienze, 90128 Palermo, Italy; enrico.viola01@unipa.it (E.V.); luca.settanni@unipa.it (L.S.); vittorio.farina@unipa.it (V.F.); raimondo.gaglio@unipa.it (R.G.)

[2] Department of Biological, Chemical and Pharmaceutical Science and Technology (STEBICEF), University of Palermo, Via Archirafi, 90123 Palermo, Italy; carla.buzzanca@unipa.it (C.B.); vita.distefano@unipa.it (V.D.S.)

[3] Centre for Sustainability and Ecological Transition, University of Palermo, Piazza Marina, 90133 Palermo, Italy

* Correspondence: ilenia.tinebra@unipa.it

Abstract: In recent years, a main goal of research has been to exploit waste from agribusiness industries as new sources of bioactive components, with a view to establishing a circular economy. Non-compliant avocado fruits, as well as avocado seeds and peels, are examples of promising raw materials due to their high nutritional yield and antioxidant profiles. This study aimed to recycle avocado food waste and by-products through dehydration to produce functional bread. For this purpose, dehydrated avocado was reduced to powder form, and bread was prepared with different percentages of the powder (5% and 10%) and compared with a control bread prepared with only semolina. The avocado pulp and by-products did not alter organoleptically after dehydration, and the milling did not affect the products' color and retained the avocado aroma. The firmness of the breads enriched with avocado powder increased due to the additional fat from the avocado, and alveolation decreased. The total phenolic content of the fortified breads was in the range of 2.408–2.656 mg GAE/g, and the antiradical activity was in the range of 35.75–38.235 mmol TEAC/100 g ($p < 0.0001$), depending on the percentage of fortification.

Keywords: avocado wastes/by-products; functional bread; lactic acid bacteria; sourdough; peels; pulp; seeds; polyphenols; antioxidant properties

Citation: Viola, E.; Buzzanca, C.; Tinebra, I.; Settanni, L.; Farina, V.; Gaglio, R.; Di Stefano, V. A Functional End-Use of Avocado (cv. Hass) Waste through Traditional Semolina Sourdough Bread Production. *Foods* **2023**, *12*, 3743. https://doi.org/10.3390/foods12203743

Academic Editors: Grazia Maria Borrelli and Donatella Bianca Maria Ficco

Received: 14 September 2023
Revised: 9 October 2023
Accepted: 10 October 2023
Published: 11 October 2023

1. Introduction

Avocado (*Persea americana* Mill.) is a subtropical/tropical fruit native to Mexico and Central America and is widely produced and consumed worldwide [1]. In recent years, avocado production has steadily increased globally due to the growing popularity and demand for the fruit [2]. The main avocado producers are Mexico (33%), the Dominican Republic (10.5%), Peru (7.8%), Indonesia (5.7%), and Colombia (5.1%) [3]. Spain and Italy are the only European countries with significant commercial production of avocados, which are cultivated, respectively, on the Andalusian Mediterranean coast, mainly in the provinces of Malaga and Granada, and in Sicily, along the Tyrrhenian coastal areas and close to Catania [4]. In Europe, avocado consumption per capita increased by an average of 180% between 2012/13 and 2018/20, with industry expectations for further increases [5]. A significant portion of this demand is driven by the young millennial generation in Europe and increased consumer interest in so-called "superfoods". Avocado fruit has great potential to meet consumers' desired requirements due to its high nutritional value, particularly its antioxidants, fiber, and low sugar content [6–8]. For these reasons, eating avocados is generally recommended for people with diabetes because it is a high-energy food [9] and can be used in a wide range of food products [10]. However, consumers' avocado preference may depend on several quality attributes [11,12]. These quality attributes refer to physical product characteristics such as freshness, color, and size and experience attributes such as taste, aroma, and stage of maturation [13,14]. External factors that most influence consumer choice include fruit weight (commercial size), peel color (green or black), absence of defects (crick side, blanch, terminal spot), and ripening stage, which is closely related to fruit firmness, given the high perishability of avocados [14,15]. For these reasons, consumers prefer to buy unripe and/or not fully ripe avocado fruit [14], avoiding fruit that is already ripe or overripe.

Avocados are mostly consumed fresh, but they are also processed to extract oil and other products, such as guacamole [16]. Therefore, several components of the fruit, including the peel and seeds, are not used and are wasted, becoming a source of environmental contamination. However, these components are rich in protein, fiber, and numerous bioactive compounds [17,18]. For instance, the seed and peel of the "Hass" avocado account for about 15% and 14% of the weight of the fruit, respectively [18–20]. This is equivalent to at least 1.6 million tons of avocado seeds and peels annually discarded worldwide [1], which adds to the global share of food waste. Among the various processing techniques that can be used to recycle both the discarded (due to overripeness) avocado fruit and its by-products is dehydration.

Drying is probably one of the oldest methods of food preservation [21] and consists of the removal of water to a final concentration, which assures microbial stability and ensures the expected shelf-life of the product [22]. In addition, this technique is the most widely used for creating powders from fresh fruits [23].

Fruit and vegetable powders can be used as intermediates in the beverage industry, functional food additives that improve the nutritional value of foods, flavoring agents, or natural coloring agents [24]. Fruit and vegetable powders also serve as ingredients for pasta, breads, dry soups, and other food recipes [25–29].

Powder quality depends largely on the drying and milling conditions as well as the composition and quality of the raw material [30,31].

Fruit and vegetable powder ingredients for dough and/or bread preparation must be strategically selected to achieve the optimal composition and physical properties and avoid adverse effects [32]. In fact, vegetable powder can potentially decrease the stability of dough because the fiber in it slows down the rate of hydration and gluten development. This depends on the amount of vegetable powder incorporated [33]. Similarly, vegetable powder may affect texture differently depending on the type of by-product. On the other hand, fruit and vegetable powders can impart coloring and stabilizing properties to the final product due to the presence of carotenoids and polyphenols [33]. This work aimed to recycle avocado waste to produce new value-added ingredients. To this end, avocado waste was dehydrated and milled. The resulting powder was then used as ingredients for processing sourdough semolina bread to functionalize this food, widely consumed daily in Southern Italy.

2. Materials and Methods

2.1. Chemicals and Reagents

Methanol, sodium carbonate, gallic acid, Folin-Ciocalteu's phenol reagent, DPPH (2,2-diphenyl-1-picrylhydrazyl), ABTS (2,2′ azino-bis (3-ethylbenzothiazoline-6-sulfonic-acid), potassium persulphate, sodium hydroxide (NaOH) and Trolox (6-hydroxy-2,5,7,8-tetramethylchroman-2-carboxylic acid) were obtained from Fluka (Buchs, Switzerland). HPLC-grade water was obtained by purifying double distilled water in a Milli-Q Gradient A10 system (Millipore, Bedford, MA, USA), 0.45 µm PTFE syringe filter (Whatman, Milan, Italy).

2.2. Production of Avocado Waste Powder (AWP) and Commercial Semolina

Avocado fruit (*Persea americana* Mill.) cv. Hass was harvested at the experimental field of the Department of Agricultural, Food, and Forestry Sciences, University of Palermo.

After being harvested, the fruits were left to ripen at a temperature of 20 ± 5 °C, and the progress of ripening was assessed by the change in epicarp color; therefore, the shade angle indicator was used, as described by Sánchez-Quezada et al. [34].

To obtain a uniform sample representative of the ripening stage, the researchers chose to pick avocado fruit at the overripe stage, i.e., fruits with hue angle values of $\geq 45 \pm 7$ h°, as determined by Sánchez-Quezada et al. [34]. After being sanitized in chlorinated water (2% *v*/*w*) for 10 min, the avocado fruits were peeled and their seeds were removed.

The different components of the avocado fruit (pulp, seeds, and peels) were separated as they required different dehydration times and temperatures. After several preliminary tests the avocado pulp and its by-products were dried as follows:

- Pulp at a temperature of 75 °C for 28 h;
- Peel and seeds at a temperature of 60 °C for 4 h.

Before dehydration, the seeds were washed with water, and the outer covering of the seeds was removed manually during washing. A tray dryer (Ausla, 1000 Watt, Milan, Italy) was used to dry the pulp and by-products.

The time/temperature binomials chosen generated a moisture content of less than 12–13%, which is the range selected to avoid microbial proliferation and to achieve rapid drying that would not lead to degradation of the bioactive components [35]. The dehydrated products obtained (pulp, seed, and peel) were separately processed into "powder" by an ultra-centrifugal mill (Fritsch, Pulverisette 14, Lainate, Italy). To obtain a powder with particle sizes between 1.5 and 2 mm from each fruit part (pulp, seeds, and peels), they were processed at 700 rpm for 10 s.

For breadmaking, a "powder mixture" (Avocado Waste Powder: AWP) comprised of pulp, seeds, and peels was used. The AWP consisted of the following percentages of dried fruit: 50% pulp, 25% seeds, and 25% peel.

A commercial semolina (Cuore Mediterraneo, Santa Giusta, Italy) was used to process the bread for this study. Its nutritional values (per 100 g) were: 12.5 g of protein; 1.5 g of fat; 0.3 g of saturated fats; 69 g of carbohydrates, and 26 g of fiber.

2.3. Determination of Color Characteristics of AWP

The color of both semolina and AWP samples was measured using a Minolta colorimeter (Chroma Meter CR-400, Konica Minolta Sensing Inc., Tokyo, Japan), and the L* (brightness), Chroma (C*), and hue angle (h°) parameters were evaluated [21]. The instrument was calibrated using a standard white plate. Chroma (C*) values and hue angles (h°) were calculated using Equations (1) and (2), respectively:

$$C^* = \left(a^2 + b^2\right)^{\frac{1}{2}} \tag{1}$$

$$h° = \arctang\left(\frac{b}{a}\right) \tag{2}$$

Using the obtained values of L*, a*, and b*, a color table was created by converting the CIEL*a*b* color space to the red/green/blue (RGB) scale through the e-paint.co website (accessed on 15 June 2023).

2.4. Hygienic Characteristics of AWP

The AWP was microbiologically analyzed for some microbial groups that are unwanted during food fermentation, as reported by Messina et al. [36]. Briefly, 10 g of AWP was first homogenized by a BagMixer® 400 stomacher (Interscience, Saint Nom, France) and then serially diluted. The diluted samples were analyzed for the following microbial groups: total mesophilic microorganisms (TMM), members of the Enterobacteriaceae family, total coliforms, and spore-forming aerobic bacteria. The analyses were performed in duplicate.

2.5. Bacterial Strains

Lactic acid bacteria (LAB) isolated from Sicilian sourdoughs and previously tested to produce semolina breads with the addition of by-product ingredients [37] were used to prepare a multiple-strain sourdough starter. The strains *Lentilactobacillus diolivorans* SD4, *Fructilactobacillus sanfranciscensis* SD22, *Levilactobacillus brevis* SD46, *Lactiplantibacillus plantarum* SD96, *Weissella cibaria* SD123, *Lactiplantibacillus pentosus* SD130, *Leuconostoc citreum* SD142, and *Leuconostoc holzapfelii* SD148, all belonging to the Culture Collection of the Agricultural Laboratory of the University of Palermo, Italy, were defrosted from $-80\,^\circ$C and cultivated in de Man-Rogosa-Sharpe medium modified as described by Lhomme et al. [38] at $30\,^\circ$C for 24 h.

2.6. Sourdough Propagation

After reactivation in a synthetic medium, all the LAB strains were propagated in sterile semolina extract (SSE) broth [39]. Commercial semolina was used to both prepare liquid SSE broth and propagate solid sourdough. The individual cultivation of LAB and the mixed cell culture representing sourdough inoculum were performed as reported by Gaglio et al. [40].

The LAB mixed culture was diluted in sterile tap water to reach a final volume of 187.5 mL. This cell suspension was added to 312.5 g of semolina to obtain a 500 g dough with a dough yield (DY = weight of the dough/weight of semolina $\times$ 100) of 160 and a cell density of about 10^6–10^7 CFU/g. The dough was then left to ferment at $28\,^\circ$C for 16 h and subjected to seven consecutive daily refreshments to generate a mature sourdough inoculum [41].

2.7. Bread Doughs and Baking Process

Bread production was carried out solely with the sourdough developed from the selected LAB strains. No baker's yeast and kitchen salt were added to evaluate the effect of AWP on the performance of LAB. The control (CTR) doughs (800 g) to be leavened before baking were processed by adding 228.6 mL of sterile tap water and 457.2 g of semolina to 114.2 g of mature sourdough (DY = 175). The experimental AWP doughs were produced with the same amount of water and sourdough, but the amount of semolina was reduced to 434.3 g and 411.4 g for the 5-AWP [containing 5% (w/w) AWP] and 10-AWP [containing 10% (w/w) AWP] trials, respectively. A planetary mixer (model XBM10S; Electrolux Professional SpA, Pordenone, Italy) was used to mix all the ingredients for 15 min with a paddle turning on Speed 4. Aliquots of 100 g per dough were transferred into trapezoidal stainless steel baking pans [42], kept at $28\,^\circ$C for 8 h, and then baked as reported by Alfonzo et al. [43]. Two technical repeats were obtained from each bread trial (performed in duplicate), and all bread baking was repeated after two weeks to obtain two independent replicates.

2.8. Acidification Process

Sourdough fermentation was monitored by pH measurement, total titratable acidity (TTA) determination, and the evolution of LAB numbers following the approach of Francesca et al. [44]. To perform LAB viable counts and that of other microbial groups relevant during dough fermentation, the sourdough and bread doughs were microbiologically evaluated to enumerate TMM, sourdough LAB, yeasts, members of the Enterobacteriaceae family, and total coliforms, as reported by Gaglio et al. [37]. All analyses were performed in duplicates.

2.9. Quality Characteristics of Breads

The breads were cooled at room temperature for approximately 30 min after baking and investigated for several quality parameters, as reported by Cirlincione et al. [45]. In particular, the following parameters were considered: weight loss (WL, %), specific volume (cm^3/g bread), firmness (N/mm^2), crust and crumb color [Lightness (L*), redness (a*) and yellowness (b*)], void fraction (%), cell density (number of cells/cm^2), and mean cell area (mm^2). The analyses were performed in duplicate.

2.10. Chemical Characterization

2.10.1. Total Phenolic Content Analysis

Total phenolic content (TPC) was determined using the optimized Folin–Ciocalteu method previously published [46]. One gram of each bread sample (CTR-Bread, bread produced with control dough; 5-AWP Bread, experimental bread enriched with 5% (*w/w*) of avocado waste powder (AWP); 10-AWP Bread, experimental bread enriched with 10% (*w/w*) of AWP) was added to 5 mL of methanol/water (80:20 *v/v*) and sonicated and filtered through Whatman 0.45 μm PTFE filters. This was followed by a reaction with the Folin-Ciocalteu reagent in the presence of sodium carbonate to form a blue-colored complex. The intensity of the color was proportional to the phenolic compounds in the sample. The resulting colorimetric reaction was measured at 765 nm using a UV-VIS spectrophotometer (Varian Cary 50, Agilent, Santa Clara, United States). The amount of TPC was calculated by interpolation from a calibration curve of gallic acid [0.001 to 0.25 mg/mL] (y = 10.945x + 0.1305, R^2 = 0.993). The results were expressed as mg gallic acid equivalents per g (mg GAE g^{-1}) of the sample.

2.10.2. Radical Scavenging Properties Evaluation, DPPH and ABTS Assay

The measurement of the powder and fortified bread samples' antiradical activity (DPPH (2,2-diphenyl-1-picrylhydrazyl) and ABTS (2,2′ azino-bis (3-ethylbenzothiazoline-6-sulfonic-acid) assays) followed a procedure previously described by Di Stefano et al. [47]. The DPPH assay was used for the in vitro evaluation of the scavenger activity toward free radicals. One g of each bread sample (AWP and semolina) was extracted with 4 mL of methanol, mixed with 3 mL of DPPH (60 μM) and placed in the dark for 30 min. Scavenging activity was monitored by spectrophotometric analysis of the absorbance at a wavelength of 517 nm with a UV-VIS spectrophotometer (Varian Cary$^®$ 50, Agilent, Santa Clara, United States) and using methanol as the blank. The results were reported as Trolox equivalent antioxidant activity and expressed as mmol Trolox equivalent (TE)/100 g of the sample. The absorbance signal was translated into antioxidant activity using Trolox as the standard and the calibration curve in the range of 5–400 μM (y = −0.0008x + 0.4036, R^2 = 0.998). All experiments were performed in triplicate. According to Re et al. [43], for the ABTS assay, the ABTS$^+$ radical cation was produced by reacting ABTS stock solution with 2.45 mM potassium persulfate; the solution turned a dark blue-green at the end of the reaction time. One gram of each sample was added to 4 mL of methanol, sonicated, and filtered through Whatman 0.45 μm PTFE filters. A calibration curve using Trolox at increasing concentrations [2.5–30 μM] was constructed. The assays were performed in triplicate.

2.11. Sensory Analysis

A panel of 17 judges, including 10 women and seven men whose ages ranged from 26 to 57, was recruited to perform a descriptive sensory analysis of the breads produced with different percentages of AWP. First, the judges were trained to acquire familiarity with bread attributes by tasting commercial semolina bread. The judges were asked to score appearance, texture, odor, and taste descriptors from among those reported by Ruisi et al. (2021) [48]. The evaluation for each attribute was expressed on a 9-point scale (1 = extremely bad; 9 = extremely good). The evaluation was performed by the judges in individual chambers following the ISO 13299 guidelines [49].

2.12. Statistical Analysis

Differences between the microbiological and physicochemical data were identified by way of one-way variance analysis, while Tukey's test was used for multiple mean comparisons (statistical significance $p < 0.05$). In addition, a hierarchical cluster analysis (HCA) was performed to group the produced breads according to their dissimilarity, as reported by Martorana et al. [50]. The data were statistically processed using XLStat software version 7.5.2 for Excel (Addinsoft, New York, NY, USA).

3. Results

3.1. Color Characteristics

From the analysis of colorimetric data (presented in Table 1), the semolina sample showed an elevated brightness (86.90 ± 1.63) compared to the AWP sample but a lower C* index. This was due to the color of the semolina, which, as can be seen from the RGB data, tended toward white. The AWP, on the other hand, showed lower values of L* but higher values of Chroma. From the values obtained and the evaluation of the color table, the powders maintained a color tending toward green. Therefore, processing the dehydrated avocados into powder did not alter their color.

Table 1. Brightness (L*), chroma (C*), and hue angle (h°*) of the CTR and AWP samples. CIELab* values were then converted to RGB. Abbreviation: AWP (avocado waste powder). Results indicate mean values ± S.D. (standard deviation). n.a. = not analyzed. Data within a column followed by different letters are significantly different according to Tukey's test.

Samples	L*	C*	h°	RGB
Semolina	86.90 ± 1.63 a	32.31 ± 0.72 b	1.35 ± 0.07	
AWP	48.50 ± 1.19 b	50.72 ± 1.90 a	1.52 ± 0.02	
p value	<0.0001	<0.0001	0.777	n.a.

3.2. Monitoring of the Fermentation Process

The growth of the selected LAB strains in SSE showed a consistent decrease in pH until the average value of 4.03 ± 0.10. The lowest pH (3.82 ± 0.18) was reached by the strains *L. pentosus* SD130 and *L. plantarum* SD96. The mixture with the eight LABs developed individually in SSE that was obtained after three propagation days was used as a liquid inoculum to produce sourdough for bread production. Soon after inoculation of semolina and tap water with the eight strains, the obtained dough showed a pH of 5.44 ± 0.21; this value decreased to 3.92 ± 0.04 in the sourdough obtained after seven days with daily refreshments. The TTA of the mature sourdough was 12.10 ± 0.07 NaOH 0.1 N/10 g. Table 2 presents the pH and TTA data for the unbaked doughs up to 8 h of leavening. A pH value of 5.5 was measured for the CTR dough at the start of fermentation.

The initial pH of the 5-AWP dough was slightly lower (5.42) than that of the 10-AWP dough (5.56) and the CTR dough. During fermentation, the pH values decreased progressively until they reached almost similar values of 4.19, 4.33, and 4.37 for the CTR, 5-AWP, and 10-AWP doughs, respectively, at the end of the monitoring period. An inverse trend was noted for the TTA, which increased linearly over time. After 8 h of fermentation, the TTA of the CTR dough (8.93 mL NaOH 0.1 N/10 g) was slightly higher than the value expressed by the AWP doughs (8.65 and 8.60 for 5-AWP and 10-AWP, respectively). The results of the plate count of the doughs are presented in Table 3. Sourdough developed from LAB selected at the seventh refreshment was characterized by 8.64 and 7.74 Log CFU/g of LAB and yeasts, respectively. Regarding the doughs leavened for bread production, LAB accounted for 7.60–7.71 Log CFU/g soon after ingredient mixing. These data were a little higher than those shown by the TMM (6.60–6.66 Log CFU/g) and confirmed that LAB from the sourdough inoculum was significantly transferred to the bread doughs. The levels of yeast immediately after production (0 h) were one order of magnitude lower than the LAB. After 8 h of fermentation, the cell densities of LAB increased in all the trials. The LAB levels of the CTR and 5-AWP doughs were almost comparable (8.94 and 8.98 Log CFU/g, respectively), while a slightly lower density (8.51 Log CFU/g) characterized the 10-AWP dough. A very limited increase in cell density was registered for the yeasts, barely overcoming 7.0 Log CFU/g for the CTR and 5-AWP doughs at the end of fermentation. Regarding hygiene indicators, although the levels of Enterobacteriaceae and total coliforms in the AWP were below the detection limit (for this reason, these results are not included in Table 3), after 8 h of fermentation, their presence was revealed in both AWP bread doughs at levels around 10^3 CFU/g. No spore-forming bacteria were detected in the AWP and corresponding bread doughs.

Table 2. Chemical parameters of doughs. Results indicate mean values $\pm$ S.D. (standard deviation) of four determinations (carried out in two technical repeats for two independent experiments). Data within a line followed by different letters are significantly different according to Tukey's test. Abbreviations: TTA, total titratable acidity; CTR, control dough; 5-AWP, experimental dough enriched with 5% (*w/w*) of avocado waste powder (AWP); 10-AWP, experimental dough enriched with 10% (*w/w*) of AWP; n.a. = not analyzed.

Time	Parameter	Samples				*p*-Value
		Sourdough	CTR	5-AWP	10-AWP	
0 h	pH	3.92 ± 0.04 b	5.50 ± 0.08 a	5.42 ± 0.05 a	5.56 ± 0.08 a	<0.0001
	TTA	12.10 ± 0.07 a	7.58 ± 0.04 b	7.45 ± 0.07 bc	7.33 ± 0.04 c	<0.0001
2 h	pH	n.a.	5.38 ± 0.03	5.34 ± 0.04	5.43 ± 0.13	0.440
	TTA	n.a.	7.63 ± 0.04	7.65 ± 0.07	7.58 ± 0.11	0.565
4 h	pH	n.a.	4.82 ± 0.06 b	4.98 ± 0.01 a	5.07 ± 0.04 a	0.001
	TTA	n.a.	8.08 ± 0.04 a	7.90 ± 0.07 b	7.83 ± 0.04 b	0.003
6 h	pH	n.a.	4.48 ± 0.06 b	4.71 ± 0.02 a	4.70 ± 0.07 a	0.003
	TTA	n.a.	8.50 ± 0.14	8.30 ± 0.07	8.28 ± 0.11	0.092
8 h	pH	n.a.	4.19 ± 0.04 b	4.33 ± 0.02 a	4.37 ± 0.02 a	0.001
	TTA	n.a.	8.93 ± 0.11 a	8.65 ± 0.07 b	8.60 ± 0.07 b	0.007

Table 3. Microbial loads of doughs. Results indicate mean values $\pm$ S.D. (standard deviation) of four plate counts (carried out in two technical repeats for two independent experiments), expressed as Log CFU/g. Data within a line followed by different letters are significantly different according to Tukey's test. Abbreviations: TMM, total mesophilic microorganisms; LAB, lactic acid bacteria; CTR, control dough; 5-AWP, experimental dough enriched with 5% (*w/w*) of avocado waste powder (AWP); 10-AWP, experimental dough enriched with 10% (*w/w*) of AWP; n.a. = not analyzed; n.d. = not detected.

Media	Time	Samples				*p*-Value
		Sourdough	CTR	5-AWP	10-AWP	
TMM	0 h	7.76 ± 0.20 a	6.66 ± 0.24 b	6.62 ± 0.13 b	6.60 ± 0.16 b	0.0001
	8 h	n.a.	7.22 ± 0.11	6.76 ± 0.31	7.03 ± 0.14	0.087
Sourdough LAB	0 h	8.64 ± 0.15 a	7.60 ± 0.33 b	7.68 ± 0.28 b	7.71 ± 0.27 b	0.004
	8 h	n.a.	8.94 ± 0.20	8.98 ± 0.19	8.51 ± 0.22	0.055
Yeasts	0 h	7.74 ± 0.25 a	6.52 ± 0.37 b	6.63 ± 0.31 b	6.39 ± 0.27 b	0.002
	8 h	n.a.	7.06 ± 0.16	7.03 ± 0.19	6.78 ± 0.22	0.225
Total coliforms	0 h	<1	<1	<1	<1	n.d.
	8 h	n.a.	<1 b	2.90 ± 0.17 a	3.02 ± 0.28 a	<0.0001
Enterobacteriaceae	0 h	<1	<1	<1	<1	n.d.
	8 h	n.a.	<1 b	3.06 ± 0.32 a	2.80 ± 0.17 a	<0.0001

3.3. Bread Quality Attributes

The characteristics of the final breads produced are summarized in Table 4.

Weight loss after baking was 11.30% in the CTR bread, while lower values were displayed by the AWP breads. The specific volume of the breads decreased with the different percentages of AWP; a value of 3.18 cm^3/g was registered for the CTR bread compared to a value of 2.74 cm^3/g for the 10-AWP bread. The addition of AWP determined a linear increase in firmness, with the highest value (0.113 N/mm^2) recorded for the 10-AWP bread. Furthermore, the addition of AWP determined a change in the color parameters of both the crust and crumb of the breads, especially for L* and a*. Both parameters decreased progressively with the AWP percentages. Negative values were registered for the crumbs of all the trials. Image analysis of the breads indicated an increase in the void fraction and cell density of the crumbs with increasing percentages of AWP and a decrease in alveolation.

Table 4. Quality attributes of bread samples. Results indicate mean values $\pm$ S.D. (standard deviation) of four determinations (carried out in two technical repeats for two independent experiments). Data within a line followed by different letters are significantly different according to Tukey's test. Abbreviations: CTR-Bread, bread produced with control dough; 5-AWP Bread, experimental bread enriched with 5% (w/w) of avocado waste powder (AWP); 10-AWP Bread, experimental bread enriched with 10% (w/w) of AWP.

Attributes	Samples			p-Value
	CTR-Bread	5-AWP Bread	10-AWP Bread	
Weight loss (%)	11.30 $\pm$ 1.37	9.48 $\pm$ 0.79	9.94 $\pm$ 1.20	0.210
Specific volume (cm^3/g bread)	3.18 $\pm$ 0.11 a	3.09 $\pm$ 0.16 a	2.74 $\pm$ 0.12 b	0.014
Firmness (N/mm^2)	0.073 $\pm$ 0.007 b	0.102 $\pm$ 0.017 a	0.113 $\pm$ 0.006 a	0.012
Crust color				
Lightness (L*)	57.07 $\pm$ 2.58 a	52.47 $\pm$ 1.98 ab	49.87 $\pm$ 2.18 b	0.021
Redness (a*)	12.16 $\pm$ 1.62 a	5.14 $\pm$ 2.70 b	5.39 $\pm$ 2.37 b	0.015
Yellowness (b*)	33.58 $\pm$ 5.52	36.58 $\pm$ 1.61	35.17 $\pm$ 0.38	0.574
Crumb color				
Lightness (L*)	71.62 $\pm$ 0.87 a	65.37 $\pm$ 2.27 b	59.01 $\pm$ 0.96 c	<0.001
Redness (a*)	-4.44 ± 0.19 c	-2.80 ± 0.19 b	-1.61 ± 0.17 a	<0.0001
Yellowness (b*)	25.68 $\pm$ 0.87	25.20 $\pm$ 1.19	24.42 $\pm$ 0.93	0.365
Void fraction (%)	34.86 $\pm$ 1.75 b	41.14 $\pm$ 0.75 a	42.96 $\pm$ 2.71 a	0.005
Cell density (n/cm^2)	58.22 $\pm$ 1.54	64.44 $\pm$ 10.62	72.59 $\pm$ 3.45	0.091
Mean cell area (mm^2)	0.70 $\pm$ 0.05	0.65 $\pm$ 0.11	0.59 $\pm$ 0.05	0.278

3.4. Chemical Characterization of Raw Materials and Bread Samples

The antioxidant activity and antiradical scavenging activity of the raw materials were measured; in particular, as shown in Table 5, high TPC was mostly highlighted in the AWP (197.775 mgGAE/g) compared to the semolina (3.676 mgGAE/100 g). The highest increase in antiradical activity was observed in the AWP, with values of 38.235 mmol TE/100 g and 35.175 mmol TE/100 g for the DPPH and ABTS assays, respectively, while the lowest was recorded for semolina (2.656 and 2.408 mmol TE/100 g for the DPPH and ABTS assays, respectively).

Table 5. Antioxidant and antiradical activity of semolina and avocado waste powder (AWP). Results indicate mean values $\pm$ S.D. Data within a column followed by different letters are significantly different according to Tukey's test. Abbreviations: AWP (avocado waste powder); TPC (total phenolic content).

Samples	TPC mgGAE/g	DPPH$_{TEAC}$ mmol TE/100 g	ABTS$_{TEAC}$ mmol TE/100 g
Semolina	3.676 $\pm$ 0.15 b	2.656 $\pm$ 0.01 b	2.408 $\pm$ 0.04 b
AWP	197.775 $\pm$ 0.27 a	38.235 $\pm$ 0.09 a	35.175 $\pm$ 0.97 a
p value	<0.0001	<0.0001	<0.0001

The same analyses were carried out on the bread samples fortified with different percentages of AWP. The addition of AWP enhanced the samples' antiradical and antioxidant activity. As Table 6 shows, 10-AWP bread had higher values of antioxidant (23.882 mgGAE/100 g) and antiradical activity (6.656 and 9.234 mmol TE/100 g for DPPH and ABTS assays, respectively) compared to the CTR bread, which was made with only semolina.

Table 6. Antioxidant and antiradical activity of fortified and control bread samples (5-AWP Bread, 10-AWP Bread and CTR-Bread). Results indicate mean values ± S.D. Abbreviations: CTR-Bread, bread produced with control dough; 5-AWP Bread, experimental bread enriched with 5% (*w/w*) of avocado waste powder (AWP); 10-AWP Bread, experimental bread enriched with 10% (*w/w*) of AWP; TPC (total phenolic content).

Samples	TPC mgGAE/g	DPPH$_{TEAC}$ mmol TE/100 g	ABTS$_{TEAC}$ mmol TE/100 g
CTR Bread	2.972 ± 0.04	2.311± 0.02	2.102 ± 0.03
5-AWP Bread	23.033 ± 0.38	8.796 ± 0.01	5.985 ± 0.013
10-AWP Bread	23.882 ± 0.09	9.234 ± 0.07	6.656 ± 0.04
p value	<0.0001	<0.0001	<0.0001

3.5. Bread Sensory Attributes

Figure 1 presents the spider plot resulting from the sensory evaluations of the CTR and AWP breads.

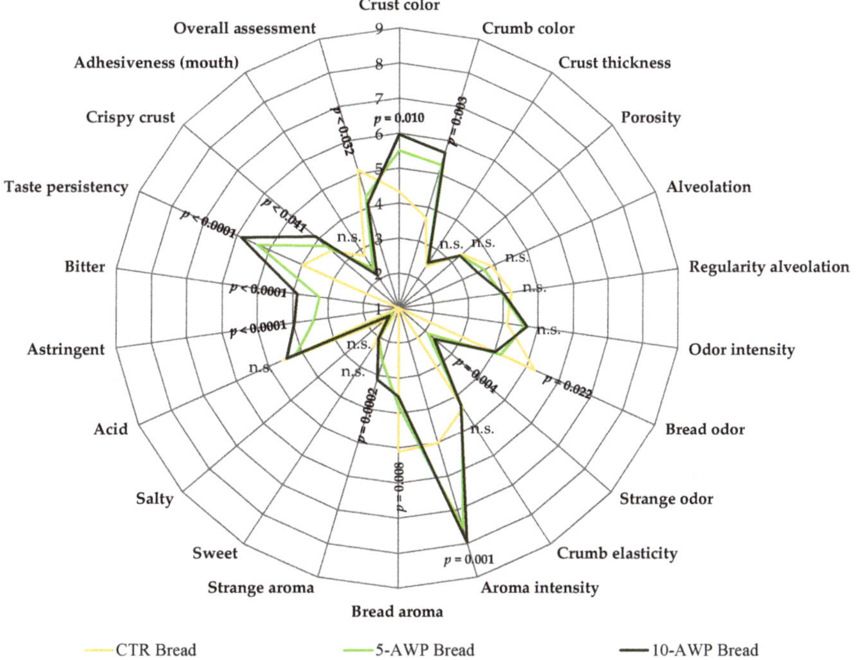

Figure 1. Spider diagrams of descriptive sensory analysis of breads. Abbreviations: CTR Bread; 5-AWP Bread, experimental bread enriched with 5% (*w/w*) of avocado waste powder (AWP); 10-AWP Bread, experimental bread enriched with 10% (*w/w*) of AWP; n.s., not significant (*p* > 0.05).

The addition of AWP, at both percentages, greatly affected the sensory characteristics of the semolina breads. The sensory traits significantly different from those of the control breads were crust and crumb color; crispy crust; bread and strange odor; astringent, bitter taste persistency; bread and strange aroma; and, especially, aroma intensity. Except for bread odor and bread aroma, which were scored at a lower level than the CTR bread, all the other traits mentioned had higher scores for the AWP breads. Regarding bread structure, although alveolation and adhesiveness are lower in AWP breads, the differences are not significant. Considering the overall assessment based on all these traits, the CTR bread received the highest scores, and between the 5-AWP bread and 10-AWP bread trials, the breads processed with 5% AWP were more appreciated.

3.6. Multivariate Analysis

The HCA clustered the breads based on their dissimilarity and relationship using a total of 37 variables, including quality attributes, antioxidant and antiradical properties, and sensory traits. The resulting cluster presented in Figure 2 shows low levels of dissimilarity (0.099%) among the breads. However, the breads enriched with AWP formed a single cluster and were clearly separated from control production.

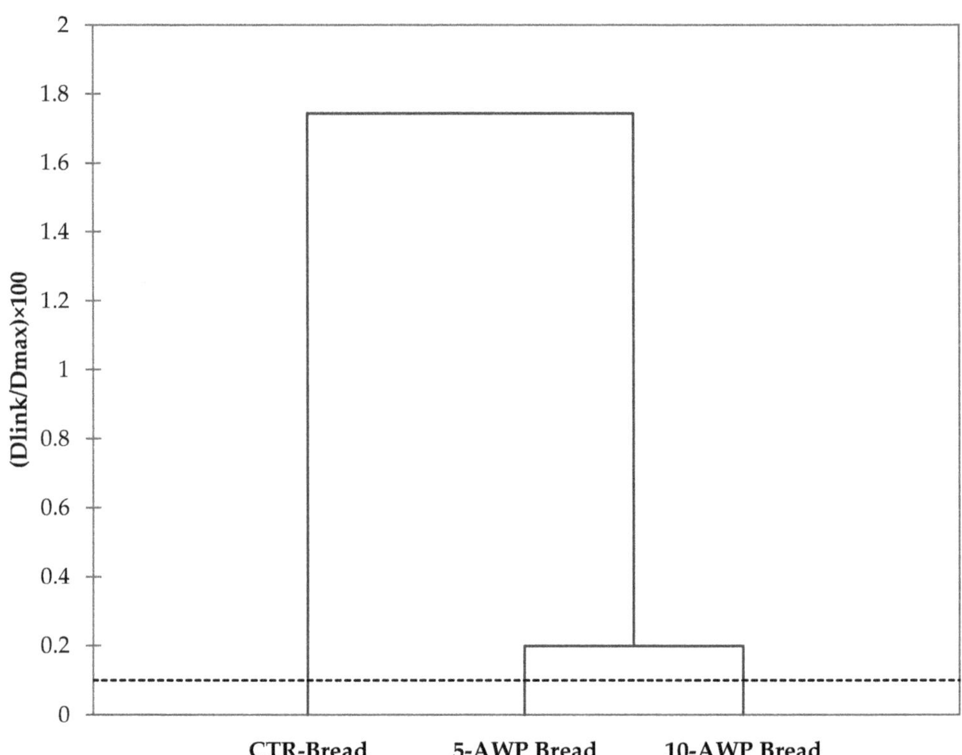

Figure 2. Dendrograms obtained from hierarchical cluster analysis based on values of quality attributes, antioxidant and antiradical properties and sensory traits of breads. Abbreviations: CTR Bread; 5-AWP Bread, experimental bread enriched with 5% (w/w) of avocado waste powder (AWP); 10-AWP Bread, experimental bread enriched with 10% (w/w) of AWP.

4. Discussion

In the production of cereal-based foods, fiber has grown in popularity as an added functional ingredient. Fiber, such as inulin, improves the rheological and technological characteristics of food, as well as its consistency, acceptability, and healthy properties. It also targets the prevention of metabolic syndromes. In addition to these improvements, when a food product is enriched with fiber, its shelf life is extended [51]. Other fiber-rich matrices, such as wheat or oat bran, have been used to replace wheat flour in baking [52]. One of the main reasons to supplement foods with dietary fiber is that it produces a wide variety of flavors that make products more palatable [53]. In a recent study, it was also shown that fortifying semolina bread with hemp seed flour improves its nutritional and antioxidant properties without significant changes in rheological properties [54]. In the work of Gómez and Martinez [32], the incorporation of fruit and vegetable by-products in baked products was evaluated to create foods with a higher fiber content. The authors highlighted a slowdown in the digestion of starch and other carbohydrates present in cereals and an improvement in rheological properties and interactions with digestive enzymes in the stomach. The improvement of antioxidant activity in fortified foods is due to an increase in the content of bioactive compounds, such as polyphenols and carotenoids [32]. Additionally, in the study of Gaglio et al. [40], the reuse of by-products for the production of fortified bread was evaluated. In particular, the authors used powdered almond skin at different percentages (5% and 10%) to produce functional products by modifying a traditional sourdough bread recipe. The final characteristics of the bread were influenced by the fortification and its percentage. The powdered almond skin positively influenced the sensory characteristics of the bread, with an increase in the intensity of the odor and the color of the crust and crumb. Thanks to the phytochemicals released by the fortified bread, an increase in the antioxidant capacity that can provide antioxidant protection at the level of human intestinal cells was also highlighted. Moreover, the microbiological parameters during fermentation were influenced by the development of coliforms due to the presence of spores after baking [37]. A recent study investigated Cava lees, another type of by-product that represents 25% of wine industry waste and is rich in antioxidant compounds and dietary fiber. This study aimed to evaluate the effect of Cava lees on microbial populations during natural leavening and bread fermentation. The results showed that the bread formulation with 5% Cava lees promoted the growth of both LAB and yeast and increased the concentration of volatile substances typically present [55].

Considering managing agro-wastes and food by-products while avoiding environmental concerns, the present work focused on valorizing the avocado production chain through the reuse of waste arising from non-compliant avocado fruits as well as avocado seeds and peels, to produce functional bread. From the results obtained in this study, it can be stated that the right time-temperature combination for the dehydration of avocado pulp and by-products, which maintained colorimetric and antioxidant characteristics after the dehydration process, was found. In particular, the grinding process did not affect or alter the organoleptic characteristics of the powder. It is important to emphasize that color is an extremely important characteristic because it makes the product attractive and acceptable, inducing consumers to purchase it; in other words, it is the first quality that guides consumers' purchase choices [56]. The dehydration process tends to alter the surface characteristics of the food and, consequently, alters both reflectivity and color properties; particularly in fruit, alterations occur at the expense of carotenoids and chlorophyll. Such alterations were not observed in this study, and as reported in the results, the powders retained a greenish color reminiscent of fresh avocado. In addition, due to the dehydration process, loss of the aromatic substances could occur [57], depending on the amount of heat energy absorbed by the product, in the form of sensible and latent heat, for the vaporization of water. The amount of this loss depends on temperature, moisture content of the food, vapor pressure, and the solubility of the volatile compounds in water vapor. However, the extent of this loss is also related to lipid oxidation reactions [58]. The oxidation of fatty acids gives rise to aldehydes, ketones, and acids that cause rancidity and off-flavors [59]. In this case, the products obtained by the dehydration protocol used did not cause such alterations, and this is deducible from the sensory evaluation of the functional bread. For bread making, the sourdough inoculum was developed from selected starter strains. They all acidified the SSE used for pH values in the range of 3.82–4.34, which are generally registered for sourdough *Lactobacillus*, *Leuconostoc*, and *Weissella* grown in this semolina-derived medium [39,60]. Looking at the TTA data for the starter strains and bread doughs, this parameter was confirmed to evolve (increase) inversely with pH [61]. The acidification process was also followed through the LAB development, and at the end of fermentation (8 h), they increased about two Log cycles in all the bread doughs, as commonly observed in sourdough bread production [62]. The LAB levels were slightly higher than the TMM levels, indicating the absolute dominance of the added strains during fermentation. Furthermore, the low TMM levels registered are a consequence of the high nutritional requirements of LAB that are not fully satisfied by principal component analysis [63,64]. Of course, during sourdough fermentation, the development of yeasts is also particularly important [65]. Yeast cell densities estimated at the end of the leavening duration were between 6.78 and

7.06 Log CFU/g, although they were not deliberately inoculated. However, yeasts develop spontaneously in sourdough [66], and the results of this study are generally found in semolina sourdough fermentation [41], even in the presence of waste/by-product addition [40,64]. Furthermore, in this research, the ratio between yeasts and LAB was optimal at 1:100, which is considered optimal for sourdough preparations [67]. The AWP was also investigated for several undesired groups such as *Salmonella* spp., *Listeria monocytogenes*, coagulase-positive staphylococci, members of Enterobacteriaceae, coliforms, *Escherichia coli*, spore-forming bacteria, and *Pseudomonas*; none of these bacterial groups exceeded the detection limit. However, after 8 h of fermentation, members of Enterobacteriaceae and total coliforms, but no *E. coli*, were detected in both AWP bread doughs. Although undesirable in dough because they compete with LAB and yeast development, the levels estimated were particularly low, and this is imputable to the fact that Enterobacteriaceae are limited in their growth by the low pH encountered in sourdough during fermentation [68]. The quality attributes of the breads were impacted by the addition of AWP. In general, a diminution in WL and the specific volume of the breads is generally reported when food waste is added [37]. The data presented in this work confirmed this trend, but only the 10-AWP breads were characterized by a WL and specific volume significantly different from those displayed by the CTR bread. The browning of the crust and crumb was due to the darker color of AWP compared to semolina. This browning is a positive sign since, as reported by Sandvik et al. [69], the dark color of bread is linked to health among consumers.

The firmness of the breads increased as a consequence of the increase in dietary fiber, as observed by Ruisi et al. [48]. Regarding the image analysis of the central slices of the breads, alveolation diminished with the AWP-enriched breads. This phenomenon was also observed with the addition of pumpkin pomace and dry tomato waste [70,71] and is due to the low percentage of gluten [72]. The antiradical and antioxidant activity was mostly highlighted in the AWP breads compared to the semolina breads. The highest value of TPC was also observed in the AWP sample, while the lowest was recorded for the semolina sample. The supplementation of AWP in bread enhanced the food's antiradical and antioxidant activity. The 10-AWP bread had higher values of antioxidant and antiradical scavenging activity compared to the CTR bread, which was made with only semolina. According to TPC values, the chemical analyses showed that the fortification of bread with AWP in different percentages increased its antiradical and antioxidant activity and organoleptic and baking qualities proportionally to the percentage of fortification used.

5. Conclusions

The overall data collected showed the excellent suitability of AWP for functional bread making. The present research found that avocado waste products that have been dehydrated and processed into powder can be successfully incorporated into leavened baked products to improve their characteristics, particularly in terms of antioxidant content. A positive relationship between the proportion of added powder and antioxidant content appeared, as well as the organoleptic and baking qualities of bread. The addition of 10% AWP produced dough with a higher antioxidant profile than the control bread. In addition, the bread produced from this dough was highly appreciated on a sensory level in terms of aroma and color. Fortified bread, therefore, has shown great potential to serve as a functional food among consumers.

Author Contributions: Conceptualization, V.F. and I.T.; methodology, V.F., I.T., V.D.S., R.G., L.S. and C.B.; formal analysis, I.T., E.V. and C.B.; data curation, C.B., I.T., R.G. and E.V.; writing—original draft preparation, I.T., C.B., R.G., E.V. and L.S.; writing—review and editing, V.F., L.S., V.D.S. and R.G.; supervision, V.F., V.D.S., R.G. and L.S.; funding acquisition, V.F. All authors have read and agreed to the published version of the manuscript.

Funding: This research was funded by the TINFRUT project, funded by PSR Sicilia 2014–2020 Programma di Sviluppo Rurale SOTTOMISURA 16.1 "Sostegno per la costituzione e la gestione dei gruppi operativi del PEI in materia di produttività e sostenibilità dell'agricoltura".

Informed Consent Statement: Informed consent was obtained from all subjects involved in the study.

Data Availability Statement: The data presented in this study are available on request from the corresponding author.

Conflicts of Interest: The authors declare no conflict of interest.

References

1. Salazar-López, N.J.; Domínguez-Avila, J.A.; Yahia, E.M.; Belmonte-Herrera, B.H.; Wall-Medrano, A.; Montalvo-González, E.; González-Aguilar, G.A. Avocado Fruit and By-Products as Potential Sources of Bioactive Compounds. *Food Res. Int.* **2020**, *138*, 109774. [CrossRef] [PubMed]
2. FAOSTAT. Available online: http://faostat.fao.org2021 (accessed on 12 June 2023).
3. FAO_FAOSTAT. Available online: http://data.un.org/Data.aspx?d=FAO&f=itemCode%3A572 (accessed on 25 July 2023).
4. Migliore, G.; Farina, V.; Tinervia, S.; Matranga, G.; Schifani, G. Consumer Interest towards Tropical Fruit: Factors Affecting Avocado Fruit Consumption in Italy. *Agric. Econ.* **2017**, *5*, 24. [CrossRef]
5. OECD. *Avocados—CBI Ministry of Foreign Affairs*; International Standards for Fruit and Vegetables; OECD: Paris, France, 2023; ISBN 978-92-64-01979-9.
6. Kopec, R.E.; Cooperstone, J.L.; Schweiggert, R.M.; Young, G.S.; Harrison, E.H.; Francis, D.M.; Clinton, S.K.; Schwartz, S.J. Avocado Consumption Enhances Human Postprandial Provitamin A Absorption and Conversion from a Novel High–β-Carotene Tomato Sauce and from Carrots. *J. Nutr.* **2014**, *144*, 1158–1166. [CrossRef] [PubMed]
7. Fulgoni, V.L.; Dreher, M.; Davenport, A.J. Avocado Consumption Is Associated with Better Diet Quality and Nutrient Intake, and Lower Metabolic Syndrome Risk in US Adults: Results from the National Health and Nutrition Examination Survey (NHANES) 2001–2008. *Nutr. J.* **2013**, *12*, 1. [CrossRef] [PubMed]
8. Araújo, R.G.; Rodriguez-Jasso, R.M.; Ruiz, H.A.; Pintado, M.M.E.; Aguilar, C.N. Avocado By-Products: Nutritional and Functional Properties. *Trends Food Sci. Technol.* **2018**, *80*, 51–60. [CrossRef]
9. Obenland, D.; Collin, S.; Sievert, J.; Negm, F.; Arpaia, M.L. Influence of Maturity and Ripening on Aroma Volatiles and Flavor in 'Hass' Avocado. *Postharvest Biol. Technol.* **2012**, *71*, 41–50. [CrossRef]
10. Migliore, G.; Farina, V.; Guccione, G.D.; Schifani, G. Quality Determinants of Avocado Fruit Consumption in Italy. *Implic. Small Farms. Calitatea* **2018**, *19*, 148–153.
11. Gellynck, X.; Kühne, B.; Van Bockstaele, F.; Van de Walle, D.; Dewettinck, K. Consumer Perception of Bread Quality. *Appetite* **2009**, *53*, 16–23. [CrossRef] [PubMed]
12. Harker, R. *Consumer Preferences and Choice of Fruit: The Role of Avocado Quality*; Plant & Food Research: Auckland, New Zealand, 2009; p. 48.
13. Badar, H.; Ariyawardana, A.; Collins, R. Capturing Consumer Preferences for Value Chain Improvements in the Mango Industry of Pakistan. *Int. Food Agribus. Manag. Rev.* **2015**, *18*, 131–148.
14. Gamble, J.; Harker, F.R.; Jaeger, S.R.; White, A.; Bava, C.; Beresford, M.; Stubbings, B.; Wohlers, M.; Hofman, P.J.; Marques, R. The Impact of Dry Matter, Ripeness and Internal Defects on Consumer Perceptions of Avocado Quality and Intentions to Purchase. *Postharvest Biol. Technol.* **2010**, *57*, 35–43. [CrossRef]
15. Bill, M.; Sivakumar, D.; Thompson, A.K.; Korsten, L. Avocado Fruit Quality Management during the Postharvest Supply Chain. *Food Rev. Int.* **2014**, *30*, 169–202. [CrossRef]
16. Kourgialas, N.N.; Dokou, Z. Water Management and Salinity Adaptation Approaches of Avocado Trees: A Review for Hot-Summer Mediterranean Climate. *Agric. Water Manag.* **2021**, *252*, 106923. [CrossRef]
17. Dalle Mulle Santos, C.; Pagno, C.H.; Haas Costa, T.M.; Jung Luvizetto Faccin, D.; Hickmann Flôres, S.; Medeiros Cardozo, N.S. Biobased Polymer Films from Avocado Oil Extraction Residue: Production and Characterization. *J. Appl. Polym. Sci.* **2016**, *133*, 43957. [CrossRef]
18. Rodríguez-Carpena, J.G.; Morcuende, D.; Estévez, M. Avocado By-Products as Inhibitors of Color Deterioration and Lipid and Protein Oxidation in Raw Porcine Patties Subjected to Chilled Storage. *Meat Sci.* **2011**, *89*, 166–173. [CrossRef] [PubMed]
19. Calderón-Oliver, M.; Escalona-Buendía, H.B.; Medina-Campos, O.N.; Pedraza-Chaverri, J.; Pedroza-Islas, R.; Ponce-Alquicira, E. Optimization of the Antioxidant and Antimicrobial Response of the Combined Effect of Nisin and Avocado Byproducts. *LWT-Food Sci. Technol.* **2016**, *65*, 46–52. [CrossRef]
20. Melgar, B.; Dias, M.I.; Ciric, A.; Sokovic, M.; Garcia-Castello, E.M.; Rodriguez-Lopez, A.D.; Barros, L.; Ferreira, I.C. Bioactive Characterization of *Persea Americana* Mill. by-Products: A Rich Source of Inherent Antioxidants. *Ind. Crops Prod.* **2018**, *111*, 212–218. [CrossRef]
21. Tinebra, I.; Passafiume, R.; Scuderi, D.; Pirrone, A.; Gaglio, R.; Palazzolo, E.; Farina, V. Effects of Tray-Drying on the Physico-chemical, Microbiological, Proximate, and Sensory Properties of White-and Red-Fleshed Loquat (*Eriobotrya japonica* Lindl.) Fruit. *Agronomy* **2022**, *12*, 540. [CrossRef]
22. Karam, M.C.; Petit, J.; Zimmer, D.; Baudelaire Djantou, E.; Scher, J. Effects of Drying and Grinding in Production of Fruit and Vegetable Powders: A Review. *J. Food Eng.* **2016**, *188*, 32–49. [CrossRef]

23. Jiang, H.; Zhang, M.; Adhikari, B. 21—Fruit and Vegetable Powders. In *Handbook of Food Powders*; Woodhead Publishing Series in Food Science, Technology and Nutrition; Bhandari, B., Bansal, N., Zhang, M., Schuck, P., Eds.; Woodhead Publishing: Sawston, UK, 2013; pp. 532–552; ISBN 978-0-85709-513-8.
24. Camire, M.E.; Dougherty, M.P.; Briggs, J.L. Functionality of Fruit Powders in Extruded Corn Breakfast Cereals. *Food Chem.* **2007**, *101*, 765–770. [CrossRef]
25. Zhang, M.; Tang, J.; Mujumdar, A.; Wang, S. Trends in Microwave-Related Drying of Fruits and Vegetables. *Trends Food Sci. Technol.* **2006**, *17*, 524–534. [CrossRef]
26. Salehi, F. Recent Applications of Powdered Fruits and Vegetables as Novel Ingredients in Biscuits: A Review. *Nutrire* **2020**, *45*, 1. [CrossRef]
27. Argyropoulos, D.; Heindl, A.; Müller, J. Assessment of Convection, Hot-air Combined with Microwave-vacuum and Freeze-drying Methods for Mushrooms with Regard to Product Quality. *Int. J. Food Sci. Technol.* **2011**, *46*, 333–342. [CrossRef]
28. Betoret, E.; Rosell, C.M. Enrichment of Bread with Fruits and Vegetables: Trends and Strategies to Increase Functionality. *Cereal Chem.* **2020**, *97*, 9–19. [CrossRef]
29. Melilli, M.G.; Di Stefano, V.; Sciacca, F.; Pagliaro, A.; Bognanni, R.; Scandurra, S.; Virzì, N.; Gentile, C.; Palumbo, M. Improvement of Fatty Acid Profile in Durum Wheat Breads Supplemented with *Portulaca oleracea* L. Qual. Trait. Purslane-Fortif. Bread. *Foods* **2020**, *9*, 764.
30. Sablani, S.S.; Andrews, P.K.; Davies, N.M.; Walters, T.; Saez, H.; Bastarrachea, L. Effects of Air and Freeze Drying on Phytochemical Content of Conventional and Organic Berries. *Dry. Technol.* **2011**, *29*, 205–216. [CrossRef]
31. Rahman, M.S.; Al-Shamsi, Q.H.; Bengtsson, G.B.; Sablani, S.S.; Al-Alawi, A. Drying Kinetics and Allicin Potential in Garlic Slices during Different Methods of Drying. *Dry. Technol.* **2009**, *27*, 467–477. [CrossRef]
32. Gómez, M.; Martinez, M.M. Fruit and Vegetable By-Products as Novel Ingredients to Improve the Nutritional Quality of Baked Goods. *Crit. Rev. Food Sci. Nutr.* **2018**, *58*, 2119–2135. [CrossRef] [PubMed]
33. Ajila, C.; Leelavathi, K.; Rao, U.P. Improvement of Dietary Fiber Content and Antioxidant Properties in Soft Dough Biscuits with the Incorporation of Mango Peel Powder. *J. Cereal Sci.* **2008**, *48*, 319–326. [CrossRef]
34. Sánchez-Quezada, V.; Campos-Vega, R.; Loarca-Piña, G. Prediction of the Physicochemical and Nutraceutical Characteristics of 'Hass' Avocado Seeds by Correlating the Physicochemical Avocado Fruit Properties According to Their Ripening State. *Plant Foods Hum. Nutr.* **2021**, *76*, 311–318. [CrossRef] [PubMed]
35. Misra, S.; Pandey, P.; Mishra, H.N. Novel Approaches for Co-Encapsulation of Probiotic Bacteria with Bioactive Compounds, Their Health Benefits and Functional Food Product Development: A Review. *Trends Food Sci. Technol.* **2021**, *109*, 340–351. [CrossRef]
36. Messina, C.M.; Gaglio, R.; Morghese, M.; Tolone, M.; Arena, R.; Moschetti, G.; Santulli, A.; Francesca, N.; Settanni, L. Microbiological Profile and Bioactive Properties of Insect Powders Used in Food and Feed Formulations. *Foods* **2019**, *8*, 400. [CrossRef]
37. Gaglio, R.; Tesoriere, L.; Maggio, A.; Viola, E.; Attanzio, A.; Frazzitta, A.; Badalamenti, N.; Bruno, M.; Franciosi, E.; Moschetti, G. Reuse of Almond By-Products: Functionalization of Traditional Semolina Sourdough Bread with Almond Skin. *Int. J. Food Microbiol.* **2023**, *395*, 110194. [CrossRef]
38. Lhomme, E.; Lattanzi, A.; Dousset, X.; Minervini, F.; De Angelis, M.; Lacaze, G.; Onno, B.; Gobbetti, M. Lactic Acid Bacterium and Yeast Microbiotas of Sixteen French Traditional Sourdoughs. *Int. J. Food Microbiol.* **2015**, *215*, 161–170. [CrossRef] [PubMed]
39. Alfonzo, A.; Urso, V.; Corona, O.; Francesca, N.; Amato, G.; Settanni, L.; Di Miceli, G. Development of a Method for the Direct Fermentation of Semolina by Selected Sourdough Lactic Acid Bacteria. *Int. J. Food Microbiol.* **2016**, *239*, 65–78. [CrossRef] [PubMed]
40. Gaglio, R.; Barbera, M.; Tesoriere, L.; Osimani, A.; Busetta, G.; Matraxia, M.; Attanzio, A.; Restivo, I.; Aquilanti, L.; Settanni, L. Sourdough "Ciabatta" Bread Enriched with Powdered Insects: Physicochemical, Microbiological, and Simulated Intestinal Digesta Functional Properties. *Innov. Food Sci. Emerg. Technol.* **2021**, *72*, 102755. [CrossRef]
41. Corona, O.; Alfonzo, A.; Ventimiglia, G.; Nasca, A.; Francesca, N.; Martorana, A.; Moschetti, G.; Settanni, L. Industrial Application of Selected Lactic Acid Bacteria Isolated from Local Semolinas for Typical Sourdough Bread Production. *Food Microbiol.* **2016**, *59*, 43–56. [CrossRef] [PubMed]
42. American Association of Cereal Chemists. *Approved Methods Committee Approved Methods of the American Association of Cereal Chemists*; AACC: St. Paul, MN, USA, 2000.
43. Alfonzo, A.; Gaglio, R.; Barbera, M.; Francesca, N.; Moschetti, G.; Settanni, L. Evaluation of the Fermentation Dynamics of Commercial Baker's Yeast in Presence of Pistachio Powder to Produce Lysine-Enriched Breads. *Fermentation* **2019**, *6*, 2. [CrossRef]
44. Francesca, N.; Gaglio, R.; Alfonzo, A.; Corona, O.; Moschetti, G.; Settanni, L. Characteristics of Sourdoughs and Baked Pizzas as Affected by Starter Culture Inoculums. *Int. J. Food Microbiol.* **2019**, *293*, 114–123. [CrossRef] [PubMed]
45. Cirlincione, F.; Venturella, G.; Gargano, M.L.; Ferraro, V.; Gaglio, R.; Francesca, N.; Rizzo, B.A.; Russo, G.; Moschetti, G.; Settanni, L. Functional Bread Supplemented with *Pleurotus eryngii* Powder: A Potential New Food for Human Health. *Int. J. Gastron. Food Sci.* **2022**, *27*, 100449. [CrossRef]
46. Bonacci, S.; Di Stefano, V.; Sciacca, F.; Buzzanca, C.; Virzì, N.; Argento, S.; Melilli, M.G. Hemp Flour Particle Size Affects the Quality and Nutritional Profile of the Enriched Functional Pasta. *Foods* **2023**, *12*, 774. [CrossRef] [PubMed]

47. Di Stefano, V.; Buzzanca, C.; Melilli, M.G.; Indelicato, S.; Mauro, M.; Vazzana, M.; Arizza, V.; Lucarini, M.; Durazzo, A.; Bongiorno, D. Polyphenol Characterization and Antioxidant Activity of Grape Seeds and Skins from Sicily: A Preliminary Study. *Sustainability* **2022**, *14*, 6702. [CrossRef]

48. Ruisi, P.; Ingraffia, R.; Urso, V.; Giambalvo, D.; Alfonzo, A.; Corona, O.; Settanni, L.; Frenda, A.S. Influence of Grain Quality, Semolinas and Baker's Yeast on Bread Made from Old Landraces and Modern Genotypes of Sicilian Durum Wheat. *Food Res. Int.* **2021**, *140*, 110029. [CrossRef] [PubMed]

49. ISO. *Sensory Analysis: Methodology: General Guidance for Establishing a Sensory Profile*; ISO: Geneva, Switzerland, 2003.

50. Martorana, A.; Alfonzo, A.; Settanni, L.; Corona, O.; La Croce, F.; Caruso, T.; Moschetti, G.; Francesca, N. An Innovative Method to Produce Green Table Olives Based on "Pied de Cuve" Technology. *Food Microbiol.* **2015**, *50*, 126–140. [CrossRef] [PubMed]

51. Ktenioudaki, A.; Gallagher, E. Recent Advances in the Development of High-Fibre Baked Products. *Trends Food Sci. Technol.* **2012**, *28*, 4–14. [CrossRef]

52. Popoola-Akinola, O.O.; Raji, T.J.; Olawoye, B. Lignocellulose, Dietary Fibre, Inulin and Their Potential Application in Food. *Heliyon* **2022**, *8*, e10459. [CrossRef]

53. Olawoye, B.; Gbadamosi, S.O. Sensory Profiling and Mapping of Gluten-free Cookies Made from Blends Cardaba Banana Flour and Starch. *J. Food Process. Preserv.* **2020**, *44*, e14643. [CrossRef]

54. Sciacca, F.; Virzì, N.; Pecchioni, N.; Melilli, M.G.; Buzzanca, C.; Bonacci, S.; Di Stefano, V. Functional End-Use of Hemp Seed Waste: Technological, Qualitative, Nutritional, and Sensorial Characterization of Fortified Bread. *Sustainability* **2023**, *15*, 12899. [CrossRef]

55. Martín-Garcia, A.; Riu-Aumatell, M.; López-Tamames, E. By-Product Revalorization: Cava Lees Can Improve the Fermentation Process and Change the Volatile Profile of Bread. *Foods* **2022**, *11*, 1361. [CrossRef] [PubMed]

56. Allès, B.; Péneau, S.; Kesse-Guyot, E.; Baudry, J.; Hercberg, S.; Méjean, C. Food Choice Motives Including Sustainability during Purchasing Are Associated with a Healthy Dietary Pattern in French Adults. *Nutr. J.* **2017**, *16*, 58. [CrossRef] [PubMed]

57. Mujumdar, A.S.; Law, C.L. Drying Technology: Trends and Applications in Postharvest Processing. *Food Bioprocess Technol.* **2010**, *3*, 843–852. [CrossRef]

58. Si, X.; Chen, Q.; Bi, J.; Wu, X.; Yi, J.; Zhou, L.; Li, Z. Comparison of Different Drying Methods on the Physical Properties, Bioactive Compounds and Antioxidant Activity of Raspberry Powders. *J. Sci. Food Agric.* **2016**, *96*, 2055–2062. [CrossRef] [PubMed]

59. Dereje, B.; Abera, S. Effect of Pretreatments and Drying Methods on the Quality of Dried Mango (*Mangifera indica* L.). *Slices. Cogent Food Agric.* **2020**, *6*, 1747961. [CrossRef]

60. Settanni, L.; Ventimiglia, G.; Alfonzo, A.; Corona, O.; Miceli, A.; Moschetti, G. An Integrated Technological Approach to the Selection of Lactic Acid Bacteria of Flour Origin for Sourdough Production. *Food Res. Int.* **2013**, *54*, 1569–1578. [CrossRef]

61. Siepmann, F.B.; de Almeida, B.S.; Ripari, V.; da Silva, B.J.; Peralta-Zamora, P.G.; Waszczynskyj, N.; Spier, M.R. Brazilian Sourdough: Microbiological, Structural, and Technological Evolution. *Eur. Food Res. Technol.* **2019**, *245*, 1583–1594. [CrossRef]

62. Suo, B.; Chen, X.; Wang, Y. Recent Research Advances of Lactic Acid Bacteria in Sourdough: Origin, Diversity, and Function. *Curr. Opin. Food Sci.* **2021**, *37*, 66–75. [CrossRef]

63. Alfonzo, A.; Miceli, C.; Nasca, A.; Franciosi, E.; Ventimiglia, G.; Di Gerlando, R.; Tuohy, K.; Francesca, N.; Moschetti, G.; Settanni, L. Monitoring of Wheat Lactic Acid Bacteria from the Field until the First Step of Dough Fermentation. *Food Microbiol.* **2017**, *62*, 256–269. [CrossRef]

64. Gaglio, R.; Alfonzo, A.; Barbera, M.; Franciosi, E.; Francesca, N.; Moschetti, G.; Settanni, L. Persistence of a Mixed Lactic Acid Bacterial Starter Culture during Lysine Fortification of Sourdough Breads by Addition of Pistachio Powder. *Food Microbiol.* **2020**, *86*, 103349. [CrossRef]

65. Carbonetto, B.; Nidelet, T.; Guezenec, S.; Perez, M.; Segond, D.; Sicard, D. Interactions between *Kazachstania humilis* Yeast Species and Lactic Acid Bacteria in Sourdough. *Microorganisms* **2020**, *8*, 240. [CrossRef] [PubMed]

66. Siepmann, F.B.; Ripari, V.; Waszczynskyj, N.; Spier, M.R. Overview of Sourdough Technology: From Production to Marketing. *Food Bioprocess Technol.* **2018**, *11*, 242–270. [CrossRef]

67. Korcari, D.; Secchiero, R.; Laureati, M.; Marti, A.; Cardone, G.; Rabitti, N.S.; Ricci, G.; Fortina, M.G. Technological Properties, Shelf Life and Consumer Preference of Spelt-Based Sourdough Bread Using Novel, Selected Starter Cultures. *LWT* **2021**, *151*, 112097. [CrossRef]

68. Dinardo, F.R.; Minervini, F.; De Angelis, M.; Gobbetti, M.; Gänzle, M.G. Dynamics of Enterobacteriaceae and Lactobacilli in Model Sourdoughs Are Driven by pH and Concentrations of Sucrose and Ferulic Acid. *LWT* **2019**, *114*, 108394. [CrossRef]

69. Purlis, E. Browning Development in Bakery Products–A Review. *J. Food Eng.* **2010**, *99*, 239–249. [CrossRef]

70. Nour, V.; Ionica, M.E.; Trandafir, I. Bread Enriched in Lycopene and Other Bioactive Compounds by Addition of Dry Tomato Waste. *J. Food Sci. Technol.* **2015**, *52*, 8260–8267. [CrossRef] [PubMed]

71. Kampuse, S.; Ozola, L.; Straumite, E.; Galoburda, R. Quality Parameters of Wheat Bread Enriched with Pumpkin (*Cucurbita moschata*) by-Products. *Acta Univ. Cibinesis Ser. E Food Technol.* **2015**, *19*. [CrossRef]

72. Rathnayake, H.; Navaratne, S.; Navaratne, C. Porous Crumb Structure of Leavened Baked Products. *Int. J. Food Sci.* **2018**, *2018*, 8187318. [CrossRef] [PubMed]

154

foods

Review

The Wheat Aleurone Layer: Optimisation of Its Benefits and Application to Bakery Products

Lucie Lebert [1], François Buche [2], Arnaud Sorin [1] and Thierry Aussenac [2,*]

[1] Foricher Les Moulins, 95400 Arnouville, France
[2] Institut Polytechnique UniLaSalle, Université d'Artois, ULR 7519, 60026 Beauvais, France
* Correspondence: thierry.aussenac@unilasalle.fr

Abstract: The wheat aleurone layer is, according to millers, the main bran fraction. It is a source of nutritionally valuable compounds, such as dietary fibres, proteins, minerals and vitamins, that may exhibit health benefits. Despite these advantages, the aleurone layer is scarce on the market, probably due to issues related to its extraction. Many processes exist with some patents, but a choice must be made between the quality and quantity of the resulting product. Nonetheless, its potential has been studied mainly in bread and pasta. While the nutritional benefits of aleurone-rich flour addition to bread agree, opposite results have been obtained concerning its effects on end-product characteristics (namely loaf volume and sensory characteristics), thus ensuing different acceptability responses from consumers. However, the observed negative effects of aleurone-rich flour on bread dough could be reduced by subjecting it to pre- or post-extracting treatments meant to either reduce the particle size of the aleurone's fibres or to change the conformation of its components.

Keywords: wheat aleurone; dietary fibre; extraction process; antioxidant; bread; arabinoxylans

Citation: Lebert, L.; Buche, F.; Sorin, A.; Aussenac, T. The Wheat Aleurone Layer: Optimisation of Its Benefits and Application to Bakery Products. *Foods* 2022, *11*, 3552. https://doi.org/10.3390/foods11223552

Academic Editors: Donatella Bianca Maria Ficco and Grazia Maria Borrelli

Received: 28 September 2022
Accepted: 3 November 2022
Published: 8 November 2022

Publisher's Note: MDPI stays neutral with regard to jurisdictional claims in published maps and institutional affiliations.

1. Introduction

Wheat is indispensable in producing many staple foods around the world, including bread, biscuits, cakes and noodles. As epidemiological studies have demonstrated that an increase in the intake of whole grain products is related to a lower incidence of cardiovascular disease, obesity, diabetes and cancer, the composition of wheat makes it a valuable asset in the diet for the prevention of chronic diseases. Consequently, in relation to the growing number of metabolic diseases, nutritional guidelines worldwide advise an increase in the consumption of whole grain products, partly as they contain fibres that many consumers lack in their diet [1,2].

These health benefits are mostly due to the presence of micronutrients, dietary fibres (DF) and bioactive components, which are mainly located in the outer layers of the grain: the bran and the aleurone layer [1,3,4]. Many researchers and industries have aimed to extract, isolate and introduce these grain fractions in food products as ingredients for added nutritional value. However, it seems that the addition of bran or fibre to wheat-flour-based products changes not only the technological properties of the end-product but also its sensory acceptance by consumers [3,5]. Moreover, the main challenge of the extraction process is improving the nutritional properties of the ingredient without impairing its technological properties during breadmaking. For instance, soluble fibres, such as water-extractible arabinoxylans (WEAX), answer this problem. The use of fibres from wheat bran has also been deeply investigated [3,6–8].

Recently, researchers have focused on the wheat aleurone layer, considered by millers to be the main bran layer. As it contains the majority of the grain's minerals and is also rich in protein, DF and bioactive components (mostly ferulic acid), the aleurone layer may be the source of many bran's reported health benefits [4,9–11]. However, despite its known nutritional and health-prevention properties, the aleurone layer is scarce on the market, whether included in cereal food products or as an ingredient. This could be related to the challenges posed by its extraction. Although multiple processes have been patented, the end product is often obtained with either low purity or yield [10,12–14]. Moreover, it seems that its incorporation into bakery products yields contrasting results, as some negative technological effects could be observed, such as a reduction in loaf volume and increased crumb hardness [15–17]. These limiting technological aspects on final food products could be improved by making specific modifications to the aleurone's components without losing any of the health benefits, prior to its incorporation into a food process, as is already carried out for wheat [18].

The objective of this review, therefore, is to provide readers with the elements of understanding to optimise the potential of the wheat aleurone layer, both from a nutritional and technological point of view. To do so, the wheat aleurone layer will first be described in terms of function and composition. Its potential nutritional and health benefits will then be presented based on clinical studies and in relation to the individual effects of its compounds. Aleurone's potential as an ingredient will also be reviewed, starting with the processes to extract it and related issues. Then, the applications of cereal-based food products performed in the literature will be investigated, with a tentative explanation of the underlying mechanisms of the observed effects. Finally, the last section will focus on the existing processes meant to optimise the aleurone layer's potential, both for its nutritional, health-related and technological benefits.

2. Aleurone Layer

2.1. Description: Histology and Functions

The aleurone layer is a tissue of wheat grain made of unicellular block-shaped cells (37–65 μm vs. 25–75 μm) [19]. Among the seven layers comprising the mature bran, the aleurone is the only one with the remaining living cells [20]. It is located between the endosperm and the nucellar epidermis (or hyaline layer), as shown in Figure 1 [10,21]. Although botanically part of the endosperm, it is considered by millers as a bran layer since it remains attached to the hyaline layer during milling [10,22]. It represents around 50% of wheat bran (or 75% w/w of its dry weight), making it the major bran layer [10,12]. Indeed, the aleurone layer is thick and can reach up to 65 μm [23]. It also corresponds to 5–8% (w/w) of the whole kernel [21].

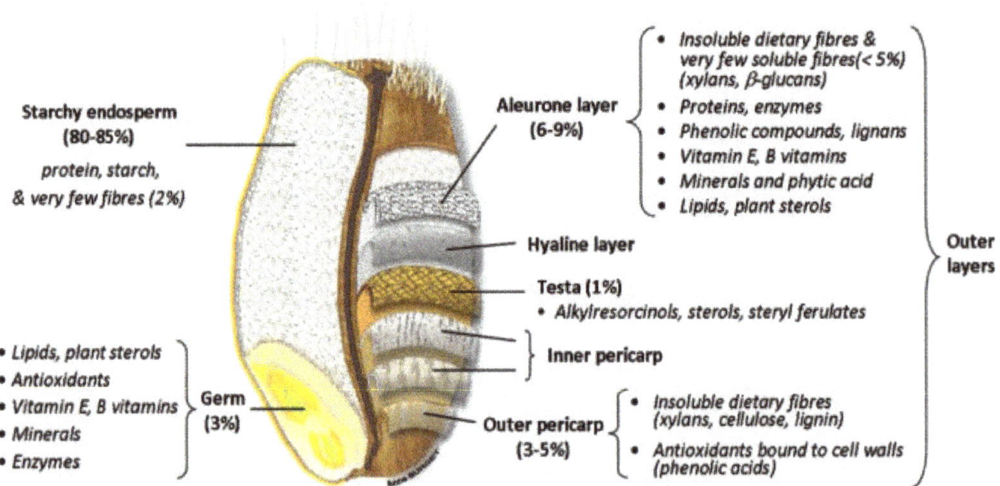

Figure 1. Wheat grain histology [10]. Adapted from Surget and Barron (2005) [21]. (Copyright for reprinting was requested and obtained through Taylor & Francis and Copyright Clearance Center. License number: 5386341293998. License date: 12 September 2022).

Although the aleurone layer is singular in wheat, it can be found multi-layered in barley, rice and oats [10].

Multiple functions in the wheat grain are allotted to the aleurone layer, namely: accumulation and transport of nutrients for seed germination, decomposition of storage materials of the endosperm for embryo growth, and protection and maintenance of caryopsis activity [24]. More specifically, its major role is during germination, where it is involved in the synthesis and release of hydrolases in the endosperm, as induced by gibberellin [11,25]. These enzymes then break down starch polymers and proteins in starchy endosperm cells, which undergo programmed cell death [22,26]. To facilitate the transfer, the aleurone's outer cell wall is degraded by endogenous hemicellulases, and only the inner resistant layer remains [25,27].

The aleurone layer is also involved in seed dormancy, induced by abscisic acid. At the same time, this hormone induces programmed cell death in endosperm cells [11]. However, the aleurone layer serves as grain storage for metabolites, minerals and amino acids [9,26,28]. It is equally involved in the regulation of water diffusion and distribution through its cell-walls [11,29].

Finally, in the crease region of the grain, some modified aleurone cells, called transfer cells, also participate in grain filling via solute uptake [30]. Due to their arabinoxylan's higher degree of arabinose substitution and lower degree of feruloylation, the transfer cells show specific cell wall hydration and porosity properties that are compatible with water diffusivity and uptake for grain filling [31].

2.2. Composition

2.2.1. Cell Wall

The aleurone cell-wall represents 35% (v/v) of the total cellular volume. It is bilayered, with a thicker outer layer (of 2 μm vs. 0.5 μm for the inner layer) and mainly composed of arabinoxylans (65%), β-glucans (29%) and phenolic acids [9], as presented in Tables 1 and 2. This high proportion of dietary fibres (DF), especially pentosans (44% w/w of total grain [32,33]), makes the aleurone's cell walls valuable from a technological and nutritional point of view, thanks to their gelling ability [6], and cancer-prevention properties [34,35].

Table 1. Carbohydrates repartition in the wheat aleurone layer, bran and whole grain.

	Aleurone Layer *		Bran * (%, *w/w*)	Whole Grain (%, *w/w*)
	Part in Tissue (%, *w/w*)	Total in Grain (%, *w/w*)		
Arabinoxylan	65 [25,36,37]	-	60–65 [32]	4–9 [36]
β-glucans	29 [36,37]	-	6 [36]	0.5–2.3 [36]
Cellulose	2–3 [32,33,36,37]	≈8 [32,33]	25–30 [32]	2–4 [32]
Glucomannan	2 [36,37]	-	-	<1 [37]
Pentosans	46 [32,33]	≈44 [32,33]	70–80 [32]	8–10 [32]
Starch	0 [32,33]	0 [32]	9–25 [3]	67–71 [32]
Total Dietary Fibre	43–48 [12]	-	40–53 [3]	13 [1]

* The aleurone layer and the bran amount respectively to 5–8% (w/w) and 5.5–9.5% (w/w) of the wheat grain [21]. All data were placed in the same unit to facilitate comparison.

Its major component, arabinoxylan (AX), is a polysaccharide made of a linear xylose chain with β-(1,4) linkages between its xylopyranosyl residues (Figure 2 [38]). These residues can either remain unsubstituted, mono-substituted at the third carbon position (C3), or with a di-substitution at both the C2 and C3 positions. Indeed, α-L-arabinofuranose residues can be esterified to the xylan backbone via α-(1,2) and α-(1,3) linkages but not in an even pattern. These arabinose residues can, in turn, be esterified by phenolic acids (mainly ferulic acid) on their primary alcohol function (O5) [9,10,39]. The feruloylation of AXs is progressive throughout grain filling [31] and takes place on average every 15 arabinofuranose residues [22]. In addition to arabinose substitution, some acetyl groups can be esterified to xylan instead [30]. The presence of these side groups confers technological value to aleurone's AXs as they can interact with AX chains (dimerization) or other molecules, such as proteins or fibres, thus strengthening the gluten network [6].

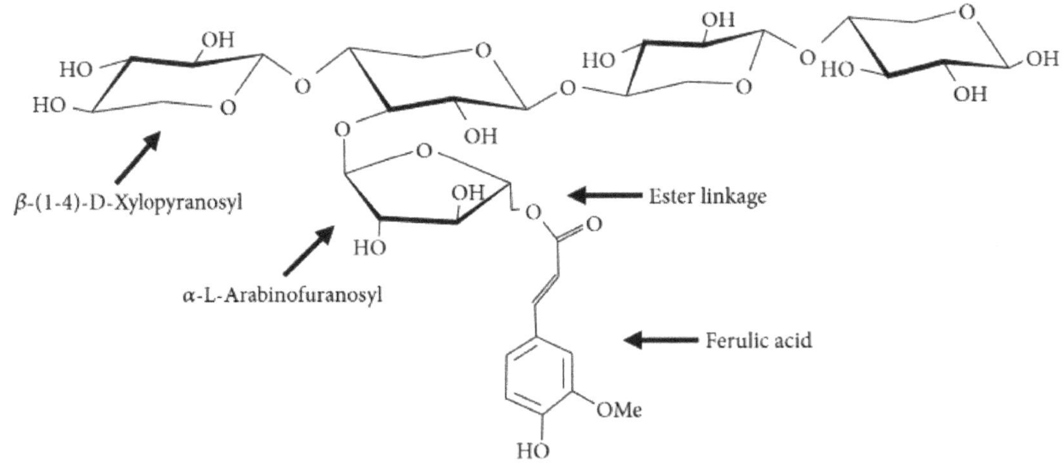

Figure 2. Arabinoxylan structure.

With a low arabinose to xylose ratio (A:X = 0.41–0.47), aleurone cell wall AXs are considered poorly cross-linked [23]. AX solubility is thus influenced since the presence of arabinose residues causes the formation of long asymmetrical polysaccharides; hence, an unsubstituted region has a tendency to aggregate (stabilisation by hydrogen bonds) and become water insoluble [6,25,40]. This insolubility increases with the presence of FA, which is shown by a high FA monomer-to-arabinose ratio (7.2–7.4) [10,23]. Only a few water-extractable AXs (WEAX) can be found at the cell-wall surface (1.5 to 2.5%), probably related to incomplete cross-linking with other components [39]. In addition, compared with other bran layers, there is less cross-linking between polysaccharides and phenolic compounds in aleurone. This might be due to the role of aleurone in the enzymatic degradation of the stored compounds, which requires facilitated movement through the cell walls [41]. Although, most health benefits related to the presence of AXs in the diet stem from the addition of WEAX, the predominance of water-unextractable AXs (WUAX) in the aleurone cell-walls can also be valued. However, it is often associated with negative rheological traits in breadmaking [39].

The second main component in wheat aleurone cell walls, β-glucan (BG), consists of a linear chain of glucose residues joined by glycosidic linkages [9]. The majority of the chain (90%) is made of cellotriosyl (DP3) and cellotetraosyl (DP4) units bonded by β-(1,3) linkages. The remaining 10% refer to β-(1,4)-linked side chains [25]. According to Jamme et al. [42], BGs are not present in the junction zone, whereas they are abundant in the inner periclinal face. As presented in Table 1, there are more BGs in aleurone cell walls compared to bran, which can be related to the layer's reported health benefits, such as the modulation of digestion [43].

Finally, phenolic acids are present in the aleurone cell walls (Table 2). They are mostly (95%) represented by ferulic acid (FA) in its trans form (90%) [44]. Their presence increases cell wall resistance by forming covalent cross-links (esterification) between AX chains through oxidative dimerization [25]. The most common dimerised forms are 5-5′-diferylate, 8-O-4′diferulate, 8-5′diferulate and 8-5′benzo diferulate [9,44]. Moreover, 92% of FA is found in bound form in the aleurone layer and accounts for 55 to 60% (w/w) of the whole

FA grain concentration according to Barron et al. [23], but can reach up to 70% (w/w) as per Brouns et al. [10]. This is highly beneficial, as FA is renowned for its antioxidant properties, which increase the antioxidant capacity of the aleurone layer [45].

Other phenolic acids include para-coumaric acid (PCA), which constitutes 67% (w/w) of the whole grain concentration and about 10% (w/w) of the total phenolics in aleurone cells [23]. It is also 3 to 5 times more concentrated in this bran layer than in the others but is not found in the endosperm [46]. Mainly in bound form (63% w/w), PCA is substituted to AX on average once every 90 arabinofuranose residues [10,22]. Sinapic and vanillic acids can also be found, mostly in conjugated form, amounting to 69 and 67%, respectively [10,23]. Traces of free syringic acid and bound flavonoids (apigenin and lutolein) have also been reported in the cell walls of the aleurone layer [10,47,48]. These phenolic compounds contribute to the total antioxidant capacity of the aleurone layer.

Table 2. Phenolic acids repartition in the wheat aleurone layer, bran and whole grain.

	Aleurone Layer * (mg/100 g dm)	Bran * (mg/100 g dm)	Whole Grain (mg/100 g dm)
Ferulic acid			
Total	628–817 [32,49]	500–1500 [3,49]	-
Monomer	798–814 [23]	-	86–87 [23]
Dimer	31–107 [23,32,49]	101 [49]	14–15 [23]
Trimer	2–15 [23,49,50]	-	3–4 [23,49]
p-Coumaric acid	15–29 [23,49,50]	13–16 [10]	1–3 [10,23,49]
Sinapic acid	6–44 [10,23,46,49]	11–28 [10]	4–8 [10,23]
p-hydroxybenzoic acid	2.8 [10]	1.9–2.2 [10,49]	0.5 [10]
Vanillic acid	2 [10]	1.6–3.5 [10,49]	0.5–2.1 [10]
Syringic acid	9 [10]	3.5–5.7 [10,49]	1.3–1.8 [10]

* The aleurone layer and the bran amount respectively to 5–8% (w/w) and 5.5–9.5% (w/w) of the wheat grain [21]. All data were placed in the same unit to facilitate comparison.

Aleurone cell wall polysaccharides (AX and BG) are in different proportions depending on their position in the bilayered cell wall. Although the thinner inner layer seems to comprise more BGs, the thicker outer layer contains more AXs. FA residues in both layers appear unchanged, although there seem to be more phenolic acids in the anticlinal than periclinal walls [25,51]. The cell walls of the transfer cells (modified aleurone cells in the crease region) contain less FA, and PCA is present in negligeable amounts. Moreover, their AXs are more substituted (A:X = 0.6), and they contain higher amounts of BGs once they mature [31]. Thus, the origin of the aleurone cell wall is related to its function and impacts on the layer's technological and health properties.

However, some minor components, such as proteins, can be found in the cell walls of the aleurone layer, cross-linked to AXs, BGs or sometimes to hydroxycinnamic acids [9,52]. They amount to about 1% of the total aleurone cell walls and show a similar amino acid composition as the proteins found in the endosperm (Table 3). These proteins can be classified as either glycine (37–86%), proline (11–39%), or serine rich (up to 23%) [52].

Compared with other bran layers, the aleurone cell-wall comprises low, if any, amounts of cellulose, glucomannan and lignin (Table 1) [10,25,44,53].

Table 3. Proteins and amino acids repartition in the wheat aleurone layer, bran and whole grain.

	Aleurone Layer *	Bran *	Whole Grain
Protein (%, w/w)			10.0–15.0 [32,33,54]
Part in tissue	≈30.0 [32,33,54–56]	15.2–16.9 [3]	-
Part in cell-wall	1.0 [36,37]	9.2 [36]	-
Total in grain	≈15.0 [32,33,54–56]	14.0 [3]	-
	AMINO ACIDS		
	(in %) [54,56]	(in g/16 g N) [54]	
Alanine	5.9	4.9	3.5
Arginine	11.1	7.0	4.6
Aspartic acid	7.9	7.2	5.0
Cysteine	-	2.0	2.2
Glutamic acid	20.9	18.6	30.6
Glycine	5.8	7.1	3.9
Histidine	3.4	2.6	2.2
Isoleucine	3.6	3.5	3.8
Leucine	6.5	6.0	6.7
Lysine	4.8	4.0	2.7
Methionine	1.6	1.6	1.7
Phenylalanine	3.8	3.9	4.6
Proline	6.3	5.9	9.8
Serine	2.9	4.5	4.8
Threonine	2.9	3.3	2.9
Tryptophan	4.0	1.6	1.2
Tyrosine	3.3	2.8	3.1
Valine	5.3	5.0	4.7

* The aleurone layer and the bran amount respectively to 5–8% (w/w) and 5.5–9.5% (w/w) of the wheat grain [21]. All data were placed in the same unit to facilitate comparison.

2.2.2. Intracellular Medium

The intracellular medium represents 70% of aleurone cell dry mass [19]. It is composed of a large nucleus and aleurone grains (or granules). The latter are inclusion bodies, or vacuolar units of two types with a spherical structure (2–4 μm in diameter), surrounded by non-polar lipid droplets. Type I inclusions comprise phytic acid crystals. In those, dihydro-phosphate traps minerals, including calcium, magnesium and zinc, by chelation, forming phytate complexes. This chelation of minerals results in a decrease in their bioaccessibility, which subsequently reduces the nutritional value of the aleurone intracellular medium unless they are released, for instance by the actions of endogenous enzymes (phytase) [9,57].

The other type of aleurone grain (type II) contains niacin (B3 vitamins) and proteins [9,10,58–60]. The proteins contained in type II inclusions represent 15 to 20% of total wheat grain proteins (dry mass) [5,61]. They are mostly storage proteins (mainly 7S globulins), despite the presence of numerous metabolic and defence enzymes [62]. The presence of lysine in these proteins is beneficial as it is a limiting amino acid in cereal products [9].

Overall, the aleurone grains provide numerous minerals accounting for 40% of total grain minerals, such as phosphate (80% in phytate form), magnesium, manganese, iron, potassium and sodium [10,11], as presented in Table 4. They are also a source of vitamins (Table 5). Indeed, thiamin and riboflavin are mostly present in this bran layer. Moreover, it is a source of lignans, especially syringaresinol, which are renowned for their antioxidant properties [10]. Significant levels of carotenoids (lutein and zeaxanthin) can also be found [63], as well as betaine and choline, which are twice as concentrated as they are in bran [64], as presented in Table 6. These bioactive components contribute to the nutritional value of the aleurone layer, and demonstrate the potential benefits of its addition to cereal products for the consumer.

Table 4. Minerals repartition in the wheat aleurone layer, bran and whole grain.

| | Aleurone Layer * | | Bran * (mg/100 g dm) | Whole Grain (mg/100 g dm) |
	Part in Tissue (mg/100 g dm)	Total in Grain (%, *w/w*)		
Calcium	73.0 [4,10]	53.3 [4]	100 [32]	30.0–70.0 [32]
Copper	12.4 [4]	67.3 [4]	0.8–1.6 [32,54]	0.4–1.5 [4,32,55]
Iron	19.0–34.0 [4,10,54]	78.6 [4]	5.0–15.0 [3,32,54]	1.8–8.0 [4,32,54]
Magnesium	600 [4,10]	86.7 [4]	500–700 [32]	100–200 [4,32]
Manganese	8.0–13.0 [4,10,54]	53.2 [4]	7.2–14.4 [3,54]	2.4–5.6 [4,54,55]
Sodium	-	-	5.0–30.0 [32]	3.0 [32]
Phosphorus	1400–3170 [4,10,54]	76.2 [4]	900–1500 [1]	218–792 [1]
Potassium	1100 [4,10]	67.6 [4]	1000–1500 [32]	350–600 [32]
Zinc	12.0 [4,10]	68.2 [4]	5.6–50 [3,32,54]	2.1–12 [4,32,54]
Total (%)	12.0 [32,33]	≈44.0 [32,33]	3.39 [1]	1.1–2.5 [1,32]

* The aleurone layer and the bran amount respectively to 5–8% (*w/w*) and 5.5–9.5% (*w/w*) of the wheat grain [21]. All data were placed in the same unit to facilitate comparison.

Table 5. Vitamins repartition in the wheat aleurone layer, bran and whole grain.

| | Aleurone Layer * | | Bran * (mg/100 g dm) | Whole Grain (mg) |
	Part in Tissue (mg/100 g dm)	Total in Grain (%, *w/w*)		
Thiamin (B1)	1.6 [36,56,65]	32 [3,49,54,66]	0.54 [25]	0.4–0.8 [32]
Riboflavin (B2)	1.0 [36,56,65]	37 [54,66]	0.39–0.75 [25]	0.1–0.2 [32]
Niacin (B3)	61.3–90.2 [36,56,65]	≈80 [3,49,54,66]	14.0–18.0 [25]	4.0–6.0 [32]
Pantothenic acid (B5)	4.51 [36,56]	41 [54,66]	2.2–3.9 [25]	1.0–2.0 [32]
Pyridoxin (B6)	3.6 [36,56,65]	≈60 [3,49,54,66]	1.0–1.3 [25]	0.5–1.0 [32]
Folate (B9)	0.2–0.8 [10]	-	0.079–0.2 [25]	0.035–0.055 [32]
Vitamin E	1.2–2.0 [10]	-	1.4 [25]	2.0–6.0 [32]

* The aleurone layer and the bran amount respectively to 5–8% (*w/w*) and 5.5–9.5% (*w/w*) of the wheat grain [21]. All data were placed in the same unit to facilitate comparison.

Table 6. Repartition of bioactive components in the wheat aleurone layer, bran and whole grain.

| | Aleurone Layer * | | Bran * | Whole Grain |
	Part in Tissue	Total in Grain		
Alkylresorcinols (mg/100 g)	≈3.0 [65,67,68]	20–400 [67]	220–400 [1,3]	42–70 [1,50,65]
Phytic acid (g/100 g)	8.4–15.6 [50,65,68]	2.2–5.2 [3]	1.2–1.4 [50,65]	1.18–1.38 [50,65]

Table 6. *Cont.*

	Aleurone Layer *		Bran *	Whole Grain
	Part in Tissue	**Total in Grain**		
Lutein (μg/100 g)	7.2–42.5 [63,68]	-	97–140 [3]	81.9 [63]
Zeaxanthin (μg/100 g)	21.2–77.6 [63,68]	-	25–219 [1]	9.0–43.8 [1,63]
Choline (mg/100 g)	1260 [54]	-	112 [1]	112 [1]

* The aleurone layer and the bran amount respectively to 5–8% (*w/w*) and 5.5–9.5% (*w/w*) of the wheat grain [21]. All data were placed in the same unit to facilitate comparison.

In addition, the aleurone intracellular medium is rich in endoplasmic reticulum, mitochondria, membrane-bound vesicles and lytic vacuoles [26]. No starch granules are present in the aleurone cells despite their proximity to the starchy endosperm [51].

In both cell walls and intracellular medium, most compounds are susceptible to great variations in concentration. Indeed, the effects of genetics, cultivar, and culture and storage conditions must be considered, as they can influence the amount of aleurone's compounds, thus having repercussions on its technological, health and nutritional properties [11,69]. For instance, coloured wheat varieties will differ greatly in their phytochemical composition [70–72].

3. Health and Nutritional Benefits

As presented, the aleurone layer is a source of many bioactive compounds that can potentially exhibit nutritional and/or health benefits for consumers. The benefits related to the layer's composition are impacted by the natural variability of the components. Moreover, as the following studies have been performed on materials of different purity and in various forms due to non-identical transformations and processes of obtention, the concluding remarks must then be approached as potential beneficial effects, as they are related to the specific conditions of materials and processes.

3.1. Digestibility and Colonic Fermentation

When ingested, the aleurone seems to have better digestibility than the other bran layers, with an overall digestibility value of 30%. This might be related to the absence of lignin in the aleurone, which is known to impair digestion [10,20]. However, this layer shows very poor digestibility in the upper digestive tract, especially when it comes to mineral absorption [57]. Furthermore, the release of minerals from the food matrix may be hindered by the presence of DFs, which cannot be digested in this part of the human gastrointestinal system [9]. Their bioaccessibility is even more decreased due to the presence of phytates, by which they are complexed [9].

In contrast, the in vitro colonic fermentation potential of aleurone seems to be better than that of wheat bran [73]. In multiple studies, the consumption of aleurone led to the stimulation of microbial activity (*Bifidobacteria dorea* and butyrate-producing *Roseburia* spp.) in the caecum and colon, thus leading to a higher yield of propionate and butyrate. These important short-chain fatty acids (SCFA) are renowned for their health benefits, namely in cancer prevention [57,74,75]. A decrease in health-detrimental bacteria, such as *Bilophila*, *Escherichia* and *Parabacteroides*, was also reported [75].

These observed effects probably arise from the presence of DFs in the aleurone layer. For instance, AXs in the layer's cell walls can exert beneficial health effects by inducing the proliferation of healthy gut microbiota and providing a substrate for increased SCFA production. Additional effects could include increased faecal bulk and viscosity, accelerated transit time and potential binding to cancer-inducing molecules [34,35]. These prebiotic effects could be accompanied by the cholesterol-lowering activity of AXs [76]. Slightly branched AXs are more easily degraded than AXs with a higher substitution degree, as they ferment more slowly or remain unfermented [57]. As the aleurone layer's AXs are poorly cross-linked, this might be beneficial when compared with other bran layers.

Nevertheless, the presence of DFs can also hinder the bioavailability of other bioactive compounds present in the aleurone layer by decreasing their bioaccessibility. Since most fibres in this layer are either only partially or not digested by the intestinal flora, most of these important compounds cannot be released for absorption [9]. This includes the absorption of minerals complexed with phytate. However, DF fermentation in the colon can increase minerals' colonic absorption by increasing the production of SCFA, in turn reducing the pH in the intestine, thus solubilising phytic complexes and releasing minerals [77].

3.2. Health Benefits of Aleurone Consumption

In various clinical studies, a diet rich in wheat aleurone has been shown to reduce the risk of cancer, cardiovascular disease, obesity and diabetes [12,78]. Symptoms such as hypertension and hyperglycaemia reduction have been reported after consumption of aleurone in the long term, which contributes to the prevention of obesity and hyperlipidaemia [79]. This can also be seen with a decreased low density lipoprotein (LDL) cholesterol level post-ingestion [80]. Moreover, aleurone ingestion seems to reduce the risk of colon cancer after fermentation in the colon by decreasing secondary bile acid production, inducing apoptosis and cell differentiation, as well as detoxification [81–84].

In human studies, a moderate intake of aleurone raised red blood cell folate levels, decreasing plasma homocysteine levels. This effect is beneficial since high plasma homocysteine is a risk factor for both cardiovascular disease and cancer via DNA damage [85–87]. Moreover, the authors concluded that the folate contained in the aleurone was then highly bioavailable and bioefficient [86].

However, Keaveney et al. [88] reported contrasting results, as the consumption of 50 g of a wheat aleurone fraction from Bühler A.G. increased the betaine plasma concentration by 2.5 times, but neither plasma tocopherols, folate nor choline levels were increased. Another study confirmed this effect of aleurone on metabolic risk factors; higher plasma betaine and related lower plasma homocysteine were observed [80]. Nonetheless, the most recent study reported that upon consumption of 27 g of aleurone daily by overweight and obese subjects, no significant changes were observed in health-related biomarkers, including plasma homocysteine and SCFA levels [89].

Studies on the effect of wheat aleurone consumption on inflammation markers have reported a decrease in pro-inflammatory tumour necrosis factor (TNF)-α (one of the most important cytokines in the immune system [90]) in LPS-stimulated U937 macrophages [91]. Moreover, a decrease in C-reactive protein (CRP) was also noticed (a biomarker of inflammation used in the prediction of coronary heart disease [92]), probably via a reduction in LDL-cholesterol levels. This could imply that aleurone consumption may change hepatic metabolism either by the action of its independent bioactive components or by their interaction [93].

Although the health benefits observed upon aleurone consumption in clinical studies are numerous, they are still hypothetical. As the studies did not involve the same test conditions and used different materials, they are thus not comparable. Moreover, it does not consider the natural variability in the composition of the aleurone layer, nor does it wholly consider the matrix effects of its consumption in a normal diet with other food products. In conclusion, the observed health benefits may arise upon aleurone consumption, but they should be referred to as potential effects.

3.3. Antioxidant Capacity

In addition to the aforementioned beneficial health effects, aleurone is also a source of many bioactive compounds with antioxidant activity. It is the bran layer with the highest total antioxidant capacity and can provoke a prolonged anti-inflammatory effect after consumption [91]. This is partly due to its high phenolic acid content and mainly allotted to FA, which accounts for 60% of the aleurone antioxidant capacity [45].

The consumption of 50 g of this milling fraction from Bühler A.G. has also been reported to increase the FA plasma concentration [94]. Furthermore, the latter phenomenon seems to be related to enhanced FA bioavailability [95], and more generally an increased phenolic acid bioavailability, which is apparently conserved even when the aleurone fractions are incorporated into bread [96].

Due to its resonance-stabilised phenoxy radical structure, FA can inhibit lipid peroxidation via superoxide scavenging [97,98]. Moreover, the presence of a methyl group in C3 enhances resonance stabilisation, thus making it a very stable antioxidant that does not initiate an oxidative chain reaction [99,100]. Although this activity is increased when FA is in its dimerised form (whereas it is mostly in monomer form in the aleurone layer, Table 2), it still exhibits better LDL inhibition than ascorbic acid, a powerful antioxidant [9,101].

However, the bioavailability of FA is decreased when in this dimerised form or if ester-linked to AX, the latter amounting to the majority of FA in the aleurone layer [45,73,102]. The low degradation of DFs by the microbiota could present better health effects than a high peak after ingestion since phenolic compounds are continuously released in the plasma [103]. Moreover, the bioavailability of such compounds could be improved by the addition of cell wall degrading enzymes that release them [104].

In addition to FA, other phenolic acids (mainly PCA and sinapic acid) also contribute to the total antioxidant capacity of the aleurone layer [12]. Bioactive components, such as phytate, phytoestrogens (including lignans), and anthocyanins, are also involved [77,105–107]. Vitamin E participates equally by quenching singlet oxygen and nitrogen-oxide radicals. More precisely, α-tocopherol works as part of an antioxidant network in breaking lipid-soluble chains [108]. Vitamin B6 also possesses antioxidant activity and thus participates in protection from oxidative stress [109]. The combined effects of these compounds present in the aleurone layers make them a valuable source of antioxidants, provided they retain this property after transformation (i.e., breadmaking).

3.4. Bioactive Components of Aleurone and Related Potential Nutritional and Health Benefits

The aforementioned effects have been reported for the consumption of aleurone as a whole, but the health benefits of the bioactive components it contains have also been investigated separately.

The presence of vitamins in the aleurone layer (Table 5) can be beneficial from both nutritional and health points of view. For instance, tocopherols are involved in the regulation of cell signalling and gene expression and are also known to delay the progress of degenerative diseases. The other type of E vitamins, tocotrienols, are involved in the prevention of neurodegeneration. They can also induce immune responses as well as lower cholesterol and prevent cancer [108]. Unlike E vitamins, folate (vitamin B9) is renowned for its biological activity in normal foetus neural tube development [11]. In addition, niacin is involved in carbohydrates and fats metabolism [110].

However, anthocyanins may exhibit anticancer properties [106,107], and lignans may lower cholesterol and show potential estrogenic activity [12,105]. Moreover, the benefits of betaine include its role as an osmolyte, remethylating total homocysteine, therapeutic agent for non-alcoholic fatty liver disease, and lipotrop [80].

Lastly, the aleurone layer contains an essential amino acid: lysine (Table 3). Its presence is beneficial for the end-product nutritional value since lysine is a limiting amino acid in cereal grains [9]. In addition, the presence of arginine in aleurone proteins can also be valorised due to its role in vascular dynamics and endothelial function, which can improve [12].

As these compounds are present in the aleurone layer, this adds to its value, provided that they remain present in the end-product and conserve their potential health-benefit properties through the process.

4. Potential of the Aleurone Layer as an Ingredient in Bread- and Cereal-Based Products

4.1. Extraction of the Aleurone Layer and Its Challenges

The aleurone layer's potential can only be revealed if it is first extracted. However, no universal process yet exists. In addition, as the grain's milling properties (friability) are dependent on its constituents, themselves related to the culture conditions and genetic background, the process must constantly adapt to the raw material. Therefore, this represents a major hurdle in the utilisation of the aleurone layer and in the exploration of its properties.

Another challenge arises from its composition. The aleurone layer, although botanically part of the endosperm, is considered by millers as belonging to the bran fraction. Since it is tightly adhered to the pericarp, it is usually removed from the endosperm during conventional milling. This tight adherence to seed coats also makes it difficult to separate the aleurone from the rest of the bran [10,12]. Consequently, most existing processes for the retrieval of the aleurone layer start with bran material. Multiple procedures have been patented [14,111–114] and two companies have mainly been known to produce and commercialise aleurone-rich flour: Bühler A.G. and Cargill Limited (through Horizon Milling with the GrainWise brand). However, it seems that most existing processes result in the obtention of aleurone-rich flour that is either not highly concentrated [13] or low yielding [14]. A summary of the composition of aleurone-rich fractions issued from the existing processes mentioned below, available in the literature, is shown in Table 7. As shown in the table, the purity of the obtained aleurone-enriched fraction depends greatly on the extraction process performed, which explains the discrepancies observed in their composition when compared to that of the pure hand-isolated aleurone layer. These differences in composition may also arise from the use of different analytical methods among publications.

Most of these processes extract the aleurone layer from bran components by dry-fractionation, a succession of mechanical or physical unitary steps [10,115,116]. Many researchers first aim to dissociate the different bran tissues, which can be conducted by grinding. They include a separation step that then enables sorting out the particles according to their size, mass, density or electrostatic properties [116]. The obtained milling fraction can thus be added to basic wheat flour to enrich it with the aleurone layer [115].

The aleurone layer is extensible, similar to the intermediate strips of wheat bran, and it has an elastoplastic rheological behaviour. Its mechanical characteristics are impacted by the degree of feruloylation of its AX, particularly by the presence of FA dehydrodimers [44]. Hence, the mechanical stress generated by dry fractionation processes first affects the aleurone cell walls, which crack, allowing the cell contents to be released [117]. According to Rosa et al. [118], the velocity of phytic acid release could thus be used as a marker to estimate aleurone cell opening.

Electrostatic separation of aleurone from other bran tissues is an interesting process since the aleurone layer presents unique electrostatic properties compared to the other strips. However, this process can be influenced by multiple parameters, such as particle size, composition, microstructure and moisture content [68].

The main advantage of physical or mechanical extraction methods is that they do not require the use of chemical products that can interact with the matrix and decrease the product's purity and phytochemicals' functionality [116]. Compared with wet processes (chemical and enzymatic treatments), they also enable higher energy efficiency [10,116]. However, the succession of unit operations may impact the antioxidant and secondary metabolites of the aleurone layer [3]. Moreover, grinding generates various particles from bran tissues of different sizes and densities, which are hard to differentiate, hence the reported end-product's low purity [116].

This type of process was used in the patented method by Stone and Minifie [14], who first used hammer-milling in wheat bran containing 34% of aleurone cells, followed by sieving, electrostatic fractionation, and a final separation through an electric field. A 95% purity of aleurone cells was obtained with a 10% yield [14,68]. Nonetheless, alternative methods exist: humidification then micro-grinding of wheat bran with a friction roller mill [114]; sequential pearling cycles in a vertical abrasive polishing machine [74]; centrifugal impact milling [68]; ultrafine grinding and electrostatic separation [119]. Different outcomes have been reported with these processes, with varying aleurone purity and yield.

However, there are limited studies on the extraction of the aleurone layer by wet processes (chemical and enzymatic treatments). For instance, the maceration of wheat bran in chemical reagents, such as organic solvents, has been tested but not in an industrial scale [10].

Nonetheless, dry and wet processes can be coupled. A patented method isolated aleurone by successive steps of cleaning, steaming, stabilising, roller-milling, sieving, fine grinding and air-classifying. However, the end-product still contained 36.5% starch, demonstrating low purity [87,120]. In addition, the patent deposited by Kvist et al. [121] subjected wheat bran to several enzymatic treatments, wet milling steps, sequential centrifugation, and ultrafiltration. Other researchers coupled successive steps of milling, sieving, air classification and centrifugation with benzene-carbon tetrachloride mixtures at laboratory scale [10].

Although many experiments have been conducted and sometimes patented to extract and isolate the aleurone layer, the challenge of measuring end-product purity has arisen. Thus, researchers have defined biochemical markers to differentiate grain parts. These biochemical markers can be used to determine the extent to which the aleurone layer is extracted from other grain components. Starch, phytate, p-coumaric acid, alkylresorcinols and FA trimer are used to estimate the proportion of the endosperm, aleurone cell content and cell-walls, intermediate layer and outer pericarp, respectively [46,65,122]. However, the relative amount of grain tissue can only be calculated when compared to the reference values. The latter were values of the same parameters from pure isolated tissues of identical wheat cultivars. Thus, it limits their utilisation for characterisation since pure fractions are obtained from hand-isolated tissues, a long and labourous process. Moreover, these markers are susceptible to natural variability among wheat cultivars due to the culture conditions and genetic background [50,123]. As an alternative to these biochemical markers, microscopy analyses can be performed to estimate the purity of the extracted fractions [115].

Table 7. Composition of aleurone-enriched fractions issued from different extracting processes in literature.

Origin of Products *	1a	1b	2	3	4
ASH (g/100 g)	9.3–13.3 [12,124,125]	≈ 10.0–13.3 [7,12,126,127]	4.1 [85]	7.2–7.34 [128,129]	3.9–5.1 [130,131]
MOISTURE (g/100 g)	-	8.0 [7]	5.4 [85,87]	8.14 [128]	6.7–8.98 [130–132]
CARBOHYDRATES					
Arabinoxylans (g/100g)	-	24.3 [127]	-	-	14.39 [131]
A:X ratio	0.62 [133]	0.35–0.46 [7,127,133]	-	-	-
β-glucans (g/100 g)	3.4 [133]	3.91–4.5 [127,133]	-	-	1.7 [132]
Cellulose (g/100 g)	10.6 [133]	6.0 [133]	-	-	-
Pentosans (g/100 g)	-	-	-	-	20.2–21.6 [134]
Starch (g/100 g)	1.9–5.8 [124,133]	2.2–5.8 [126,127,133]	36.5 [85,87]	2.5–12.75 [128,129]	33.46 [130]
Total Dietary Fibers (g/100 g)	39.7–60.0 [12,124,125,133]	39.7–49.2 [7,12, 126,127,133]	15.4 [85,87]	43.36 [128]	27.90–44.3 [130–132]
Soluble Dietary Fibers (g/100 g)	4.1 [125]	-	-	3.07 [128]	-
Insoluble Dietary Fibers (g/100 g)	50.0 [125]	-	-	40.15 [128]	-
MINERALS AND TRACE ELEMENTS					
Total (g/100 g)	5.8–9.8 [124,125]	7.0–9.8 [7,126,127]	6.5 [85,87]	4.5–6.33 [128,129]	4.1–4.7 [134]
Calcium (Ca) (mg/100 g)	76.2 [125]	93 [10]	-	-	-
Copper (Cu) (mg/100 g)	-	-	-	1.35 [128]	-
Iron (Fe) (mg/100 g)	21.3 [125]	26 [10]	-	13.93 [128]	-
Magnesium (Mg) (mg/100 g)	690–800 [12,125]	850–1030 [10,12]	-	770 [128]	-

Table 7. *Cont.*

Origin of Products *	1a	1b	2	3	4
Manganese (Mn) (mg/100 g)	-	-	-	12.7 [128]	-
Sodium (Na) (mg/100 g)	1.7 [125]	-	-	-	-
Phosphorus (P) (mg/100 g)	1900 [125]	2540 [10]	-	1857 [128]	-
Potassium (K) (mg/100 g)	1900 [125]	2250 [10]	-	1780 [128]	-
Zinc (Zn) (mg/100 g)	11.4 [125]	14.0 [10]	-	12.05 [128]	-
PHENOLIC COMPOUNDS					
Total phenolic acids (mg/100 g)	-	-	-	-	457 [132]
Total hydroxycinnamic acids					
Free (mg/100 g)	1.28 [135]	1.22 [135]	-	-	-
Bound (mg/100 g)	47.58 [135]	60.65 [135]			
p-Coumaric acid					
Total (mg/100 g)	-	16.0 [126]			
Free (mg/100 g)	-	1.0–1.5 [126]	-	-	-
Bound (mg/100 g)	0.60 [135]	0.99–1.0 [126,135]			
Conjugated (mg/100 g)	-	0 [126]			
Sinapic acid—bound form (mg/100 g)	0.46 [135]	0.53 [135]	-	-	-
Alkylresorcinols (mg/100 g)	1107 [135]	993.24 [135]	-	-	138 [132]
Flavonoids (mg/100 g)	9.65 [135]	8.95 [135]	-	-	-
Lignans (mg/100 g)	6300 [133]	4700 [133]	-	-	-
Phytic acid (mg/100 g)	6900 [125]	-	2360 [85,87]	-	-
PROTEINS					
Total (g/100 g)	13.3–18.0 [12,124,125]	21.0–22.2 [7,12,126,127]	23.6 [85,87]	15.2–17.46 [128,129]	16.5–21 [130–132]
VITAMINS					
Total (mg/100 g)	>29.0 [125]	30.0 [126]	-	-	-
Thiamin (B1) (mg/100 g)	0.87–1.6 [12,125]	1.1–1.4 [10,12,126]	-	1.26 [128]	-
Riboflavin (B2) (mg/100 g)	0.3 [125]	0.2–0.3 [10,126]	-	0.32 [128]	-
Niacin (B3) (mg/100 g)	24.0 [125]	21.0–32.9 [10,126]	-	22.87 [128]	-
Pantothenic acid (B5) (mg/100 g)	-	5.0 [126]	-	1.87 [128]	-
Pyridoxin (B6) (mg/100 g)	0.3 [125]	1.3–1.4 [10,126]	-	2.52 [128]	-
Folate (B9) (mg/100 g)	0.8 [125]	0.1–0.2 [10,126]	-	0.2 [128]	-
Tocopherols and tocotrienols (E) (mg/100 g)	2.0 [125]	0.8–1.2 [10,126]	-	-	-

* Origin of products, as listed below. **1a**: Bühler A.G. (55–70% aleurone purity); **1b**: Bühler A.G. (75–90% aleurone purity)—patented method [111]; **2**: Goodman Fielder Milling and Baking Pty. Ltd. (90% aleurone-rich flour with 10% of waxy maize starch); **3**: Cargill Limited and Horizon Milling (Grainwise); **4**: Jiaxing Zhishifang Food Science and Technology Co. (Shandong, China), 14% wb. All data were placed in the same unit to facilitate comparison.

4.2. Application to Breadmaking

4.2.1. Aleurone Bread Nutritional Profile

According to past reviews and experiments, there are many benefits to incorporating an aleurone-rich flour into bread and bakery products, starting with an improved nutritional profile of the end-products. This amelioration is related to increased DF and protein (mainly albumin and globulin) content at the expense of readily digestible carbohydrates [13,15,119,131,136]. The enhancement of minerals, including phosphate, magnesium, manganese and iron, and bioactive compounds such as phenolic acids, antioxidants, phytoestrogens and sterols, also increase the value of the obtained end-products [16,136]. This improved composition confers the end product a nutritional profile similar to that of whole wheat products [128], while equally making it a good source of fibre [12]. However, the nutritional benefits of aleurone-rich products are accompanied by increased phytate content, which is known for its antinutritional effect [57].

4.2.2. Aleurone Bread Dough Characteristics

Despite these beneficial nutritional properties, the incorporation of the aleurone layer for breadmaking leads to changes during dough formation, which affects the sensory attributes of the end-product. With its high DF content (Tables 1 and 7), the aleurone layer impairs dough hydration properties. The AXs and BGs contained by the aleurone layer compete for water with the proteins forming the gluten network, thus increasing the water absorption of the dough and retarding the dough development time [136,137]. The water retention capacity is also affected by fibres that take up a large amount of water (3.5 to 6.3 times their weight for WEAX and 6.7 to 9.9 for WUAX) by binding through hydroxyl groups, resulting in a longer mixing stability due to the alteration of the gluten structure [6,131,136].

In addition, the presence of these fibres has a diluting effect on the starch granules. Damaged starch content is then decreased as well as the falling number. The latter effect is further reduced by the increase in α-amylase activity in the presence of calcium. Indeed, this metalloenzyme requires calcium for its performance, which is provided by the aleurone layer (Tables 4 and 7). In addition, these observed properties seem to increase in relation to the aleurone-rich flour dosage [136].

The effect of aleurone incorporation on starch also influences the pasting properties of dough. Multiple studies have shown a decrease in peak viscosity, as well as in the retrogradation of dough [131,132]. This might not only stem from the presence of fibres that interfere with starch granule swelling and increased amylase activity but also from the combined effect of other aleurone constituents. For instance, the presence of fat and FA can also impact pasting properties, in addition to an already low starch content [132,138].

Nevertheless, the aleurone dough exhibits higher Rapid Visco Analyzer (RVA) parameters, revealing a strong gel ability greater than that of whole wheat flour [131]. According to Bucsella et al. [136], this could be due to the swelling of fibres, which form a strong gel despite the lower starch content. This gel is described as being more resistant to heat and mechanical stress.

The gluten network can also be strengthened following the addition of aleurone-rich flour to bread dough (up to 40%). However, according to Mixolab (Chopin) measurements, the dough development time is increased due to the presence of fibres that compete for water and hinder gluten network formation by intercalating between the proteins, resulting in a more heat-stable and stress-resistant dough [131].

This increase in dough stability is also depicted by firm elastic-like behaviour due to the stronger gluten complex [131,136]. The increased protein content (albumin) and the strengthening effect of AX binding to gluten via the oxidative dimerization of FA also contribute to these observed effects [139]. Instrumentally, this translates into an increase in dough stability and break time, as well as delayed weakening [136].

Despite the aforementioned beneficial traits observed due to the aleurone components, most of them are dose dependent. Excessive addition of aleurone-rich flour to the dough can lead to deleterious effects on dough rheology.

4.2.3. End-Product: Aleurone Bread Characteristics

The addition of aleurone-rich flour to bread-making has an impact on end-product quality, although the results of the researcher's findings are contradictory. This may be related to the aleurone-enrichment level, the purity of this material as well as the bread formulation process, compiled in Table 8.

Table 8. Overview of existing aleurone-enriched bread formulation processes in literature.

Aleurone-Rich Flour Used	Enrichment Level (%, *w/w*)	Basic Flour Type and Protein Content (%)		Bread Formulation Process	Specificities	Reference
Table 7, 1a/1b	20	White flour	11–13	Straight dough	Addition of vital wheat gluten, high fructose corn syrup, and dough conditioner	[12,128]
				Sponge dough	Addition of vital wheat gluten, high fructose corn syrup, ascorbic acid, dough conditioners, mono- and diglycerides	
Table 7, 1a/1b	40 and 75	Wheat flour	11.86	ICC Standard Method 131	-	[13,15]
Table 7, 1a/1b	25 and 50	Wheat flour	12.9	ICC Standard Method 131	-	[16,17]
Table 7, 4	10, 20, 30 and 40	Wheat flour	14.16	-	-	[131]
Table 7, 1a/1b	15, 40, 75 and 100	Conventional bread wheat flour	15.24	Sourdough (MSZ method, 1989)	-	[136]
54.11% purity	18	Wheat flour	-	GB/T 358969-2018 method	Hemicellulase addition (0–60 mg/kg of flour)	[140]

* Basic flour type as described in the corresponding articles, defined as commercial flour.

Some report a decrease in loaf volume, accompanied by reduced height and increased weight [13,15–17]. For instance, Bagdi et al. [15] observed a diminution of 27% of the specific volume for a bread prepared with 100% of aleurone-rich flour compared to a control white bread, as well as a reduction in loaf height of 13%. Using the same breadmaking process (ICC Standard Method 131), Bartalné-Berceli et al. [16] obtained a height decrease of 15 and a 7.2% weight increase with a bread containing 25% of aleurone-rich flour compared with a control white bread. These tendencies are further incremented with a higher aleurone flour input (50%), where almost half of the height was decreased and 3.6% of the weight was increased compared with the control.

Other studies have observed a higher loaf volume than white bread upon aleurone incorporation or have found insignificant changes. The texture in these experiments was also reported to be softer than white bread, which means that the crumb was less dense [12,136,140]. Indeed, Tian et al. [140] described an increase of 40.91% in bread specific volume using aleurone-rich flour (modified GB/T 35869-2018 procedure with 54.11% of aleurone layer content). However, this beneficial effect could be attributed to the presence of hemicellulases (at 40 mg/kg) that enable the formation of WEAX from WUAX. Similar results were obtained from breads made with a sourdough preparation (MSZ-6369-8:1988) incorporating an aleurone fraction, even though the observed volume increase was not significantly different from that of the control bread [136]. Breads made from straight dough and sponge dough processes with additives containing aleurone-rich flour (20%) also show this beneficial trait [12].

Overall, it seems that these beneficial effects could be related to the presence of hemicellulose-degrading enzymes—either endogenous (sourdough) or exogenous (as an additive)—or additives or a special breadmaking process, each enabling the revelation of the aleurone layer's full technological potential.

Unlike for the volume and texture of the breads, the appearance of the end-product is equivocal: the crumb colour (measured by colorimetry) is darker than white bread, even brownish, which can be a limiting factor for some consumers. However, it is still lighter than whole wheat products [13,15,57].

As for the taste of the bread, diverging results also occur. Whereas some report a flavour similar to that of white bread, especially when a long fermentation process takes place [12,128]; others describe a bread that is more bitter and sour, even rye-like [13,15]. In addition to the last finding, Amrein et al. [57] outlined a gritty mouth-feel, which is a limiting factor for the consumers of the study. However, the smell of the products is reported to be more intense and sour [13,15].

Overall, bread made of aleurone-rich flour in different proportions showed similar properties to that of white breads but with the nutritional profile of whole wheat breads. More thorough experiments on the relevance of aleurone addition in cereal products compared to other wheat kernel layers should be conducted, as the diversity of breadmaking methods and starting materials is great in the existing studies. Nevertheless, the results are still contradictory and lead to either a decrease or increase in end-product consumer acceptability. The use of special breadmaking technologies or additives, such as cell wall degrading enzymes, could thus reveal the aleurone layer's full technological potential.

4.2.4. Underlying Mechanisms

Many of the adverse or positive technological effects due to the addition of aleurone-rich flour to bakery products stem from its unique composition and, more specifically, its high protein and DF content. Indeed, studies investigating the effect of DF, AX and bran incorporation into bakery products showed similar properties to those described in aleurone-enriched products.

Most experiments on this subject describe that the addition of fibres to bread dough increases dough development time, water absorption and strength. However, it also seems to weaken the dough's tolerance to mixing and fermentation [141]. This results for most studies in a reduction in loaf volume [5,141–144], an increase in crumb firmness [142–144], and a darkened crumb appearance [141–144].

Hypotheses exist to explain the mechanisms underlying these results, which corroborate those of aleurone-rich products. First, fibres with their high water binding capacity might compete for water with starch and gluten, thus keeping wheat proteins from sufficient hydration for the formation of the gluten network [5,141,142,145–149]. Another explanation is that fibres dilute gluten, thus affecting its gas-holding capacity [5,141,142,145–149]. Nonetheless, this impairment in gas retention that causes a loss of loaf volume could also be due to the shortened and lowered resistance to dough extension upon DF addition, which increases the concentration of soluble cell wall materials and disrupts the gluten network [142,150].

In addition to the previous mechanisms of action that hint at a physical mode of action, a chemical hypothesis also exists that states that FA linked to DFs could mediate AX–AX and AX–protein cross-linking (through FA–tyrosine linkages), thus impacting gluten properties [146–149]. This would be possible in the presence of oxidants or enzymes (such as laccase and peroxidases) that provoke the dimerization of FA, thus creating covalent linkages between AX chains. Moreover, this dimerization increases the water retention capacity of AXs, which directly affects the gluten network [6].

More specifically, studies conducted on the addition of AXs to bakery products could be helpful in understanding the mechanisms underlying the aleurone-enriched bread properties, since they represent a large part of this layer. Similar to the addition of general DFs, an increase in the water absorption of the dough is reported due to the high water retention capacity of AXs, which increases dough consistency [151]. According to Berger and Ducroo [6], to reach the same dough consistency as the control dough on the Brabender farinograph, 0.5 to 2% of additional water should be incremented per percent of AX supplemented.

As for the negative effects on gluten network formation due to AX addition, they could stem from the steric hindrance of the increased batter viscosity that limits components mobility, thus decreasing the formation of gluten aggregates and starch entrapment in its matrix [151]. Nonetheless, the observed effects are not as important as the extent of their addition and their molecular size, but most importantly, depend on the breadmaking quality of the initial flour used for the experiments [152].

Furthermore, the water extractability of AXs is also a determining factor in the adverse effects it causes on bread and bakery products. For instance, water-unextractable AXs (WUAX) seem to generate more deleterious side effects upon their addition than water-extractable AXs (WEAX). This might explain the contradictory results with aleurone applications since it mainly contains WUAXs, which can be transformed into WEAXs during breadmaking.

The use of WUAXs in bakery products is often reported with breads of lower volume, coarser crumb and higher firmness [39]. To explain this phenomenon, there are three hypotheses: (i) WUAXs form physical barriers for wheat proteins during dough development [39]; (ii) these AXs form intrusions in the gas cells during fermentation [39,153]; (iii) the WUAXs compete for water with the gluten network, thus impairing its formation and leading to a fracture effect that increases dough resistance to extension [6,7,39]. The latter hypothesis is believed to be more accurate since a correction of dough hydration (2% per percentage of AXs added) improves its extensibility [6].

In contrast, the use of WEAXs in bread dough yields contradictory results, even though they are mostly beneficial. Globally, a finer and more homogenous breadcrumb is depicted, with a bread that is softer [39]. The loaf volume is also impacted, but contradictory results are obtained. The observed higher volume is usually obtained with the use of high molecular weight WEAXs [39,153,154]. The underlying mechanisms explaining these effects include an increase in liquid film stability and thus in dough aqueous phase viscosity [7,39]. Moreover, WEAXs of higher molecular weight can form a secondary network, weaker than gluten, which enforces the latter and stabilises it through the dimerization of FA and by physical entanglement, either of gluten or between WEAXs [39]. By increasing the dough's gas retention capacity, the resulting breads become higher [6]. According to [6], the higher the WEAXs molecular weight, the highest beneficial effect is observed.

4.3. Application to Other Food Products

Although the incorporation of aleurone-rich flour into bakery products have mostly been studied, its application to other product categories exists. For instance, Cargill Limited [128] developed cereal flakes and extruded snacks with 35% of aleurone-rich flour, as well as high-protein bars containing 20%. Ready-to-eat cereals enriched in aleurone were also studied by Byrne [155].

Other applications entail aleurone-enriched pasta and noodles. Whereas the former was described as healthier than wholemeal spaghetti due to higher protein, fat, and DF content, its consumer acceptance was decreased. Although it showed improved quality characteristics (lower water uptake, higher cooked pasta firmness, higher tensile strength and lower stickiness), the darker, more intense, bitter and sour taste of the pasta influenced consumer appreciation of the product [138]. The second used a combination of aleurone-rich flour and transglutaminase, which resulted in noodles with less cooking loss and the best sensory evaluation when compared to traditional noodles [130].

Finally, Yang et al. [134] used aleurone-enriched flour and cell wall degrading enzymes for the production of Chinese buns. They found that the action of enzymes promoted the WEAX content while also increasing the water availability to the gluten-forming proteins, resulting in softer dough, especially when enzyme activities were combined.

5. Optimization of the Aleurone Layer's Potential

Despite the aleurone layer's nutritional and potential health-benefit properties, its use on the market is scarce. This might be related to the hurdles in its extraction, as well as the rheological issues it has with the end-product. To overcome the latter and obtain a product with improved nutritional traits without alteration of its technological properties, aleurone-rich flour can be subjected to different processes, whether physical, chemical or biological. By modifying the aleurone constituents and, most specifically, its DFs, which are mainly responsible for the observed adverse effects, the technological potential of the aleurone layer could thus be optimised. Moreover, these processes could also enhance the health benefits associated with the intake of this bioactive milling fraction.

Many physical treatments, whether thermal, non-thermal, dry or wet, focus on the particle size reduction in the material, as it can lead to many beneficial effects. It includes an improved antioxidant and bile acid-binding capacity, a greater bioavailability of phenolic compounds and vitamin E, a higher production of colonic SCFA, and a faster digestion rate, which can increase transit time and decrease faecal bulking [13,19]. For this purpose, different techniques can be performed. Whereas milling refers to the process of separating the endosperm (known as white flour) from the bran (outer layers and germ) [156], grinding uses shear stress and compression to reduce the particle size [157]. Both can be used to obtain a particle size below that of aleurone cells (around 50 μm) to release their content [117, 158].

The results of those studies concluded that as the particle size of the material decreased, phytate extractability was enhanced, and phenolic acids were released [117], thus improving the mineral bioaccessibility of aleurone [158]. The hydration properties (namely water holding and binding capacity) of the milling fraction are also reduced, which subsequently negatively affects its fermentability [127]. In addition, an increase in conjugated and free FA post-grinding has been observed, as well as the release of aleurone intracellular compounds (soluble proteins, vitamin E and phytic acid). The combined effects of improved bioaccessibility of antioxidant compounds, as well as the greater exposure of phenolic moieties, result in an enhanced antioxidant capacity of the modified aleurone ingredient [9, 118,126,127,159].

In addition to physical treatments, the use of chemical processes to modify, inter alia, the solubility of fibres is of equal interest for the improvement of the aleurone layer's technological potential. Their benefits lie in the fact that they are the only ones that can provoke a polymerisation of the materials. However, the final product can be of low purity, with a high degree of hydrolysis and modified functional groups [160,161].

Experiments conducted by Bagdi et al. [162] revealed that hydroxyl radical treatment, •OH oxidation and cross-linking of AXs extracted from aleurone-rich flour modified its bile acid-binding capacity and could even enhance the cholesterol-lowering effect of AXs. Besides these improved health effects, the use of an alkaline treatment on the aleurone's WUAXs could be beneficial, as it was reported to maintain the fibre's molecular weight while increasing its water solubility [154]. Moreover, as carboxymethylation of wheat bran enhanced its health properties (increased total antioxidant capacity, total reducing power, Fe^{2+} chelating capacity and DPPH radical scavenging capacity), the use of this chemical mean of treatment on the aleurone layer could equally raise its value [163].

Nonetheless, most of the treatments performed on aleurone-rich flour in the literature have been conducted by biochemical means. Rhodes and Stone [124] studied the effect of combined methods, namely ultra-fine grinding and enzymatic treatment with xylanase and feruloyl esterase. Upon those treatments, changes in the aleurone layer structure were observed, as well as an increase in free FA. Moreover, associated beneficial health effects, such as reduced mouse body weight and improved glucose metabolism, led the authors to conclude that a partial depolymerisation of the wheat aleurone cell wall could be favourable for their metabolism.

These conclusions are consistent with those from Rosa et al. [127], which upon xylanase and feruloyl esterase treatment of aleurone, described a release in FA, both in free and conjugated forms. Furthermore, although SCFA production was not improved, the fast metabolization of FA by the colonic microbiota promoted the production of FA colonic metabolites.

Moreover, these findings are complementary to those from Rosa, et al. [126], whom upon enzymatic treatment of an aleurone-rich fraction, also found an increase in the release of cell-wall bound phenolic acids in conjugated and free forms. However, this was also associated with an increase in the antioxidant capacity of the fraction due to both the released components and the increase in their bioaccessibility.

However, experiments conducted by Vangsøe et al. [133] demonstrated that the enzyme susceptibility to the aleurone cell wall AXs was correlated to its arabinose-to-xylose ratio. More specifically, it seems that xylanase activity was enhanced in the presence of AXs with lower substitution degrees, which are mainly found in the cell walls of the aleurone layer.

Overall, xylanase has mainly been used for the transformation of the aleurone layer in previous experiments. As those enzymes are capable of hydrolysing WUAXs, known for their negative technological effects, this explains their success. Nevertheless, complete hydrolysis should not be performed, as the dough's stickiness increases, and the crust darkens. Instead, the release of WUAX-hydrolysed fragments of high molecular weight is preferred. By increasing the dough's viscosity, decreasing the fermentation gas migration rate and thus improving its retention, the dough shows an improved tolerance to fermentation and oven baking. The final products are reported to increase in volume from 15 up to 30% [6].

In conclusion, only a few modification processes were performed on the aleurone layer despite the large range of known physical, chemical and biological processes [18,157]. As the transformation of wheat bran has been extensively studied, it would be beneficial to rely on the obtained results to undertake new experiments focusing on improving the milling layer's technological and health benefits in bakery products.

However, the experiments were performed on starting materials of different purity, which explains the disparity in the results. It would then be interesting to work on the aleurone-extracting process to make it efficient and reproduceable. Little else can be accomplished on this part besides aiming to erase the traits that are not beneficial to the rheology of the end-products.

6. Conclusions

The aleurone layer is a major bran component that exhibits numerous nutritional and potential health benefits. Multiple processes exist and have been patented to extract this layer, but it seems that a choice must be made between the extraction yield and the purity of the fraction. Although its addition to bakery products has been studied and claimed to be beneficial from a nutritional point of view, some technological negative effects still arise, such as a decrease in loaf volume. These are partly due to the presence of a large number of DFs, mainly water-unextractable. However, those side effects could be reduced, and the aleurone-rich flour ingredient's functionality improved following some transformation processes preceding its addition to a dough matrix.

Despite all the known benefits of the aleurone layer, as proved in the experiments conducted by researchers, only a few applications exist nowadays with even fewer producers and furnishers. This may be related to the challenges of its extraction—more specifically, to the fact that no universal process exists. Even though it may be created, the natural variability of the grain's constituents and the impact of the environment would still require this process to be adapted to each raw material, as the grain's technological properties would change in accordance with its composition. In addition, the negative rheological traits observed upon aleurone addition into bakery products may also explain the absence of its use on the market, unless its presence is not communicated as such.

The appeal of the aleurone layer may also be dwindled by its more expensive extraction process when compared with wholegrain flours and other by-milling products. The latter can be nutritionally relevant for the consumer and would be easier to obtain. However, as demonstrated in this review, the health benefits obtained upon aleurone consumption may be higher (i.e., antioxidant capacity, reduced risk of cancer, etc.). In addition, the presence of the outer parts of the grain could increase the concentration of pesticides residues and impair the technological and organoleptic properties of the cereal end-products, thus favouring the use of the innermost part of the bran which shows those undesirable effects to a lesser extent [18].

Perhaps investigating other types of applications, as well as transformations pre- and post-extraction of the milling fraction, could instigate increased interest in this nutritionally beneficial aleurone layer. The lack of an actual market would also make it an opportunity for a company to renew its interest in the aleurone as an ingredient, not unlike bran fibres, provided its rheological properties are improved. Investigating the limits of aleurone layer incorporation to the formulation of other bakery products than bread in a design of experiments format could be interesting.

Author Contributions: Conceptualization, F.B., A.S. and T.A.; writing—original draft preparation, L.L.; writing—review and editing, L.L., F.B., A.S. and T.A.; supervision, A.S. and T.A.; project administration, F.B. and A.S.; funding acquisition, A.S. All authors have read and agreed to the published version of the manuscript.

Funding: This research was funded by Foricher les Moulins, grant number CIFRE 2021/1733.

Institutional Review Board Statement: Not applicable.

Informed Consent Statement: Not applicable.

Data Availability Statement: Data is contained within the article.

Conflicts of Interest: The authors declare no conflict of interest.

References

1. Fardet, A. New hypotheses for the health-protective mechanisms of whole-grain cereals: What is beyond fibre? *Nutr. Res. Rev.* **2010**, *23*, 65–134. [CrossRef] [PubMed]
2. Björck, I.; Östman, E.; Kristensen, M.; Mateo Anson, N.; Price, R.K.; Haenen, G.R.M.M.; Havenaar, R.; Bach Knudsen, K.E.; Frid, A.; Mykkänen, H.; et al. Cereal grains for nutrition and health benefits: Overview of results from in vitro, animal and human studies in the HEALTHGRAIN project. *Trends Food Sci. Technol.* **2012**, *25*, 87–100. [CrossRef]
3. Chalamacharla, R.B.; Harsha, K.; Sheik, K.B.; Viswanatha, C.K. Wheat Bran-Composition and Nutritional Quality: A Review. *Adv. Biotechnol. Microbiol.* **2018**, *9*, 7.
4. O'Dell, B.L.; De Boland, A.R.; Koirtyohann, S.R. Distribution of phytate and nutritionally important elements among the morphological components of cereal grains. *J. Agric. Food Chem.* **1972**, *20*, 718–723. [CrossRef]
5. Curti, E.; Carini, E.; Bonacini, G.; Tribuzio, G.; Vittadini, E. Effect of the addition of bran fractions on bread properties. *J. Cereal Sci.* **2013**, *57*, 325–332. [CrossRef]
6. Berger, M.; Ducroo, P. Arabinoxylanes et arabinoxylanases. *Ind. Des Céréales* **2004**, *56*, 3–14.
7. Bonnand-Ducasse, M.; Della Valle, G.; Lefebvre, J.; Saulnier, L. Effect of wheat dietary fibres on bread dough development and rheological properties. *J. Cereal Sci.* **2010**, *52*, 200–206. [CrossRef]
8. Arufe, S.; Chiron, H.; Doré, J.; Savary-Auzeloux, I.; Saulnier, L.; Della Valle, G. Processing & rheological properties of wheat flour dough and bread containing high levels of soluble dietary fibres blends. *Food Res. Int.* **2017**, *97*, 123–132. [CrossRef]
9. Antoine, C.; Lullien, V.; Abécassis, J.; Rouau, X. Intérêt nutritionnel de la couche à aleurone du grain de blé. *Sci. Des Aliment.* **2002**, *22*, 545–556. [CrossRef]
10. Brouns, F.; Hemery, Y.; Price, R.; Anson, N.M. Wheat aleurone: Separation, composition, health aspects, and potential food use. *Crit. Rev. Food Sci. Nutr.* **2012**, *52*, 553–568. [CrossRef]
11. Meziani, S.; Nadaud, I.; Tasleem-Tahir, A.; Nurit, E.; Benguella, R.; Branlard, G. Wheat aleurone layer: A site enriched with nutrients and bioactive molecules with potential nutritional opportunities for breeding. *J. Cereal Sci.* **2021**, *100*, 103225. [CrossRef]
12. Atwell, B.; Reding, W.; Earling, J.; Kanter, M.; Snow, K. Aleurone: Processing, Nutrition, Product Development, and Marketing. In *Whole Grains and Health*; Blackwell Publishing Ltd.: Oxford, UK, 2007; pp. 123–136. ISBN 978-0-470-27760-7.
13. Bagdi, A. *Utilization of Wheat Aleurone-Rich Flour in Pasta and Bread Production and the Oxidative Modification of Its Arabinoxylan Fraction*; Budapest University of Technology and Economic: Budapest, Hungary, 2016.
14. Stone, B.A.; Minifie, J. Recovery of Aleurone Cells from Wheat Bran. U.S. Patent US4746073A, 24 May 1988.
15. Bagdi, A.; Tóth, B.; Lőrincz, R.; Szendi, S.; Gere, A.; Kókai, Z.; Sipos, L.; Tömösközi, S. Effect of aleurone-rich flour on composition, baking, textural, and sensory properties of bread. *LWT Food Sci. Technol.* **2016**, *65*, 762–769. [CrossRef]

16. Bartalné-Berceli, M.; Izsó, E.; Gergely, S.; Salgó, A. Development and application of novel additives in bread-making. *Czech J. Food Sci.* **2018**, *36*, 470–475. [CrossRef]
17. Bartalné-Berceli, M.; Izsó, E.; Gergely, S.; Salgó, A. Bread Quality Improvement with Special Novel Additives. *Int. J. Biol. Biomol. Agric. Food Biotechnol. Eng.* **2015**, *9*, 5.
18. Gómez, M.; Gutkoski, L.C.; Bravo-Núñez, Á. Understanding whole-wheat flour and its effect in breads: A review. *Compr. Rev. Food Sci. Food Saf.* **2020**, *19*, 3241–3265. [CrossRef] [PubMed]
19. Rosa-Sibakov, N.; Poutanen, K.; Micard, V. How does wheat grain, bran and aleurone structure impact their nutritional and technological properties? *Trends Food Sci. Technol.* **2015**, *41*, 118–134. [CrossRef]
20. Laubin, B.B.; Lullien-Pellerin, V.V.; Nadaud, I.; Gaillard-Martinie, B.B.; Chambon, C.C.; Branland, G.G. Isolation of the wheat aleurone layer for 2D electrophoresis and proteomics analysis. *J. Cereal Sci.* **2008**, *48*, 709–714. [CrossRef]
21. Surget, A.; Barron, C. Histologie du grain de blé. *Ind. Des Céréales* **2005**, *145*, 3–7.
22. Rhodes, D.I.; Sadek, M.; Stone, B.A. Hydroxycinnamic Acids in Walls of Wheat Aleurone Cells. *J. Cereal Sci.* **2002**, *36*, 67–81. [CrossRef]
23. Barron, C.; Surget, A.; Rouau, X. Relative amounts of tissues in mature wheat (*Triticum aestivum* L.) grain and their carbohydrate and phenolic acid composition. *J. Cereal Sci.* **2007**, *45*, 88–96. [CrossRef]
24. Xiong, F.; Yu, X.-R.; Zhou, L.; Wang, Z.; Wang, F.; Xiong, A.-S. Structural development of aleurone and its function in common wheat. *Mol. Biol. Rep.* **2013**, *40*, 6785–6792. [CrossRef] [PubMed]
25. Burton, R.A.; Fincher, G.B. Evolution and development of cell walls in cereal grains. *Front. Plant Sci.* **2014**, *5*, 456. [CrossRef] [PubMed]
26. Becraft, P.W. Aleurone Cell Development. *Plant Cell Monogr.* **2007**, *8*, 45–56. [CrossRef]
27. Bacic, A.; Stone, B.A. Isolation and Ultrastructure of Aleurone Cell Walls From Wheat and Barley. *Funct. Plant Biol.* **1981**, *8*, 453–474. [CrossRef]
28. Meziani, S.; Nadaud, I.; Gaillard-Martinie, B.; Chambon, C.; Benali, M.; Branland, G. Proteomic analysis of the mature kernel aleurone layer in common and durum wheat. *J. Cereal Sci.* **2012**, *55*, 323–330. [CrossRef]
29. Saulnier, L.; Sado, P.-E.; Branland, G.; Charmet, G.; Guillon, F. Wheat arabinoxylans: Exploiting variation in amount and composition to develop enhanced varieties. *J. Cereal Sci.* **2007**, *46*, 261–281. [CrossRef]
30. Saulnier, L.; Guillon, F.; Chateigner-Boutin, A.-L. Cell wall deposition and metabolism in wheat grain. *J. Cereal Sci.* **2012**, *56*, 91–108. [CrossRef]
31. Robert, P.; Jamme, F.; Barron, C.; Bouchet, B.; Saulnier, L.; Dumas, P.; Guillon, F. Change in wall composition of transfer and aleurone cells during wheat grain development. *Planta* **2011**, *233*, 393–406. [CrossRef]
32. Feillet, P. *Le Grain de blé: Composition et Utilisation*; INRA: Paris, France, 2000; ISBN 2-7380-0896-8.
33. Lopez, H.W.; Adam, A.; Leenhardt, F.; Scalbert, A.; Remesy, C. Maîtrise de la valeur nutritionnelle du pain. *Ind. Des Céréales* **2001**, *124*, 15–20.
34. Andersson, R.; Andersson, A.; Åman, P. Chapter 16—Molecular Weight Distributions of Water-Extractable B-Glucan and Arabinoxylan. In *Healthgrain Methods*; Shewry, P.R., Ward, J.L., Eds.; American Associate of Cereal Chemists International; AACC International Press: Washington, DC, USA, 2009; pp. 203–216. ISBN 978-1-891127-70-0.
35. Zeng, H.; Lazarova, D.L.; Bordonaro, M. Mechanisms linking dietary fiber, gut microbiota and colon cancer prevention. *World J. Gastrointest. Oncol.* **2014**, *6*, 41–51. [CrossRef]
36. Fincher, G.B.; Stone, B.A. CEREALS | Chemistry of Nonstarch Polysaccharides. In *Encyclopedia of Grain Science*; Wrigley, C., Ed.; Elsevier: Oxford, UK, 2004; pp. 206–223. ISBN 978-0-12-765490-4.
37. Stone, B.; Morell, M.K. Chapter 9—Carbohydrates. In *Wheat*, 4th ed.; Khan, K., Shewry, P.R., Eds.; American Associate of Cereal Chemists International; AACC International Press: Washington, DC, USA, 2009; pp. 299–362. ISBN 978-1-891127-55-7.
38. Mendez-Encinas, M.A.; Carvajal-Millan, E.; Rascon-Chu, A.; Astiazaran-Garcia, H.F.; Valencia-Rivera, D.E. Ferulated Arabinoxylans and Their Gels: Functional Properties and Potential Application as Antioxidant and Anticancer Agent. *Oxidative Med. Cell. Longev.* **2018**, *2018*, e2314759. [CrossRef] [PubMed]
39. Courtin, C.M.; Delcour, J.A. Arabinoxylans and Endoxylanases in Wheat Flour Bread-making. *J. Cereal Sci.* **2002**, *35*, 225–243. [CrossRef]
40. Döring, C.; Jekle, M.; Becker, T. Technological and Analytical Methods for Arabinoxylan Quantification from Cereals. *Crit. Rev. Food Sci. Nutr.* **2016**, *56*, 999–1011. [CrossRef] [PubMed]
41. Parker, M.L.; Ng, A.; Waldron, K.W. The phenolic acid and polysaccharide composition of cell walls of bran layers of mature wheat (*Triticum aestivum* L. cv. Avalon) grains. *J. Sci. Food Agric.* **2005**, *85*, 2539–2547. [CrossRef]
42. Jamme, F.; Robert, P.; Bouchet, B.; Saulnier, L.; Dumas, P.; Guillon, F. Aleurone cell walls of wheat grain: High spatial resolution investigation using synchrotron infrared microspectroscopy. *Appl. Spectrosc.* **2008**, *62*, 895–900. [CrossRef]
43. Collins, H.M.; Burton, R.A.; Topping, D.L.; Liao, M.-L.; Bacic, A.; Fincher, G.B. REVIEW: Variability in Fine Structures of Noncellulosic Cell Wall Polysaccharides from Cereal Grains: Potential Importance in Human Health and Nutrition. *Cereal Chem.* **2010**, *87*, 272–282. [CrossRef]
44. Antoine, C.; Peyron, S.; Mabille, F.; Lapierre, C.; Bouchet, B.; Abecassis, J.; Rouau, X. Individual Contribution of Grain Outer Layers and Their Cell Wall Structure to the Mechanical Properties of Wheat Bran. *J. Agric. Food Chem.* **2003**, *51*, 2026–2033. [CrossRef]

45. Mateo Anson, N.; van den Berg, R.; Havenaar, R.; Bast, A.; Haenen, G.R.M.M. Ferulic acid from aleurone determines the antioxidant potency of wheat grain (*Triticum aestivum* L.). *J. Agric. Food Chem.* **2008**, *56*, 5589–5594. [CrossRef]
46. Antoine, C.; Peyron, S.; Lullien, V.; Abécassis, J.; Rouau, X. Wheat bran tissue fractionation using biochemical markers. *J. Cereal Sci.* **2004**, *39*, 387–393. [CrossRef]
47. Van Hung, P.; Maeda, T.; Miyatake, K.; Morita, N. Total phenolic compounds and antioxidant capacity of wheat graded flours by polishing method. *Food Res. Int.* **2009**, *42*, 185–190. [CrossRef]
48. Žilić, S. Phenolic Compounds of Wheat. Their Content, Antioxidant Capacity and Bioaccessibility. *MOJ Food Process. Technol.* **2016**, *2*, 37. [CrossRef]
49. Piironen, V.; Lampi, A.-M.; Ekholm, P.; Salmenkallio-Marttila, M.; Liukkonen, K.-H. Chapter 7—Micronutrients and Phytochemicals in Wheat Grain. In *Wheat*, 4th ed.; Khan, K., Shewry, P.R., Eds.; American Associate of Cereal Chemists International; AACC International Press: Washington, DC, USA, 2009; pp. 179–222. ISBN 978-1-891127-55-7.
50. Barron, C.; Samson, M.F.; Lullien-Pellerin, V.; Rouau, X. Wheat grain tissue proportions in milling fractions using biochemical marker measurements: Application to different wheat cultivars. *J. Cereal Sci.* **2011**, *53*, 306–311. [CrossRef]
51. Jääskeläinen, A.-S.; Holopainen-Mantila, U.; Tamminen, T.; Vuorinen, T. Endosperm and aleurone cell structure in barley and wheat as studied by optical and Raman microscopy. *J. Cereal Sci.* **2013**, *57*, 543–550. [CrossRef]
52. Rhodes, D.I.; Stone, B.A. Proteins in Walls of Wheat Aleurone Cells. *J. Cereal Sci.* **2002**, *36*, 83–101. [CrossRef]
53. Bacic, A.; Stone, B.A. Chemistry and Organization of Aleurone Cell Wall Components From Wheat and Barley. *Funct. Plant Biol.* **1981**, *8*, 475–495. [CrossRef]
54. Pomeranz, Y. *Wheat: Chemistry and Technology*, 3rd ed.; Pomeranz, Y., Ed.; AACC Monograph Series; American Association of Cereal Chemists: St. Paul, MN, USA, 1988; Volume I, ISBN 978-0-913250-65-5.
55. Wieser, H.; Koehler, P.; Scherf, K.A. Chapter 2—Chemical Composition of Wheat Grains. In *Wheat-An Exceptional Crop*; Wieser, H., Koehler, P., Scherf, K.A., Eds.; Woodhead Publishing: Cambridge, UK, 2020; pp. 13–45. ISBN 978-0-12-821715-3.
56. Shewry, P.R.; D'Ovidio, R.; Lafiandra, D.; Jenkins, J.A.; Mills, E.N.C.; Békés, F. Chapter 8—Wheat Grain Proteins. In *Wheat*, 4th ed.; Khan, K., Shewry, P.R., Eds.; American Associate of Cereal Chemists International; AACC International Press: Washington, DC, USA, 2009; pp. 223–298. ISBN 978-1-891127-55-7.
57. Amrein, T.M.; Gränicher, P.; Arrigoni, E.; Amadò, R. In vitro digestibility and colonic fermentability of aleurone isolated from wheat bran. *LWT-Food Sci. Technol.* **2003**, *36*, 451–460. [CrossRef]
58. Buttrose, M. Ultrastructure of the Developing Aleurone Cells of Wheat Grain. *Aust. J. Biol. Sci.* **1963**, *16*, 768. [CrossRef]
59. Marion, D.; Dubreil, L.; Douliez, J.-P. Functionality of lipids and lipid-protein interactions in cereal-derived food products. *Ol. Corps Gras Lipides* **2003**, *10*, 47–56. [CrossRef]
60. Morrison, I.N.; Kuo, J.; O'Brien, T.P. Histochemistry and fine structure of developing wheat aleurone cells. *Planta* **1975**, *123*, 105–116. [CrossRef]
61. Yan, X.; Ye, R.; Chen, Y. Blasting extrusion processing: The increase of soluble dietary fiber content and extraction of soluble-fiber polysaccharides from wheat bran. *Food Chem.* **2015**, *180*, 106–115. [CrossRef]
62. Arte, E.; Huang, X.; Nordlund, E.; Katina, K. Biochemical characterization and technofunctional properties of bioprocessed wheat bran protein isolates. *Food Chem.* **2019**, *289*, 103–111. [CrossRef] [PubMed]
63. Ndolo, V.U.; Beta, T. Distribution of carotenoids in endosperm, germ, and aleurone fractions of cereal grain kernels. *Food Chem.* **2013**, *139*, 663–671. [CrossRef] [PubMed]
64. Graham, S.F.; Hollis, J.H.; Migaud, M.; Browne, R.A. Analysis of Betaine and Choline Contents of Aleurone, Bran, and Flour Fractions of Wheat (*Triticum aestivum* L.) Using 1H Nuclear Magnetic Resonance (NMR) Spectroscopy. *J. Agric. Food Chem.* **2009**, *57*, 1948–1951. [CrossRef] [PubMed]
65. Hemery, Y.; Lullien-Pellerin, V.; Rouau, X.; Abecassis, J.; Samson, M.-F.; Åman, P.; von Reding, W.; Spoerndli, C.; Barron, C. Biochemical markers: Efficient tools for the assessment of wheat grain tissue proportions in milling fractions. *J. Cereal Sci.* **2009**, *49*, 55–64. [CrossRef]
66. Rosentrater, K.A.; Evers, A.D. Chapter 4—Chemical Components and Nutrition. In *Kent's Technology of Cereals*, 5th ed.; Rosentrater, K.A., Evers, A.D., Eds.; Woodhead Publishing Series in Food Science, Technology and Nutrition; Woodhead Publishing: Cambridge, UK, 2018; pp. 267–368. ISBN 978-0-08-100529-3.
67. Landberg, R.; Kamal-Eldin, A.; Salmenkallio-Marttila, M.; Rouau, X.X.; Aman, P. Localization of alkylresorcinols in wheat, rye and barley kernels. *J. Cereal Sci.* **2008**, *48*, 401–406. [CrossRef]
68. Chen, Z.; Zha, B.; Wang, L.; Wang, R.; Chen, Z.; Tian, Y. Dissociation of aleurone cell cluster from wheat bran by centrifugal impact milling. *Food Res. Int.* **2013**, *54*, 63–71. [CrossRef]
69. Batifoulier, F.; Verny, M.-A.; Chanliaud, E.; Rémésy, C.; Demigné, C. Variability of B vitamin concentrations in wheat grain, milling fractions and bread products. *Eur. J. Agron.* **2006**, *25*, 163–169. [CrossRef]
70. Saini, P.; Kumar, N.; Kumar, S.; Mwaurah, P.W.; Panghal, A.; Attkan, A.K.; Singh, V.K.; Garg, M.K.; Singh, V. Bioactive compounds, nutritional benefits and food applications of colored wheat: A comprehensive review. *Crit. Rev. Food Sci. Nutr.* **2021**, *61*, 3197–3210. [CrossRef]
71. Garg, M.; Kaur, S.; Sharma, A.; Kumari, A.; Tiwari, V.; Sharma, S.; Kapoor, P.; Sheoran, B.; Goyal, A.; Krishania, M. Rising Demand for Healthy Foods-Anthocyanin Biofortified Colored Wheat Is a New Research Trend. *Front. Nutr.* **2022**, *9*, 913. [CrossRef]

72. Zheng, Q.; Li, B.; Li, H.; Li, Z. Utilization of blue-grained character in wheat breeding derived from Thinopyrum poticum. *J. Genet. Genom.* **2009**, *36*, 575–580. [CrossRef]
73. Mateo Anson, N.; van den Berg, R.; Havenaar, R.; Bast, A.; Haenen, G.R.M.M. Bioavailability of ferulic acid is determined by its bioaccessibility. *J. Cereal Sci.* **2009**, *49*, 296–300. [CrossRef]
74. Bach Knudsen, K.E.; Steenfeldt, S.; Børsting, C.F.; Eggum, B.O. The nutritive value of decorticated mill fractions of wheat. 1. Chemical composition of raw and enzyme treated fractions and balance experiments with rats. *Anim. Feed Sci. Technol.* **1995**, *52*, 205–225. [CrossRef]
75. D'hoe, K.; Conterno, L.; Fava, F.; Falony, G.; Vieira-Silva, S.; Vermeiren, J.; Tuohy, K.; Raes, J. Prebiotic Wheat Bran Fractions Induce Specific Microbiota Changes. *Front. Microbiol.* **2018**, *9*, 31. [CrossRef]
76. Demuth, T.; Betschart, J.; Nyström, L. Structural modifications to water-soluble wheat bran arabinoxylan through milling and extrusion. *Carbohydr. Polym.* **2020**, *240*, 116328. [CrossRef] [PubMed]
77. Schlemmer, U.; Frølich, W.; Prieto, R.M.; Grases, F. Phytate in foods and significance for humans: Food sources, intake, processing, bioavailability, protective role and analysis. *Mol. Nutr. Food Res.* **2009**, *53*, S330–S375. [CrossRef] [PubMed]
78. Harris, P.J.; Ferguson, L.R. Dietary fibres may protect or enhance carcinogenesis. *Mutat. Res. Genet. Toxicol. Environ. Mutagen.* **1999**, *443*, 95–110. [CrossRef]
79. Sagara, M.; Mori, M.; Mori, H.; Tsuchikura, S.; Yamori, Y. Effect of dietary wheat aleurone on blood pressure and blood glucose and its mechanisms in obese spontaneously hypertensive rats: Preliminary report on comparison with a soy diet (Clinical and Experimental Pharmacology and Physiology. *Clin. Exp. Pharmacol. Physiol.* **2007**, *34*, S37–S39. [CrossRef]
80. Price, R.K.; Keaveney, E.M.; Hamill, L.L.; Wallace, J.M.W.; Ward, M.; Ueland, P.M.; McNulty, H.; Strain, J.J.; Parker, M.J.; Welch, R.W. Consumption of wheat aleurone-rich foods increases fasting plasma betaine and modestly decreases fasting homocysteine and LDL-cholesterol in adults. *J. Nutr.* **2010**, *140*, 2153–2157. [CrossRef]
81. Borowicki, A.; Stein, K.; Scharlau, D.; Glei, M. Fermentation Supernatants of Wheat (Triticum aestivum L.) Aleurone Beneficially Modulate Cancer Progression in Human Colon Cells. *J. Agric. Food Chem.* **2010**, *58*, 2001–2007. [CrossRef]
82. Borowicki, A.; Stein, K.; Scharlau, D.; Scheu, K.; Brenner-Weiss, G.; Obst, U.; Hollmann, J.; Lindhauer, M.; Wachter, N.; Glei, M. Fermented wheat aleurone inhibits growth and induces apoptosis in human HT29 colon adenocarcinoma cells. *Br. J. Nutr.* **2010**, *103*, 360–369. [CrossRef]
83. Borowicki, A.; Michelmann, A.; Stein, K.; Scharlau, D.; Scheu, K.; Obst, U.; Glei, M. Fermented Wheat Aleurone Enriched With Probiotic Strains LGG and Bb12 Modulates Markers of Tumor Progression in Human Colon Cells. *Nutr. Cancer* **2011**, *63*, 151–160. [CrossRef] [PubMed]
84. Stein, K.; Borowicki, A.; Scharlau, D.; Glei, M. Fermented wheat aleurone induces enzymes involved in detoxification of carcinogens and in antioxidative defence in human colon cells. *Br. J. Nutr.* **2010**, *104*, 1101–1111. [CrossRef] [PubMed]
85. Fenech, M.; Noakes, M.; Clifton, P.; Topping, D. Aleurone flour is a rich source of bioavailable folate in humans. *Am. Soc. Nutr. Sci.* **1999**, *129*, 1114–1119. [CrossRef] [PubMed]
86. Fenech, M.; Noakes, M.; Clifton, P.; Topping, D. Aleurone flour increases red-cell folate and lowers plasma homocyst(e)ine substantially in man. *Br. J. Nutr.* **2005**, *93*, 353–360. [CrossRef]
87. McIntosh, G.H.; Royle, P.J.; Pointing, G. Wheat Aleurone Flour Increases Cecal β-Glucuronidase Activity and Butyrate Concentration and Reduces Colon Adenoma Burden in Azoxymethane-Treated Rats. *J. Nutr.* **2001**, *131*, 127–131. [CrossRef]
88. Keaveney, E.M.; Hamill, L.L.; Price, R.K.; Wallace, J.M.W.; McNulty, H.; Ward, M.; Strain, J.J.; Ueland, P.M.; Scott, J.M.; Molloy, A.M.; et al. Evaluation of the uptake of bioactive components from wheat-bran and wheat-aleurone fractions in healthy adults. *Proc. Nutr. Soc.* **2009**, *67*, E244. [CrossRef]
89. Fava, F.; Ulaszewska, M.M.; Scholz, M.; Stanstrup, J.; Nissen, L.; Mattivi, F.; Vermeiren, J.; Bosscher, D.; Pedrolli, C.; Tuohy, K.M. Impact of wheat aleurone on biomarkers of cardiovascular disease, gut microbiota and metabolites in adults with high body mass index: A double-blind, placebo-controlled, randomized clinical trial. *Eur. J. Nutr.* **2022**, *61*, 2651–2671. [CrossRef]
90. Mahdavi Sharif, P.; Jabbari, P.; Razi, S.; Keshavarz-Fathi, M.; Rezaei, N. Importance of TNF-alpha and its alterations in the development of cancers. *Cytokine* **2020**, *130*, 155066. [CrossRef]
91. Mateo Anson, N.; Havenaar, R.; Bast, A.; Haenen, G.R.M.M. Antioxidant and anti-inflammatory capacity of bioaccessible compounds from wheat fractions after gastrointestinal digestion. *J. Cereal Sci.* **2010**, *51*, 110–114. [CrossRef]
92. Shrivastava, A.K.; Singh, H.V.; Raizada, A.; Singh, S.K. C-reactive protein, inflammation and coronary heart disease. *Egypt. Heart J.* **2015**, *67*, 89–97. [CrossRef]
93. Price, R.K.; Wallace, J.M.W.; Hamill, L.L.; Keaveney, E.M.; Strain, J.J.; Parker, M.J.; Welch, R.W. Evaluation of the effect of wheat aleurone-rich foods on markers of antioxidant status, inflammation and endothelial function in apparently healthy men and women. *Br. J. Nutr.* **2012**, *108*, 1644–1651. [CrossRef] [PubMed]
94. Hamill, L.L.; Keaveney, E.M.; Price, R.K.; Wallace, J.M.W.; Strain, J.J.; Welch, R.W. Absorption of ferulic acid in human subjects after consumption of wheat-bran and wheat-aleurone fractions. *Proc. Nutr. Soc.* **2008**, *67*, E255. [CrossRef]
95. Bresciani, L.; Scazzina, F.; Leonardi, R.; Dall'Aglio, E.; Newell, M.; Dall'Asta, M.; Melegari, C.; Ray, S.; Brighenti, F.; Del Rio, D. Bioavailability and metabolism of phenolic compounds from wholegrain wheat and aleurone-rich wheat bread. *Mol. Nutr. Food Res.* **2016**, *60*, 2343–2354. [CrossRef] [PubMed]

96. Dall'Asta, M.; Bresciani, L.; Calani, L.; Cossu, M.; Martini, D.; Melegari, C.; Del Rio, D.; Pellegrini, N.; Brighenti, F.; Scazzina, F. In Vitro Bioaccessibility of Phenolic Acids from a Commercial Aleurone-Enriched Bread Compared to a Whole Grain Bread. *Nutrients* **2016**, *8*, 42. [CrossRef] [PubMed]

97. Graf, E. Antioxidant potential of ferulic acid. *Free Radic. Biol. Med.* **1992**, *13*, 435–448. [CrossRef]

98. Toda, S.; Kimura, M.; Ohnishi, M. Effects of Phenolcarboxylic Acids on Superoxide Anion and Lipid Peroxidation Induced by Superoxide Anion. *Planta Med.* **1991**, *57*, 8–10. [CrossRef]

99. Rietjens, S.J.; Bast, A.; de Vente, J.; Haenen, G.R.M.M. The olive oil antioxidant hydroxytyrosol efficiently protects against the oxidative stress-induced impairment of the NO• response of isolated rat aorta. *Am. J. Physiol.-Heart Circ. Physiol.* **2007**, *292*, H1931–H1936. [CrossRef]

100. Zhao, Z.; Moghadasian, M.H. Chemistry, natural sources, dietary intake and pharmacokinetic properties of ferulic acid: A review. *Food Chem.* **2008**, *109*, 691–702. [CrossRef]

101. Castelluccio, C.; Bolwell, G.P.; Gerrish, C.; Rice-Evans, C. Differential distribution of ferulic acid to the major plasma constituents in relation to its potential as an antioxidant. *Biochem. J.* **1996**, *316*, 691–694. [CrossRef]

102. Andreasen, M.F.; Kroon, P.A.; Williamson, G.; Garcia-Conesa, M.-T. Esterase Activity Able To Hydrolyze Dietary Antioxidant Hydroxycinnamates Is Distributed along the Intestine of Mammals. *J. Agric. Food Chem.* **2001**, *49*, 5679–5684. [CrossRef]

103. Vitaglione, P.; Napolitano, A.; Fogliano, V. Cereal dietary fibre: A natural functional ingredient to deliver phenolic compounds into the gut. *Trends Food Sci. Technol.* **2008**, *19*, 451–463. [CrossRef]

104. Mateo Anson, N.; Aura, A.; Selinheimo, E.; Mattila, I.; Poutanen, K.; van den Berg, R.; Havenaar, R.; Bast, A.; Haenen, G.R.M.M. Bioprocessing of Wheat Bran in Whole Wheat Bread Increases the Bioavailability of Phenolic Acids in Men and Exerts Antiinflammatory Effects ex Vivo. *J. Nutr.* **2011**, *141*, 137–143. [CrossRef] [PubMed]

105. Heinonen, S.; Nurmi, T.; Liukkonen, K.; Poutanen, K.; Wähälä, K.; Deyama, T.; Nishibe, S.; Adlercreutz, H. In Vitro Metabolism of Plant Lignans: New Precursors of Mammalian Lignans Enterolactone and Enterodiol. *J. Agric. Food Chem.* **2001**, *49*, 3178–3186. [CrossRef]

106. Kamei, H.; Kojima, T.; Hasegawa, M.; Koide, T.; Umeda, T.; Yukawa, T.; Terabe, K. Suppression of Tumor Cell Growth by Anthocyanins In Vitro. *Cancer Investig.* **1995**, *13*, 590–594. [CrossRef] [PubMed]

107. Zhao, C.; Giusti, M.M.; Malik, M.; Moyer, M.P.; Magnuson, B.A. Effects of Commercial Anthocyanin-Rich Extracts on Colonic Cancer and Nontumorigenic Colonic Cell Growth. *J. Agric. Food Chem.* **2004**, *52*, 6122–6128. [CrossRef]

108. Lampi, A.-M.; Nurmi, T.; Ollilainen, V.; Piironen, V. Tocopherols and Tocotrienols in Wheat Genotypes in the HEALTHGRAIN Diversity Screen. *J. Agric. Food Chem.* **2008**, *56*, 9716–9721. [CrossRef] [PubMed]

109. Mooney, S.; Leuendorf, J.-E.; Hendrickson, C.; Hellmann, H. Vitamin B6: A Long Known Compound of Surprising Complexity. *Molecules* **2009**, *14*, 329–351. [CrossRef]

110. Ndolo, V.U.; Fulcher, R.G.; Beta, T. Application of LC–MS–MS to identify niacin in aleurone layers of yellow corn, barley and wheat kernels. *J. Cereal Sci.* **2015**, *65*, 88–95. [CrossRef]

111. Bohm, A.; Buri, R. Composition Comprising Aleurone, Method of Administering the Same, and Method of Manufacturing the Same. U.S. Patent US20040258776A1, 26 May 2004.

112. Bohm, A.; Kratzer, A. Method for Isolating Aleurone Particles. U.S. Patent US7780101B2, 24 August 2008.

113. Bohm, A.; Bogoni, C.; Behrens, R.; Otto, T. Method for the Extraction of Aleurone from Bran. U.S. Patent US20030175384A1, 4 October 2011.

114. Laux, J.; Von Reding, W.; Widmer, G. Aleurone Product and Corresponding Production Method. Patent WO2004073416A1, 13 January 2004.

115. Barron, C.; Abecassis, J.; Chaurand, M.; Lullien-Pellerin, V.; Mabille, F.; Rouau, X.; Sadouti, A.; Samson, M.-F. Accès à des molécules d'intérêt par fractionnement par voie sèche. *Innov. Agron.* **2012**, *19*, 51–62.

116. Hemery, Y.; Rouau, X.; Lullien-Pellerin, V.; Barron, C.; Abecassis, J. Dry processes to develop wheat fractions and products with enhanced nutritional quality. *J. Cereal Sci.* **2007**, *46*, 327–347. [CrossRef]

117. Antoine, C.; Lullien-Pellerin, V.; Abecassis, J.; Rouau, X. Effect of wheat bran ball-milling on fragmentation and marker extractability of the aleurone layer. *J. Cereal Sci.* **2004**, *40*, 275–282. [CrossRef]

118. Rosa, N.N.; Barron, C.; Gaiani, C.; Dufour, C.; Micard, V. Ultra-fine grinding increases the antioxidant capacity of wheat bran. *J. Cereal Sci.* **2013**, *57*, 84–90. [CrossRef]

119. Hemery, Y.; Holopainen, U.; Lampi, A.-M.; Lehtinen, P.; Nurmi, T.; Piironen, V.; Edelmann, M.; Rouau, X. Potential of dry fractionation of wheat bran for the development of food ingredients, part II: Electrostatic separation of particles. *J. Cereal Sci.* **2011**, *53*, 9–18. [CrossRef]

120. Tupper, R.J.; Pointing, G.; Westcott, S.J. Bran Products and Methods for Production Thereof. Patent WO2001021012A1, 15 September 2001.

121. Kvist, S.; Carlsson, T.; Lawther, J.M.; DeCastro, F.B. Process for the Fractionation of Cereal Brans. U.S. Patent US7709033B2, 4 May 2010.

122. Hemery, Y.; Rouau, X.; Dragan, C.; Bilici, M.; Beleca, R.; Dascalescu, L. Electrostatic properties of wheat bran and its constitutive layers: Influence of particle size, composition, and moisture content. *J. Food Eng.* **2009**, *93*, 114–124. [CrossRef]

123. Andersson, A.A.M.; Kamal-Eldin, A.; Fraś, A.; Boros, D.; Åman, P. Alkylresorcinols in Wheat Varieties in the HEALTHGRAIN Diversity Screen. *J. Agric. Food Chem.* **2008**, *56*, 9722–9725. [CrossRef] [PubMed]

124. Pekkinen, J.; Rosa, N.N.; Savolainen, O.-I.; Keski-Rahkonen, P.; Mykkänen, H.; Poutanen, K.; Micard, V.; Hanhineva, K. Disintegration of wheat aleurone structure has an impact on the bioavailability of phenolic compounds and other phytochemicals as evidenced by altered urinary metabolite profile of diet-induced obese mice. *Nutr. Metab.* **2014**, *11*, 1. [CrossRef]
125. Boshuizen, B.; Moreno de Vega, C.V.; De Maré, L.; de Meeûs, C.; de Oliveira, J.E.; Hosotani, G.; Gansemans, Y.; Deforce, D.; Van Nieuwerburgh, F.; Delesalle, C. Effects of Aleurone Supplementation on Glucose-Insulin Metabolism and Gut Microbiome in Untrained Healthy Horses. *Front. Vet. Sci.* **2021**, *8*, 642809. [CrossRef]
126. Rosa, N.N.; Dufour, C.; Lullien-Pellerin, V.; Micard, V. Exposure or release of ferulic acid from wheat aleurone: Impact on its antioxidant capacity. *Food Chem.* **2013**, *141*, 2355–2362. [CrossRef]
127. Rosa, N.N.; Aura, A.-M.; Saulnier, L.; Holopainen-Mantila, U.; Poutanen, K.; Micard, V. Effects of Disintegration on in Vitro Fermentation and Conversion Patterns of Wheat Aleurone in a Metabolical Colon Model. *J. Agric. Food Chem.* **2013**, *61*, 5805–5816. [CrossRef]
128. Wellnitz, J. The Ingredient 'Wheat Aleurone' and Nutritional Benefits. In *International Nutrition & Health Industry Expo*, 3rd ed.; Whole Grain Forum: Beijing, China, 2011.
129. Nkhata, L.; Izydorczyk, M.; Beta, T. Effect of Water-Extractable Arabinoxylans from Wheat Aleurone and Bran on Lipid Peroxidation and Factors Influencing their Antioxidant Capacity. *Bioact. Carbohydr. Diet. Fibre* **2017**, *10*, 20–26. [CrossRef]
130. Yang, Y.-L.; Guan, E.-Q.; Li, M.; Li, N.; Bian, K.; Wang, T.; Lu, C.; Chen, M.; Xu, F. Effect of transglutaminase on the quality and protein characteristics of aleurone-riched fine dried noodles. *LWT* **2022**, *154*, 112584. [CrossRef]
131. Li, X.; Hu, H.; Xu, F.; Liu, Z.; Zhang, L.; Zhang, H. Effects of aleurone-rich fraction on the hydration and rheological properties attributes of wheat dough. *Int. J. Food Sci. Technol.* **2018**, *54*, 1777–1786. [CrossRef]
132. Xu, M.; Hou, G.G.; Ma, F.; Ding, J.; Deng, L.; Kahraman, O.; Niu, M.; Trivettea, K.; Lee, B.; Wu, L.; et al. Evaluation of aleurone flour on dough, textural, and nutritional properties of instant fried noodles. *LWT* **2020**, *126*, 109294. [CrossRef]
133. Vangsøe, C.T.; Sørensen, J.F.; Bach Knudsen, K.E. Aleurone cells are the primary contributor to arabinoxylan oligosaccharide production from wheat bran after treatment with cell wall-degrading enzymes. *Int. J. Food Sci. Technol.* **2019**, *54*, 2847–2853. [CrossRef]
134. Yang, M.; Li, N.; Wang, A.; Tong, L.; Wang, L.; Yue, Y.; Yao, J.; Zhou, S.; Liu, L. Evaluation of rheological properties, microstructure and water mobility in buns dough enriched in aleurone flour modified by enzyme combinations. *Int. J. Food Sci. Technol.* **2021**, *56*, 5913–5922. [CrossRef]
135. Martín-García, B.; Gómez-Caravaca, A.M.; Marconi, E.; Verardo, V. Distribution of free and bound phenolic compounds, and alkylresorcinols in wheat aleurone enriched fractions. *Food Res. Int.* **2021**, *140*, 109816. [CrossRef]
136. Bucsella, B.; Molnár, D.; Harasztos, A.H.; Tömösközi, S. Comparison of the rheological and end-product properties of an industrial aleurone-rich wheat flour, whole grain wheat and rye flour. *J. Cereal Sci.* **2016**, *69*, 40–48. [CrossRef]
137. Hemdane, S.; Leys, S.; Jacobs, P.J.; Dornez, E.; Delcour, J.A.; Courtin, C.M. Wheat milling by-products and their impact on bread making. *Food Chem.* **2015**, *187*, 280–289. [CrossRef]
138. Bagdi, A.; Szabó, F.; Gere, A.; Kókai, Z.; Sipos, L.; Tömösközi, S. Effect of aleurone-rich flour on composition, cooking, textural, and sensory properties of pasta. *LWT-Food Sci. Technol.* **2014**, *59*, 996–1002. [CrossRef]
139. Noort, M.W.J.; van Haaster, D.; Hemery, Y.; Schols, H.A.; Hamer, R.J. The effect of particle size of wheat bran fractions on bread quality–Evidence for fibre–protein interactions. *J. Cereal Sci.* **2010**, *52*, 59–64. [CrossRef]
140. Tian, B.; Zhou, C.; Li, D.; Pei, J.; Guo, A.; Liu, S.; Li, H. Monitoring the Effects of Hemicellulase on the Different Proofing Stages of Wheat Aleurone-Rich Bread Dough and Bread Quality. *Foods* **2021**, *10*, 2427. [CrossRef] [PubMed]
141. Zhou, Y.; Dhital, S.; Zhao, C.; Ye, F.; Chen, J.; Zhao, G. Dietary fiber-gluten protein interaction in wheat flour dough: Analysis, consequences and proposed mechanisms. *Food Hydrocoll.* **2021**, *111*, 106203. [CrossRef]
142. Sivam, A.S.; Sun-Waterhouse, D.; Quek, S.; Perera, C.O. Properties of Bread Dough with Added Fiber Polysaccharides and Phenolic Antioxidants: A Review. *J. Food Sci.* **2010**, *75*, R163–R174. [CrossRef] [PubMed]
143. Sudha, M.L.; Vetrimani, R.; Leelavathi, K. Influence of fibre from different cereals on the rheological characteristics of wheat flour dough and on biscuit quality. *Food Chem.* **2007**, *100*, 1365–1370. [CrossRef]
144. Wang, M.; Hamer, R.J.; van Vliet, T.; Oudgenoeg, G. Interaction of Water Extractable Pentosans with Gluten Protein: Effect on Dough Properties and Gluten Quality. *J. Cereal Sci.* **2002**, *36*, 25–37. [CrossRef]
145. Anil, M. Using of hazelnut testa as a source of dietary fiber in breadmaking. *J. Food Eng.* **2007**, *80*, 61–67. [CrossRef]
146. Wang, M.; Hamer, R.J.; van Vliet, T.; Gruppen, H.; Marseille, H.; Weegels, P.L. Effect of Water Unextractable Solids on Gluten Formation and Properties: Mechanistic Considerations. *J. Cereal Sci.* **2003**, *37*, 55–64. [CrossRef]
147. Wang, M.; Oudgenoeg, G.; van Vliet, T.; Hamer, R.J. Interaction of water unextractable solids with gluten protein: Effect on dough properties and gluten quality. *J. Cereal Sci.* **2003**, *38*, 95–104. [CrossRef]
148. Wang, M.; Van Vliet, T.; Hamer, R.J. Evidence that pentosans and xylanase affect the re-agglomeration of the gluten network. *J. Cereal Sci.* **2004**, *39*, 341–349. [CrossRef]
149. Wang, M.; van Vliet, T.; Hamer, R.J. How gluten properties are affected by pentosans. *J. Cereal Sci.* **2004**, *39*, 395–402. [CrossRef]
150. Collar, C.; Santos, E.; Rosell, C.M. Assessment of the rheological profile of fibre-enriched bread doughs by response surface methodology. *J. Food Eng.* **2007**, *78*, 820–826. [CrossRef]

151. Frederix, S.A.; Van Hoeymissen, K.E.; Courtin, C.M.; Delcour, J.A. Water-Extractable and Water-Unextractable Arabinoxylans Affect Gluten Agglomeration Behavior during Wheat Flour Gluten−Starch Separation. *J. Agric. Food Chem.* **2004**, *52*, 7950–7956. [CrossRef] [PubMed]
152. Biliaderis, C.G.; Izydorczyk, M.S.; Rattan, O. Effect of arabinoxylans on bread-making quality of wheat flours. *Food Chem.* **1995**, *53*, 165–171. [CrossRef]
153. Autio, K. Effects of cell wall components on the functionality of wheat gluten. *Biotechnol. Adv.* **2006**, *24*, 633–635. [CrossRef] [PubMed]
154. Courtin, C.M.; Delcour, J.A. Relative Activity of Endoxylanases Towards Water-extractable and Water-unextractable Arabinoxylan. *J. Cereal Sci.* **2001**, *33*, 301–312. [CrossRef]
155. Byrne, J. Baked Goods with Aleurone Fractions may Boost Heart Health. Available online: https://www.bakeryandsnacks.com/Article/2010/05/07/Baked-goods-with-aleurone-fractions-may-boost-heart-health (accessed on 5 April 2022).
156. Shewry, P.R.; Wan, Y.; Hawkesford, M.J.; Tosi, P. Spatial distribution of functional components in the starchy endosperm of wheat grains. *J. Cereal Sci.* **2020**, *91*, 102869. [CrossRef]
157. Mert, I.D. The applications of microfluidization in cereals and cereal-based products: An overview. *Crit. Rev. Food Sci. Nutr.* **2020**, *60*, 1007–1024. [CrossRef]
158. Aslam, M.F.; Ellis, P.R.; Berry, S.E.; Latunde-Dada, G.O.; Sharp, P.A. Enhancing mineral bioavailability from cereals: Current strategies and future perspectives. *Nutr. Bull.* **2018**, *43*, 184–188. [CrossRef]
159. Hemery, Y.M.; Anson, N.M.; Havenaar, R.; Haenen, G.R.M.M.; Noort, M.W.J.; Rouau, X. Dry-fractionation of wheat bran increases the bioaccessibility of phenolic acids in breads made from processed bran fractions. *Food Res. Int.* **2010**, *43*, 1429–1438. [CrossRef]
160. Gan, J.; Xie, L.; Peng, G.; Xie, J.; Chen, Y.; Yu, Q. Systematic review on modification methods of dietary fiber. *Food Hydrocoll.* **2021**, *119*, 106872. [CrossRef]
161. Arzami, A.N.; Ho, T.M.; Mikkonen, K.S. Valorization of cereal by-product hemicelluloses: Fractionation and purity considerations. *Food Res. Int.* **2021**, *151*, 110818. [CrossRef] [PubMed]
162. Bagdi, A.; Tömösközi, S.; Nyström, L. Structural and functional characterization of oxidized feruloylated arabinoxylan from wheat. *Food Hydrocoll.* **2017**, *63*, 219–225. [CrossRef]
163. Zhang, M.-Y.; Liao, A.-M.; Thakur, K.; Huang, J.-H.; Zhang, J.-G.; Wei, Z.-J. Modification of wheat bran insoluble dietary fiber with carboxymethylation, complex enzymatic hydrolysis and ultrafine comminution. *Food Chem.* **2019**, *297*, 124983. [CrossRef] [PubMed]

MDPI

St. Alban-Anlage 66

4052 Basel

Switzerland

www.mdpi.com

Foods Editorial Office

E-mail: foods@mdpi.com

www.mdpi.com/journal/foods